# Information Technology and Development Series

## Volume 2

### Informatics and Industrial Development

# Information Technology and Development Series

**Volume 1:** **Information Technology: Impact on the Way of Life**
A Selection of Papers from the EEC Conference on The Information Society, held in Dublin, Ireland, 18–20 November 1981.

*Editors*: Liam Bannon, Ursula Barry and Olav Holst

Conference organized by FAST Programme, Commission of the European Communities and National Board for Science and Technology, Ireland.

xii + 381 pp; 17 Tables; 8 Figures.

**Paperback:** ISBN 0 907567 35 5; **Library Edition:** ISBN 0 907567 34 7
Published: September 1982

**Volume 2:** **Informatics and Industrial Development**
Proceedings of the International Conference on Policies for Information Processing for Developing Countries, 9–13 March 1981, Trinity College, Dublin, Ireland.

*Editor*: F.G. Foster

Conference organized by Systems Development Programme, Trinity College Dublin, and National Board for Science and Technology, Ireland.

Conference sponsored by IBI, UNIDO and ITM.

xiv + 453 pp; 31 Tables; 43 Figures.

**Paperback:** ISBN 0 907567 29 0; **Library Edition:** ISBN 0 907567 28 2
Published: November 1982

# Informatics and Industrial Development

Proceedings of the International Conference on Policies for Information Processing for Developing Countries
9–13 March 1981, Trinity College Dublin, Ireland

*Edited by*
**F.G. Foster**

*Published by*
TYCOOLY INTERNATIONAL PUBLISHING LIMITED
DUBLIN

*Published by*:
Tycooly International Publishing Ltd.,
6 Crofton Terrace,
Dun Laoghaire,
Co. Dublin, Ireland
Telephone: (+353-1) 800245/6
Telex: 30547 SHCN EI

First edition 1982



Typeset by Intercontinental Photocomposition Ltd., Dublin
Printed in Ireland by Irish Elsevier Printers Ltd., Shannon

ISBN 0 907 567 28 2 Hardback
ISBN 0 907 567 29 0 Softcover

## *Editorial Committee*

F. Gordon Foster, B.A., M.A., D.Phil.,
*Trinity College Dublin*

Richard J. Nolan, B. Comm., M.Econ.Sc., M.Sc., Ph.D.,
*Department of the Public Service, Ireland*

Tevfik Dalgic, B.S., M.B.A., Ph.D.,
*Dublin Institute of Technology*

John Moriarty, B.E., M.A.,
*Trinity College Dublin*

Diarmuid Murphy, M.Sc., Ph.D.,
*National Board for Science and Technology, Ireland*

## *Conference Committee*

F. Gordon Foster, *Chairman*
Diarmuid Murphy, *Co-Chairman*
Dermot O'Doherty, *Co-Ordinator*
Tevfik Dalgic, *Secretary*
F. Piera Gomez, *IBI*
Roch T. de Mautort, *UNIDO*
Didier Lecerf, *ITM*
Robert Cochran
Peter Daly
Brendan Finucane
John Moriarty
Richard J. Nolan
Lilly Sheehan, *Conference Officer*

# ACKNOWLEDGEMENTS

***Patron***
The Minister for Foreign Affairs of Ireland

***Organised by***
Systems Development Programme, Trinity College Dublin
National Board for Science and Technology, Ireland

***Sponsored by***
Intergovernmental Bureau for Informatics (IBI)
United Nations Industrial Development Organization (UNIDO)
Informatique pour les Tiers Mondes (ITM)

***The Conference organisers acknowledge support from***
European Research Office of the US Army
US Air Force European Office of Aerospace Research and Development
Allied Irish Banks
Bank of Ireland
Industrial Development Authority (IDA), Ireland
IBM Ireland Limited

# Contents

# Foreword I

The International Conference on Informatics and Industrial Development, held in Dublin from 9 to 13 March 1981, was the first international meeting on this subject and had the great merit of raising a series of issues concerning both the applications of informatics to industry and the development of the informatics industry itself. The need for further in-depth research was also stressed.

The keynote of the debates concerned the effective transfer of technology to developing countries as an important element in the whole development process, considering all the relevant aspects and showing to the organizations involved the importance of this issue and the need to emphasize and expand their activities of assistance to developing countries.

The conference dealt with subjects of great importance to developing countries and will surely constitute one of the main contributions to the preparatory work of the IBI SPIN II Conference on Strategies and Policies for Informatics.

**F.A. Bernasconi**
*Director General*
Intergovernmental Bureau for Informatics (IBI)

*Rome, 1982*

# Foreword II

Information technology is in a state of dynamic growth and the Dublin Conference on Informatics and Industrial Development, held 9–13 March 1981, has highlighted the importance developing countries attach to this development which is to have substantial influence on their industrial development strategies. It has also been noted in the Conference that action must be initiated by the developing countries to build the technical capacity to benefit from the recent advances in this area. Accordingly, UNIDO intends to pursue a programme of action in this field.

A signatory to the Declaration of Mexico on Informatics, Development and Peace in June 1981, UNIDO will co-operate with the Intergovernmental Bureau for Informatics and other UN organizations on the forthcoming World Conference on Strategies and Policies for Informatics (SPIN II). These proceedings will, I hope, provide valuable material for the preparatory activities for this Conference.

**Abd-el Rahman Khane**
*Executive Director*
United Nations Industrial Development
Organization (UNIDO)

*Vienna, May 1982*

# Preface

THE INTERNATIONAL CONFERENCE on Informatics and Industrial Development, held in Trinity College, Dublin, 9–13 March 1981 was attended by some 130 participants, coming from 45 countries (mostly developing countries) and from 9 international organizations. There was also a balance in the speakers between industrialised and developing countries. As may be seen from the list of participants (Appendix), these included many prominent authorities, so that the deliberations reflect the most current thinking on the impact of informatics on the industrialisation of the developing countries.

Sponsored jointly by the Intergovernmental Bureau for Informatics (IBI), the United Nations Industrial Development Organisation (UNIDO) and Informatique pour les Tiers Mondes (ITM) and organised by the Systems Development Programme of Trinity College Dublin and the Irish National Board for Science and Technology, it was felt that the selected theme was timely and, in view of Ireland's very recently developed strategy for informatics technology and her now rapidly growing electronics industry, that the location was appropriate.

Within the industrialised countries there has been a growing consciousness of the potentialities of informatics, a sector that will soon account for over half their GNP. Meanwhile three quarters of the globe, the developing nations, have only a tiny share of the technological capacity and industrial potential in this vital field. It is hoped that these Proceedings will contribute to the development of operational strategies for the informatisation of the Third World.

The organisers thank the Minister for Foreign Affairs of Ireland for his patronage of the Conference. The organisers are very appreciative of the support received from the international and Irish sponsors and of the lively personal interest demonstrated by their representatives. They also wish to thank all those who participated so fully and enthusiastically in the Conference and hope that they can be involved as fully as possible in whatever follow-up actions are eventually identified and decided upon.

**F. Gordon Foster**
*Conference Chairman*

**Diarmuid Murphy**
*Conference Co-Chairman*

# Introduction

## What is Informatics?

Rather than start with a definition let us first consider the objective of this book. Over the last quarter of a century a new, powerful and integrative technology, based on microelectronics, has been created that has revolutionised the handling of information, and this technology is still in the process of rapid development. We are here concerned specifically with its impact on industry, with special reference to industrial development in the Third World.

Let us set this in perspective. In the first phase of the development of the new information handling technology, starting in the late 1940s, computers were used, almost exclusively, for scientific calculation. The impact was felt in the natural sciences, where new advances were made possible by the then, relatively, huge increase in computing power that became for the first time available. A decade later, and continuing to the present, applications of computers began to be made as integral components of organisational systems for information and control purposes in the pursuit of productivity and economic growth. In the early days only the largest organisations could afford to operate a computer. More recently, dramatic decreases, both in unit cost and in actual size of components, coupled with increased flexibility in handling have brought computing power to even very small scale enterprises.

The general trend does not stop there. In the industrialised countries the impact of computerisation, or informatisation as it is also being called, is now felt not just within organisations but by society itself. The systems supporting modern society are increasingly computer-based and this is affecting the ways in which society is evolving. It is being said that the final phase of informatisation is now just commencing, in which the impact will be on the individual for whom information technology is creating new potentialities for self-realisation.

The informatisation of industry may therefore be regarded as but one aspect of a broader socio-economic development. It is convenient to use the term *informatics* to refer to this whole phenomenon that is largely based on the information handling technology. The term also connotes the analysis and formulation of national and international strategies and policies for informatisation.

As in the case of other technological frontiers, information technology has a massive dual potential both for destruction and for peaceful uses. Other technologies have extended man's muscular power and manipulative abilities. The deep significance of the integrated information handling technology of computers, communications and control is that it extends his nervous system and his brain. Only man's will and creative imagination can restrict the pervasiveness of its applications in the support of economic, social and personal development.

## Informatics and the Third World

Turning to the problems of the Third World which pose such a tremendous challenge, and which only radically innovative and unconventional approaches have any hope of touching, can we propose that this revolutionary new technology really does provide a short-cut to socio-economic development, or is it just another mystification? To examine this question in depth many complex issues must be confronted. For the Third World, industrial development is a priority. Studies are revealing that informatics will profoundly affect the productive infrastructure and the international division of labour. Policy options for industrialisation are narrowed. A re-conceptualisation of development strategies is required. Developing countries must not merely reproduce what the industrialised North has done but base their actions on an assessment of long-term comparative advantages.

## Scope of the Book

In this book an attempt is made to cover all the issues in an examination of the opportunities and threats, constraints and the necessary national infrastructures relating to the informatisation of industry in developing countries.

Perhaps the most significant contrast that can be drawn between the industrialised North and the developing South is that between information-rich and information-poor communities. Information is seen to be a 'commodity' of crucial importance in development. Technology transfer is therefore a basic issue. National, regional and international organisations have a key role to play in this area and their activities are described. This leads on to a presentation of the variety of ways of making information for industrial development available to agencies and enterprises requiring it. The distinction is made between the needs of those engaged in research and development (R & D) and in industrial planning and investment. Only knowledge relevant to the needs and the level of understanding of the user will be utilised. It is not sufficient just to set up information systems: a determined and sustained effort must be mounted to 'market' information.

Since the informatic hardware industry now ranks in size with energy and transportation, it must itself be a subject of study as a sector of industry. How they can participate in this industry is therefore an issue of major interest to policy makers in developing countries. The policy options, preconditions and constraints relating to an indigenisation strategy are therefore examined.

The necessary infrastructures to support organic and self-sustaining growth are examined, with particular reference to education and training, telecommunications, research and development (R & D) and institutional arrangements.

The potential economic effects on productivity and employment are of major concern and this issue is taken up in regard to both the informatic hardware industry and informatisation of other sectors.

A number of surveys and case studies are presented, both of advanced applications and applications in smaller enterprises. These studies, ranging from real-time production control to the use of personal computers but with the main emphasis on applications in developing countries, provide an opportunity to gauge the pulse of actual industrial development.

The conclusions and recommendations of the four workshop sessions that were held during the Conference are presented and the Conference opening and closing addresses comprise the first and last sections of the book.

## Summary

What emerges may be briefly summarised as follows. Informatics has immense potential but in developing countries severe structural and cultural constraints hinder the effective implementation of even the simplest applications. Developing countries are heterogeneous and the circumstances of each will need to be separately considered. Specific pilot projects and case studies and the diffusion of information on them are urgently required.

The problems are particularly acute because information technology is the prime example of a systems technology which only works effectively when many interacting components, people and hardware, are brought into harmonious relationship.

Mechanistic 'solutions' to poorly understood problems must be avoided. Education and training have the highest priority and an effective informatics policy will be people-oriented: one that seeks to improve existing skills rather than replace them.

The question remains whether fundamentally new insights are still required for the promotion of economic growth. The concept of an organising principle referred to as "social intelligence" in the paper by Professor Dedijer must surely be pointing in the right direction.

# Introductory Statements

## Address by the Minister for Foreign Affairs of Ireland, Mr. Brian Lenihan

I WOULD LIKE to congratulate the organisers of the conference, the National Board for Science and Technology (NBST) and the Systems Development Programme of Trinity College Dublin. The sponsorship has been undertaken by the United Nations Industrial Development Organisation and the Intergovernmental Bureau for Informatics. I am associated with the Conference as patron and in this capacity congratulate the organisers and the sponsors.

I hope that this Conference will be just the first in a series under the general title of "Dublin Conference on Development" to be devoted to a variety of development issues of interest to the developing world. In a highly technological world where not only access to and acquisition of technical and other information, but also its management and effective utilisation are important factors in development, it is essential that developing countries get into a position where they can develop their industrial capabilities based on sound knowledge of what is available and how it can be applied effectively, also how it can be applied rationally to meet their own development needs.

On a different level the conference is also significant in that it provides an excellent example of the kind of concrete follow-up action which can and indeed should be taken in response to resolutions which have been adopted at various important conferences held during the past few years within the UN framework. Reference might be made in this connection to two conferences held in 1979—the UN Conference on Science and Technology for Development, UNCSTD, and the Intergovernmental Conference on Scientific and Technological Information for Development, UNISIST, held under the aegis of UNESCO. Both these conferences highlighted in particular the need for better access for developing countries to information sources and technical know-how, as well as for greater international co-operation in the exchange and transfer of information as a means to assisting developing countries to realise their development potential. I am confident that this conference will succeed in a concrete and practical way towards meeting some of these objectives.

A message by the executive director of UNIDO, Dr. Abd-el Rahman Khane, to me to mark the conference, expresses his hope that the conference will identify concrete ways and means by which the wealth of information available in the world can be harnessed for the accelerated industrialisation of the developing world. UNIDO, as you are aware, has been involved since its inception as the United Nations organisation responsible for co-ordinating assistance within the UN system in the field of industrial development, in industrial and technological information activities, and this particular function has been further strengthened by the establishment a few years ago of a scientific and technological information bank within UNIDO. Thus UNIDO looks forward, as we all do, to the results of the conference as an input to further efforts in this field both by UNIDO itself and by others at national and international levels.

There is one area which I feel deserves mention in being essential to the effective utilisation of informatics, and that is the one of human resources development. This particular area, when coupled with the development of appropriate physical infrastructures, is an area, in my view, of vital importance, not only in relation to the field of informatics and industrial development but to national development as a whole. For this reason, we here in Ireland, where we have our own development co-operation efforts, have always placed strong emphasis on the drawing and transfer of technical expertise from our own pool of experience, as well as training the human resources potential of those countries to which we lend help.

In conclusion I would like again to thank the organisations, to whom I have referred, responsible for organising this conference, and to thank the UNIDO and the IBI as well who have been associated with them in this matter. It is an excellent sign of the proper approach in my view, to these matters, that we have the state sponsored agency of the NBST working in conjunction with one of our major educational and university establishments, Trinity College, and their Systems Development Programme, in taking this initiative. It is a most useful forum, in my view, for a practical in depth examination of the issues involved and will almost certainly result in the practical identification of concrete action for the future. That is the main thrust of this conference, a thrust that I hope will bring together much of the information to secure a degree of co-ordination that can give positive recommendations for action in the future. Therefore in association with Dr. Khane, Executive Director of UNIDO, I wish you every success in your deliberations.

**Professor Bernasconi, Director General, Intergovernmental Bureau for Informatics** (Delivered in his absence by Mr. F. Piera Gomez)

Informatics, handicraft yesterday, science today, must be considered as one of the essential tools that man has at his disposal to reach levels that otherwise would be unattainable. Informatics is a new language that must be mastered. It is necessary to learn how to speak, read and write it, as it has shown itself to be a critical factor in the development process. Hence the need for all countries to learn this language, and IBI is conducting this familiarisation campaign which is too important to be left only to the technicians. In parts of the world, society is in the midst of a measure of transition recasting it into an age of cybernetic dimensions. In the less developed parts of the world major transitions are also occurring, taking a slower pace to work out alternative futures. In this new area computers will be even more closely united and integrated into things thus affecting and transforming much of the technological fabric of society, changing many aspects of human behaviour through the use of informatics and thereby plunging society deeper into a further evolution. This area comes about through the incorporation of microcomputer-like devices into machines making the resulting devices increasingly intelligent. In turn, society will speed even more rapidly into this new age through the rapid adoption of these devices. The new intellectual framework now developing to complement and supplement the mechanistic and rationalistic doctrines of the analytic modes of thought are deductions of teleology and synthesis. This concept includes the frameworks of choice, goals, the anticipatory sciences, participatory government, cybernetics, systems, etc.

The primary mode of thinking points towards the diffusion of sciences, arts and humanities, together with technology, sociology, economics, politics and culture, and quickly evolves into a multidisciplinary, interdisciplinary, and trans-disciplinary system. A common threat to most however, is that informatisation, to a great extent, will take over the task of physical labour. The new work roles for people in this new age appear to be diverted away from laborious and hazardous tasks, tasks of monotony and rote, tasks which chain people to machines and the like. Efforts to make machines more humanistic and to amplify human capabilities are strong trends occurring at present and will certainly evolve and expand in the future. The extent to which people/machine symbiosis will evolve is almost unlimited in a service oriented future as machines become more intelligent. As a result the national definition of work will undergo considerable change and most people will have a multiplicity of careers throughout their lifetime, requiring a life-long approach and dedication to education. The concepts of life-long learning will of necessity be merged and incorporated into new forms of work, both work and leisure time activities. And so education may well become an integral part of everyone's daily life style.

This conference is the first that has been convened to deal with an integral approach, with the promise of industrial development through informatics and of informatics, two subsets of a larger set of objectives. IBI is already working on the preparation of the Second Inter-Governmental Conference on the Strategies and Policies for Informatics, SPIN II, which will be held in Havana in the Spring of 1983. Its scope is the informatisation of emerging nations. The results of the conference which is starting today will be of great importance in the preparation of SPIN II and of its objectives. These are the reasons why the IBI has actively supported this conference.

**Dr. G.S. Gouri, UNIDO**

It is indeed a great pleasure and a privilege for us in UNIDO to be associated with this most important timely conference. It has been said that we are in the midst of the second industrial revolution. The technological breakthroughs that have occurred and that will be occurring in the next five or ten years in a sense will greatly transform the present industrial structure, the present distribution and location of industries. It has also been said that this rate of change, in fact, will be so quick that it is very necessary to prepare oneself and to see that there is not only a partnership in this great change but also to meet the challenge that is in front of us. It is in this connection that we have been greatly honoured to be associated with this conference as microelectronics is the subject of a major focus in our technology programme.

When studying and monitoring the technology breakthroughs, which UNIDO is currently doing, in the field of genetic engineering, microelectronics, petrochemicals and other related technologies, machine tools, lighter than air technologies, it is seen that the field of microelectronics, in a sense, cuts across all these and is not only interacting but is contributing to all these developments. Consequently the entire question of the application of microelectronics, informatics and information, is of great importance. I am sure that this conference will indeed answer a number of questions. One question, of course, to be considered is the rate of application and the policies of developing countries to the adoption of this technology. Should the discussion be closely related to the aggravation of employment aspects in the developing countries if microelectronics are introduced into the economy? What is the trade-off between productivity benefits in the short and long-term and the unemployment effects? What are the possibilities of increasing employment in the downstream industries, such as consumer products, audio industry, simple process control and so on? These, I am sure, will receive very detailed attention by all of you.

## KEYNOTE ADDRESS: PRESENT SIGNIFICANCE AND FUTURE PROSPECTS OF INFORMATICS

**Dr. I.H. Abdel Rahman,** Special Advisor to the President of Egypt

### Introduction

Recent scientifically based major technological developments have been more concentrated in fewer countries, and within those countries, in even fewer institutions. Informatics is no exception in this respect. The concentration of development, as a result of intensive research and high risk investment, results in efforts to promote the technology in the form of new product development, servicing its users, and licensing to other manufacturers. The terms and conditions of promotion vary according to the competitive situation, market potential, and speed of obsolescence and speed of innovation. The rapid developments in computers, since they emerged commercially in the early 1950's have led to successive 'generations' of these computers with considerable improvements in cost, efficiency and capacity. With thc development of integrated circuits and microprocessors, further technological advances in the coming years may be mainly in the exploitation of those important core units and the equipment built around them. These pieces of equipment include memories, inputs, outputs, terminals, as well as the telecommunication infrastructure to assemble and distribute information between widely separated points. It has been generally recognised that developments in hardware have been accomplished relatively ahead of the software development necessary for optimum utilisation, and hence the marketed hardware may not exactly correspond to the appropriate real possible utilisation, resulting in less benefit accruing from the information content carried by the hardware facilities. The marketing techniques of offering attractive rental rates, facilities for training, software supplies and stimulated upgrading have helped to develop the informatics industry, and expand its economic size to become the fastest growing industry during the last few years and, in absolute magnitude of sales, to reach a very advanced order in the world in the next few years, if it has not already done so.

The whole world seems to need a breathing space of a few years to consolidate the rapid core technological advances already realised, and to examine more systematically the impact of, and prospects for its widespread and persuasive applications in a variety of fields and countries. The marketing drives of the sellers are not likely to relent, since they are driven under competition by the urgency to recuperate the heavy past investments in the shortest possible time, before their wares become themselves out-paced by the innovations realised through the succeeding current investments. This is a feverish situation, in which not only business corporations have been drawn, but also governments and public institutions. No one has time to wait until the dust of the storm settles, so as to see clearly. These remarks are preliminary and general in nature, and subsequent more specialised speakers at this conference will undoubtedly confirm or refute the general import of this statement.

### Hardware Industry

Next we may, however, examine for a moment whether the developing countries, and for this matter, all countries with no major hardware informatic sector, will have a chance for a share in this growing industry, which requires world wide markets. Historical precedents of diffusion of manufacturing capacities of products of other technologies may be helpful but not conclusive in this case. We would not be concerned here with the period between innovation and application, which has been extremely short in the case of information equipment, as compared, for example, to nuclear energy, radar, or the relevant branches of bio-technology. We are, rather, contemplating the possible spreading of the manufacturing capacities of the different components related to informatics and telematics in other developed and developing countries, especially those in which marketing potential exists.

For argument's sake, let us assume that concerned Governments may establish, in consultation with industry, a number of policies and guidelines, which would lead to securing for their countries a fair share of industry, compatible with their industrial capacities and the size of their prospective market. The possible advances in informatics may be applied to allow such decentralization of production without loss of quality or efficiency. The recipient countries cannot deny themselves the benefits of the new informatic applications, and should not, on the other hand, lose the opportunity of having a reasonable share of experience and competence. Only the relatively advanced developing countries are likely to have a chance to succeed in this direction. This is a second question, which may be discussed in the next few days in the conference. Historical experience with other technologies such as aviation, motor car industry, and communication equipment may be helpful, but due attention should be given to the specific characteristics of the informatics hardware industry.

## Software Industry

Hardware now represents about half the costs of an informatics project, excluding the normally available channels of telecommunication. This share is likely to be reduced in the next few years to about one third or less. There are two main reasons for these expected changes, namely an expected increase of software costs, and the further reduction of competitive hardware prices, due to more streamlined techniques of production and larger output. Actually software may be considered to be relatively lagging, and hence it may be a good area for further expansion and improvement. Here is a chance for the non-manufacturing countries, to invest in training and software development, first for their own use, and second for the use of other neighbouring countries. It was reported a few years ago that the United Kingdom was a good source of software, with a share much larger than in the case of hardware. Some of the large producers tie software partially or totally to design features in the hardware they produce. Such a tendency should be discouraged. Eventually the users in all countries may come together and express their preference for certain codes of manufacture and software development. A feedback signal from the market may be sufficient to warn against such monopolistic tendencies which are starting to appear.

### Trained Personnel

A major part of informatics is based on intellectual technology. Consequently, manpower with the necessary training is a crucial element of paramount importance, especially for those preparing and using software and their collaborators. These persons include programme designers, systems programmers, analysts, data base designers, controllers and administrators, in addition to the more basic specialisations in operations research, mathematical logic and scientific management. There is a high demand for trained people everywhere, a large personnel turnover, much head hunting, and a very short period of obsolescence. Yet the developing countries may seriously consider establishing, at a relatively low cost and large benefit, a well qualified and agile corps of trained software personnel. This may be one of the few areas in which they can advance appreciably quickly and cheaply and hence may deserve priority attention. These considerations may be examined in establishing a national policy for the procurement and licensing of informatic projects. Special training facilities will be required apart from normal education and university centres. The cost of the highly skilled personnel required for informatic projects in developing countries may reach up to 80 percent of the total costs of installation (excluding the running costs during the lifetime of the facility) as compared with the average of 25 percent mentioned before. Because of heavy handed marketing procedures, the developing countries are likely to over invest in hardware, and thus underutilise the potential capacities they acquire. The suppliers, in their turn, may oversell packages of software which are not fully tested, and hence the additional time and cost of de-bugging and streamlining such packages, over and above the normal requirements of maintenance and repair.

### Employment Effects of Informatics

Mechanical and electric equipment, from the point of view of manning skills, amplifies the power of the muscles of man. Automation did reduce the need for continuous attention and judgement by operators in industry and other activities. It is now said that the new information revolution takes over some (if not all) of the thinking requirements. If so the level of skill of operators in manufacturing industries may be reduced, with increasing dependence on the 'thinking' machine. The machine will be much faster, and more precise than the human operator it replaces. Hence a fully electronically equipped production facility, would need a different mix of floor skills, and a smaller number of operators.

Not all aspects of human judgement and control could be easily programmed and transferred to the electronic mind. Two results may consequently be noted. The first is the possibility of displacing cheap human labour by electronically controlled programmed machines. The second is the reduced need for skills on the shop floor, and the creation of the need for other skills in the management and design areas. The first trend would erode the comparative advantage of the developing countries in some of their important labour intensive industries such as textiles and leatherware. The second trend will afford a partial short cut to the developing countries in acquiring the skilled labour force for certain other industries. There is certainly some over-simplification in presenting these two trends side by side. The real choices may be more complex. Industrial and development planners seem now to be more preoccupied with the detrimental effect on the developing countries of the first trend in which the comparative advantage of abundant cheap labour is eroded or lost. It may be worth while for them to consider the possibility of the second trend, especially since it supports the previous recommendation to the developing countries to intensify their efforts in the area of selective skilled manpower as the spearhead of a more appropriate development.

### Decentralisation

If precision and reproducibility could be better assured at low cost by informatic programming and control in certain industrial manufacturing operations, then fixed overhead costs per unit product may be reduced in comparison to total costs, and thus the advantages of large scale production could be realised for smaller size units. Production could be achieved in a large number of small or medium size units as economically as in a smaller number of large units. This type of decentralisation if it were assured, may help the developing countries, where size of the market is small, to establish industries which would otherwise be uneconomic to establish. The possible higher costs of procurement and distribution of smaller quantities of inputs and output may be balanced by lower costs of transport especially if the product is designed for local consumption. These cases will not herald a new era of small and medium scale local community industries, but to a certain extent, and in specific cases, may be bringing industry to the people, in place of the huge flux of the last two centuries in which people flocked to industry.

This movement will naturally encourage the development of mini-computers and decentralised informatic systems, with possible connection to major centres. Macro-frames still have certain advantages, and may exist simultaneously with medium and mini-computers.

### Better Information for Industry

We may consider three major successive steps in transformation industry, namely the supply of raw materials, energy and other material inputs, the series of electrical, mechanical, physical and chemical transformation operations, and lastly the distribution of the output product to appropriate users. Industry can benefit from improvements in any one of these three steps. We are apt, when speaking about industrialisation to think foremost about the second step, which is accomplished within the factory by organised labour, appropriate equipment and technology and under the supervision of efficient management. In the feasibility study of industrial projects, all of the three steps are considered. Better information can however help in every one of these three steps, through the information sources and channels, which would be more accessible at cost in most cases, in an information-rich society. The developing countries may therefore be placed in a better position to know about sources of raw materials and openings in export markets for their manufactured goods. They could also choose better technologies and identify more suitable partners. These merits will at the same time be made available to the business community of the advanced countries, especially the transnational corporations with their already complex technology and machinery trade across national boundaries. New centres of technological data would emerge, from which one would find answers to specific questions about details and economics of alternative technologies required in working out feasibility studies and solving industrial and trade problems. The work of industrial and business consultancies will therefore be changed, but not disposed of altogether, by the improvement of data collection and retrieval as just described. This looks like a double-edged weapon, which works for or against the developing countries, but it can be particularly useful in fostering co-operation between developing countries, or stimulating development within the same country between different industries. One needs to judge the practicability of such situations and their possible relevance in the near future or ultimately. The UNIDO approach in the project INTIB addresses itself to this category of information and we will hear more about it at this conference. New equipment and procedures will undoubtedly displace a number of workers, but may also create new jobs somewhere else in the industry, or outside it, or lead to a greater volume of activity, and hence gain full employment in other sectors of the society. The whole issue of the employment effects of informatics will be taken up later in this conference. The organised labour movement in the advanced countries seems to be living cautiously so far with the spread of information technology.

## Rational Use of Resources

It may be noted that mass information gathering for social, economic, and administrative macro-purposes is a phenomenon less than 50 years old in most countries. It may have originated in, or been stimulated by, the needs of mobilisation of resources, and the management of shortages associated with the two world wars, and the intervening efforts of economic recovery from the severe depressions of the 1930s. Later mass collection of data about all aspects of individual, corporate and national life mushroomed practically in all countries for social and economic reasons related to the establishment of policies and plans of development, and in general for the management of the modern "nation state". Yet, almost invariably, scholars and decision makers alike feel that whenever a subject is examined, data are found wanting, and they call for new data collection. We may not be able as yet to ascertain to what extent informatics will improve this paradoxical situation. Certain aspects of the question have attracted public attention, such as the danger of breach of individual privacy and freedom. There is also the increased flow of data across borders, not equally in all directions. It has been said, for example, that basic data flow out of the developing countries, while policy and decision data flow into them. What concerns us most here is the direct information about individual consumer demand which is a major criterion (and theoretically the ultimate determining factor) in the supply of goods in general and manufactured goods in particular: how this demand changes, to what extent it is modulated by production and distribution lobbies, and how it could be ascertained in information rich and information poor societies. Basic demand can be derived from biological needs of survival, or long established needs of social coherence. But the erratic and irregular demand on appliances and gadgets including important ones such as motor cars and international travel, have led to major social and economic changes, which are likely to continue to happen in the future in magnitudes and directions which are not predictable with a sufficient degree of accuracy. In this manner, the demand for industrial goods, the availability of resources, technology, energy and purchasing power for their acquisition, and other important planning data for industry, are still being handled by rudimentary methods of extrapolation and projection. In the serious cases, when demand does not come up to supply already established or predicted, arm twisting methods are used to modulate demand and make "business" successful. To what extent would the information revolution coming (or has it actually come?) help in this basic question of the rational use of resources and the ordering of priorities of needs at national and international levels? In short can we better answer the question, "what industrialisation, and for whom?" Electronic data processing manufacturers, and multinational corporations will certainly not be the ones to determine the answer for future generations. The currently bogged-down North-South dialogue about trade, resources and energy, may eventually add the subject to its agenda, namely the worldwide management and manipulation of the king consumer, who became a slave to undetermined masters for no worthy causes.

In addition to the codes governing transfer of technology, which are being examined internationally, and about the conduct of transnational corporations, a third effort may well be required in regard to demand stimulation and manipulation of consumption.

## CONCLUSION

THESE ARE some incoherent reflections about the subject matter of this conference, in which I give expression to my ignorance and hope for further clarification in the course of the discussions that follow. These reflections relate to the new information machines, which receive data, process those data according to certain instructions given to them by their designers and operators, feed data in and out of memory and finally produce outputs, which can be simple organised data, or instructions transmitted through special devices to manipulate other machines. The programming of the new universal information machines is expanding and developing both quantitatively and qualitatively, to emulate in certain aspects the human mind, which in itself and in this context, can be described as a super information machine. The output of the processor, if appropriately programmed, could make choices between possible alternatives, and react to signals generated during its processing of external data, as gathered and supplied to it by special sensors working continuously to serve it. The input data can be received by the machine not necessarily through cards and tapes, but by more direct methods, and likewise the output.

The more obvious phenomena of human growth and daily body functioning are in themselves subject to programming and control. The data gathered by the senses, added to the data accumulated by the memory, and assisted by external data in written or spoken form feed into the appropriate parts of the body-machine and produce outward signs of movements, expressions and manifestations of intelligence, in addition to the routine and careful attendance to the internal management of the involuntary body functions necessary for survival.

The informatic aspect of the human individual is shared to a limited extent by all living animals and plants, which grow and function throughout their life cycles. We know of the existence of some chemical and biological agents of control and programming of the actions and of functioning, growth and decay of animals and plants. But not enough yet.

We may observe that the only physical connection between the parents and the offspring is the fertilised ovum (in higher creatures). This tiny living particle seems to contain all the instructions necessary for its succeeding evolution into an embryo, then into a living individual which grows into maturity with specific 'inherited' characters, related to those of the parents.

We have not yet been able to make information machines that grow and reproduce. We cannot imagine how nature could have developed the necessary information systems and programmes for cell division and organ specialisation, and to encapsulate all that within a small living unit transmitted from parents to offspring.

The working of the human mind, and the information and controlled programming of heredity, growth, and regular functioning, as compared with human achievements in informatics, however marvellous, give us every reason to be modest and reflective.

SECTION 1

# Informatics and Industrial Development—Another Mystification or a Short Cut to Industrialisation?

The broad subject of informatics is defined and discussed and its role in industrial development, with particular reference to the Third World, is considered. The critical importance for development of the 'commodity' which is information is discussed in some detail. The remarkable advances during the last two decades in microelectronics are described and it is argued that the information technology revolutionized by these advances has vitally important implications for the speeding up of socio-economic development in Third World countries.

# An Overview of Informatics and Its Impact on Development

***Ibrahim M. Sharkas***
*General Organization for Industrialization, Cairo, Egypt*

***Sayed Abdel Kader***
*General Organization for Industrialization, Cairo, Egypt*

***Salah El-Din Aidarous***
*Ain Shams University, Cairo, Egypt*

## INTRODUCTION TO INFORMATICS

FOR THOUSANDS OF YEARS, man has had limited amounts of information available and the processing of it did not pose a great problem. With the passing of time the amount of information available grew slowly but steadily. The industrial revolution in the last century and the launching of research programmes in the first half of the present century, as well as the modification of economic, social and political conditions around the world, have resulted in a tremendous increase in the amount of information to be handled.

The computer was invented in the fourth decade of this century, in the beginning purely as a calculating machine. As the technology for storing data was developed, vast amounts of information could be processed in a very systematic and rational way.

More recently, advances in telecommunications have resulted in unexpected transmission capabilities of computer data through networks. As a consequence, it clearly appears that man can and should restructure information in order to profit more from it, and to increase the speed of the social and economic development of nations.

This awareness of the value of information leads to the uncovering of specific properties and characteristics of information, to the perception of information as a physical entity, to its identification as a natural resource and as a form of energy available to mankind waiting to be exploited.

This new approach to information suggests the need for a new word which expresses this new concept. Therefore the first definition of informatics was prepared by the French Academy in 1966 and read as follows: "Informatics is the science of the systematic and effective treatment, especially by automated machines, of information seen as the medium for human knowledge and for communication in technical, economic and social contexts."

The concept of informatics, even while maintaining the validity of the original definition, has been evolving. Informatics is now considered as covering a wider area than was originally envisaged. Without disregarding the validity of the original concept, new definitions which better express the extension of the concept are being provided, such as the following: "Informatics is the discipline that studies the information phenomenon, the systems of information and the processing, transfer and utilization of information, mainly, but not necessarily, with computers and telecommunications systems as the tools, for the benefit of mankind."

## Associated Concepts of Informatics

From these concepts of informatics it is necessary to further analyze the concept of information and to comment on the role of the computer and other associated concepts such as computer science, computing sciences and information science.

Information may have, for the purpose of clearer definition of informatics, several meanings, but for any of them the relationship between information and data should be distinguished. Data can be considered as "units of information", and information as "a structured and organized set of data." The same piece of information can be, according to the level, the case and the use which is made of it, either data or information.

Information can also be considered as sets of data which can be transmitted by means of a signal and having a meaning by themselves. In other aspects, information can be considered as the amount of knowledge covered by a given number of signs stored by some means and which are able to be transferred.

All the above concepts are, as stated, meaningful to informatics.

The electronic computer as a tool of informatics is one of the most important elements for helping to solve the information-handling problems of modern society. When a country reaches a certain stage of development, a range of major and complex information-handling operations arise in government, insurance, banking, etc. Computers have an essential role in enabling these operations to be satisfactorily carried out. At present, computer hardware, due to technological progress, is decreasing in cost. The development of mini- and micro-computers is expanding potential areas of application into entirely new fields. Computer Science studies the theoretical, practical and technological aspects of the computer. Information Science deals with the handling of information in libraries, archives and documentation services.

### Related Disciplines to Informatics

The following disciplines are directly related to informatics. *Data Processing*, which attempts to find out how to process information systematically and rationally by collecting, organizing, codifying, sorting, combining and comparing data. *The Theory of Systems*, which includes the study and analysis of systems consisting of sets of objects involving relationship between these objects, and their attributes, and aimed at an end. This theory also includes the classification, simulation and design of systems. *Simulation*, which includes the construction of discrete and continuous models used in economics and in process control. *Operations Research*, which may be considered as a set of mathematical methods of analysis of organizational phenomena for optimization. *The Theory of Information Structures*, which involves the study of information, analyzing its structures and thus allowing a better understanding of such information. *Communications and Network Theory*, which studies the physical structures that make possible the transfer of information between two or more points and the properties of networks. *Organisation Theory*, which studies the organisational and hierarchical structures and communications channels of administrative and industrial bodies; it is a decisive factor in the design of information systems. These disciplines have a great influence on the development of informatics, and informatics is in turn becoming a driving force in their development.

## ECONOMIC IMPACT OF INFORMATICS

THE ECONOMIC IMPACT of informatics gives rise to a unique phenomenon because of both the wide range of its field of application and the degree to which it affects this field, and also because of the concentration of its technology in a very few centres of power.

The technology of informatics is a costly exercise both from the point of view of research, development and manufacturing and from that of acquisition of hardware, software and consultancy in systems. In particular, the total turnover of trade in the technology of informatics is becoming (as expressed in many publications) the third highest in the world. However, about 90 per cent of the whole trade is concentrated in a very few multinational companies (just one company has about two thirds of the total world market), all with headquarters located in a single highly developed country. It is interesting to note that the remaining 10 per cent is being produced either partially under licence or inspired by the technology of a highly developed country.

On a first analysis, it may be argued that the case is not unique and that the characteristics of informatics are similar to those of, for instance, aeronautics. However, informatics has a much more important impact than others because, it is affecting the daily life of men, the structure of society, the transfer of information between countries and between peoples and their governments and *vice versa*, and is sustaining the development of all other technologies (aeronautics included). It is a substantial tool for building up the structures through which social and economic development should inevitably pass.

Without informatics the whole campaign for the New Economic Order would be just an intellectual exercise which could polarize the enthusiasm of people and enhance the expectations of developing countries, but would be void in terms of feasibility because of the eventual inability of developing countries to handle the results of the new circumstances which are arising from this campaign.

The aforementioned observations lead one to draw two conclusions, one related to research, development, and manufacturing of hardware, software and systems and the other to the use of informatics for the socioeconomic development of developing countries.

## RESEARCH, DEVELOPMENT AND PRODUCTION OF INFORMATICS PRODUCTS

From the point of view of research, development and manufacturing of hardware, software and systems, developing societies (and even more advanced ones) can hardly, in the short and medium term, have a significant participation in the highly monopolised hardware industry. This apparently irreversible phenomenon might be partially counter-balanced by, for instance, developing regional efforts aimed at less sophisticated products which may fulfil the particular requirements of the region and/or by the development of mini- and micro-computers and of more or less intelligent peripherals. However, this is still doubtful unless a full comprehension of the political implications derived from this technological economic phenomenon is achieved by high level governmental decision makers. As regards software and services production, this opens a way to countries by means of which at least part of the high cost involved may be shared by a national effort. It is sufficient to point out that the cost of software and services is already of the order of 40 per cent of the overall cost. Development of indigenous capability in this sector of informatics technology may give countries the possibility of a higher participation and a decreased dependency. Software and services can be developed in all countries, but they do need the training of qualified manpower which involves the formulation of adequate educational programmes.

## INFORMATICS IN SOCIO-ECONOMIC DEVELOPMENT

FROM THE POINT OF VIEW of the use of Informatics for the socio-economic development of developing countries, it is possible to assert that informatics is both a powerful means of increasing productivity and a crucial tool in the process of development.

**Productivity:** As a main factor in the overall increase of productivity in public administration, industry, agriculture and services or tertiary sector, informatics has an essential role within the economical field. This essential role of informatics appears more evident when the tertiary sector is specifically considered because it is a sector in which it is difficult to increase productivity unless informatics is widely utilized.

**Development:** Informatics is a necessary element in the process of development because informatics is a basic tool for planning and planning is the main means of action for a country to develop coherently and rationally.

Actually, for planning, a large amount of data is required which can only be processed through informatics. Informatics also permits the assembly, for the testing of different situations and hypotheses, of simulation models required in the planning process. Finally, the control of the implementation of development programmes and projects, which is also necessary in the process of planning, is widely facilitated by the utilization of informatics.

Another aspect for which informatics is fundamental for development lies in the fact that one of the elements which characterize developing countries (even if they have a high income per capita) is the lack of managerial structures within government and the lack of qualified people at all levels. To create or improve the managerial structure, and to minimise the lack of qualified people, a wide use should be made of informatics.

## POLITICAL IMPACT OF INFORMATICS

TO MASTER INFORMATION does not mean just to know how to collect it, but to have the capacity to process it and obtain fruitful results.

As well as the monopoly of hardware and the quasi-monopoly of software, there exists in the world a new threat, namely the fact that only the richer countries and those most powerful in technical organisations (which is the least harmful situation) are building up large data banks, and developing countries are the producers of these data. Developing countries are thus appearing as the producers of raw materials, but they find themselves, in the best hypothesis, in a situation in which they have to pay to receive information, that is to say the manufactured product derived from their own raw materials.

Similarly, there is the important problem related to data banks situated beyond the border of a country, and containing information the use of which is absolutely necessary to it.

This proves that it is not only by the exploitation of their natural resources that developing countries may be despoiled, but also through the use of their information, either provided by themselves, as is the case of information provided by government offices, or directly picked up by highly developed countries, as in the case of satellites. This information is being used by countries which have the capability to process it, giving them the possibility to define political strategies and economic policies to be applied in their own interest over the regions and/or countries from which the information is collected.

Developed and developing countries should be prepared for the new situation that is being created by the conjunction of satellite technology with teleprocessing computer networks. The symbiosis of these technologies will deeply affect all existing structures. To transform this symbiosis into a positive phenomenon for developing countries, a strong political action should be taken at international level as soon as possible to protect these developing countries from actions that may be undertaken by other countries without their consent or even with their consent but without a complete awareness of their implications.

At the national level, the impact of informatics on the structure of states and governments can be investigated through systems theory, taking into consideration that countries and societies are 'living systems' whose governments act as the decisional control sub-system. Any decisional control sub-system must be able to react to the environment and its changes at a speed compatible with that of the change in the environment, since otherwise part or all of the living systems would be destroyed. In other words, in the context of the "living system society" the governments (other decisional control sub-systems), through the decision making process, must be able to react to the changes of the political community at a speed compatible with that at which those changes take place, otherwise there is the danger that political instability and social upheaval may appear, and that the existing organisation of the community might be destroyed. However, for this, it is necessary that the feedback process to the government be well designed.

The amount of information that the feedback process provides to governments is so vast at present that only through the use of informatics is the political decision maker able to cope with it and to extract from it the necessary elements to make the right decisions. It has been said that information is power and whoever has information has power.

Informatics has a leading role as an interface between the living system (community) and its control sub-system (government), and a proper use of its potential as a two-way channel to link governed with governing may lead to an effective structure for a better life.

## APPLICATIONS OF INFORMATICS

THE FIELD OF INFORMATICS concerns all levels of activity of a country, and decisions and initiatives can result from the complex interaction of a wide variety of interested bodies, e.g. governmental authority, intergovernmental organizations, the private sector, professional organizations, and the public, which can voice its opinion either through its related representatives or through individual associations. The means for setting up a national policy could thus be established within this complex system.

The various areas of application of informatics must be distinguished by their own characteristics. They can be grouped in three main areas:

(1) Scientific, medical and engineering computation and application in the education and research fields.
(2) Industrial automation.
(3) Administration and business-oriented applications.

Particular attention should be given to applications corresponding to the needs of developing countries, such as:

(1) Processing of basic statistical data.
(2) Computerization of national accounting, of tax collection and customs income.
(3) Economical modelling, aids to planning of economic and social development.
(4) Informatics applied to issues of employment and to public health services.
(5) Assistance to agrarian reform.
(6) Public welfare and social security.

The positive results obtained through the implementation of informatics are reduced by some new constraints that informatics has created in terms of data collection, of organizational problems, of centralized processing, which sometimes renders a service which is unsatisfactory from the users' point. To what extent the evolution of technology permits the progressive reduction of these constraints and brings new advantages is an important issue. As informatics requires important investments which must be financed by the end-user, the implementation of new applications, especially in developing countries, should preferably be progressive and preceded by cost-efficiency analysis; the conditions and effects of computer utilization to accelerate education should be carefully considered; the international use of data channels requires study and preliminary agreement; it is necessary to organise international collaboration on planning, standards and software exchanges.

## CONCLUSION

It would be rather naive and even dangerous, however, to assume that development of informatics is an aim in itself. Informatics is nothing but a tool, and it is useless to increase its efficiency and power without bearing in mind its basic scope which is to help countries to better solve the fundamental problems they are facing now; such as how to conduct the national economy, the optimization of the use of resources for the promotion of social and economic development—all addressed to the welfare of mankind.

Informatics is obviously not the only way to tackle such problems, but it strongly contributes to improving the approach to their solutions. As a mental revolution, it represents a factor of the adaptation of the new to the rapid changes of society and the world.

All these advantages of informatics, however, imply that it be properly used lest they lose their effectiveness. The increase of overall productivity requires that the cost of informatics and its environment be less than its profitability, even considered in the broad sense of the term. The training of manpower is expensive, and must also be profitable. The economic significance of informatics is meaningless if it is not a driving force of progress, but a waste of part of the GNP.

Informatics should make it possible to realize what is wrong in the information processing procedure of a firm or an administration, and to improve the situation. However, one should not destroy everything systematically so as to make better use of the informatics tool; the latter is a service to society and not the opposite. It is far from obvious, however, that informatics is properly used in a spontaneous manner. On the contrary, it can easily be badly used.

It is absolutely necessary and urgent for developing countries to acquire the capability of handling and processing information for their own purposes, in the same manner as the more advanced countries. In this way the interface at political level between countries may be more balanced and equitable.

# The Socio-Economic Significance of Information Technology for Developing Countries

*R. Narasimhan*
*Tata Institute of Fundamental Research, India*

## INDUSTRIAL DEVELOPMENT AND STRUCTURAL CHANGE

THE INDUSTRIAL TRANSFORMATION of the Western societies is a phenomenon that is less than 200 years old. In spite of its recent history, we still do not fully understand the sociological, psychological and economic aspects of the mechanism that gave rise to and sustained this transformation. It is hardly necessary to emphasize that a systematic understanding of this historical process is of immediate value to the developing countries (comprising some two-thirds of the world's current population) which are trying to repeat this historical process and transform their own traditional societies into modern ones. While it may be fashionable for some in the West to decry and condemn the "dehumanising" aspects of the industrial culture, there is no denying the fact that the secular and modern world view that is an integral part of the industrial culture is indispensable to the socio-economic advancement of the developing countries. Hence, it is important for us to become aware of and to try to understand the socio-psychological and economic significance of the technologies that underlie the industrial culture of the West—in particular the significance of those technologies that were instrumental in bringing about the industrial transformation of the Western world. In this paper I shall restrict myself to a consideration, in this spirit, of the socio-economic significance of information technology.

Figure 1 summarises the main stages of the industrial and technological change in the Western countries starting from the first industrial revolution in England in the late 18th century to the so-called second industrial revolution in the post-World War II years. It is a well-documented socio-economic fact by now that industrialisation and economic development have been accompanied in all developed countries by a structural shift of labour from the primary and secondary production sectors to the tertiary or service sector. Although no detailed efforts seem to have been made to give a theoretical reason for this shift, sociologists like Bell (1973) have popularised the view that post-industrial societies will be knowledge-based societies. A more systematic analysis of the service sector would show that the 'commodity' on which this sector acts is 'information'. The increasing importance of information to control and manage the socio-economic structure of 'advancing' societies is the primary reason for this structural shift of labour to the service sector.

Before the 19th century, in the West, there were traditionally only three professions: medicine, law and religion. An important change that was brought about in the socio-economic set-up through progressive industrialisation was the

**Figure 1. Stages in Industrialization**

| | |
|---|---|
| Stage I<br>1790–1840 | Start of the first industrial revolution. Primarily agricultural and small artisan groups move into coal-mining and textile centres and become urbanised. |
| Stage II<br>1840–1870 | Improvements in industrial productivity through the use of steam power and partial automation, especially in the textile industry. Beginnings of transportation revolution through the use of steam locomotives and steam ships. |
| Stage III<br>1870–1900 | The gradual replacement of steampower in industries by electric power. The increasing impact of chemical industry on textile manufacture. |
| Stage IV<br>1900–1945 | Transportation revolution based on I.C. engines. Growth of petro-chemical and pharmaceutical industries. Beginnings of electronics revolution. |
| Stage V<br>1945–1980 | Major transformations in the electronics industry. First stage of information technology revolution. Beginnings of the second industrial revolution? |

introduction of a variety of new occupational categories and the gradually increased specialisation and professionalisation of these occupations. Some of the earliest such professional occupations that came into being were accountancy and civil engineering. Figure 2 lists a cross-section of the varieties of occupations that support a modern industrialised society.

Talcott Parsons (1968) has pointed out that "professionalisation" of an occupational category is based on three core criteria:

(i) there is a requirement of formal technical training accompanied by institutionalised modes of validation of both the adequacy of the training and the competence of the trained individuals
(ii) the training acquired must include a recognisably intellectual component
(iii) the training acquired (based in an essential way on the intellectual base) should lead to skills which can be put to social use in an institutionalised setting.

These core criteria at once make it evident that "information consciousness" is an essential aspect of professionalised activity. The major part of the information- or knowledge-oriented professions are to be found in the tertiary sector of the economy as illustrated in Figure 2. The all-encompassing nature of the tertiary sector, as also its basic importance to the primary and secondary production sectors in an advanced industrial economy, are clearly brought out in Figure 3. In the early stages of industrialisation many of the service occupations are to be found "in-house" within a corporation or an industrial establishment. But as industrialisation advances, the service occupations tend to become more and more specialised and professionalised and get transformed into industries in their own right. Each such service industry, then, is itself serviced by the entire spectrum of services indicated in Figure 3. This highly interpenetrating nature of interdependence of industries makes the economies of highly industrialised societies extremely complex and difficult to analyse and study.

## INFORMATION TECHNOLOGY AND THE SERVICE SECTOR

An analysis of Figure 2 would show that technology plays two kinds of roles in an industrialised economy. On the one hand technology enlarges the spheres of industrialised activity: it introduces new processes yielding new end-products which are enjoyed as consumer goods or which re-enter the production process as producer goods. Chemical, electrical and electronics technologies are preeminent examples of this *extensive* role of technology. On the other hand, technology increases the efficiency of a productive process already deployed in the field, thus resulting in increased productivity. The role of steam technology in the early stages of the first industrial revolution, and later of electrical technology in the introduction of electric motors, are exemplary illustrations of this *intensive* role of technology.

**Figure 2. Occupational Categories**

SOCIAL OVERHEAD SERVICES

**Transport**
road, rail, air, sea, inland waterways
**Communication**
post, telegraph, telephone, telex, data transmission
**Education**
**Police, Defence, Justice**
**Welfare**
public health, hospitals, social service institutions
**Government Administration**

BUSINESS SERVICES

**Trade**
wholesale, retail, estate agency
**Finance**
banking, insurance, brokerage
**Professional & Technical**
accountancy, engineering, management, consultancy, testing & repair, maintenance, advertising, promotional

PERSONAL SERVICES

**Domestic**
**Artisan**
barbers, cleaners, tailors, plumbers, etc., gardeners, catering staff, etc. in restaurants, hotels
**Professional**
medicine, law, finance, design & decor, training & tutoring

COMMUNITY & COOPERATIVE SERVICES

**Religion**
**Professional Associations**
trade unions, cooperatives, clubs, other social groups

CULTURE, SPORTS & RECREATION

**Entertainment**
radio, TV, theatre, cinema, concerts
**Sport & Travel**
**Information**
newspapers, periodicals, books, libraries
**Culture**
museums, galleries

**Figure 3. The Service Sector Industries**

| CONSTRUCTION | MANAGEMENT | FINANCE |
|---|---|---|
| Buildings<br>Water Supply<br>Sewerage<br>Roads<br>Rail Tracks<br>Canals & Waterways<br><br>1, 3, 5 | Planning & Research<br>Administration<br>Marketing<br>  Advertising<br>  Other Promotion<br><br>Sales<br>  Wholesale<br>  Retail<br>1, 2, 4 | Banks<br>Insurance<br>Real Estate<br>Stock Exchange<br><br>2, 3, 4 |
| **TRANSPORTATION & STORAGE** | | **COMMUNICATIONS** |
| Rail, Road, Sea, Air<br>Inland Waterways Vehicle<br>  Operation<br>  Maintenance<br>Warehousing<br>1, 2, 3, 4 | Primary<br>PRODUCTION<br>Secondary | Post, Telegraph, Telex<br>Telephones, Digital Data<br>Equipment<br>  Operations<br>  Maintenance<br><br>3, 6, 7 |
| **RECREATION & TOURISM** | **SOCIAL OVERHEADS** | **COMMUNITY SERVICES** |
| Cinema, TV, Radio<br>Audio, Sports, Racing<br>Clubs, Hotels, Restaurants<br>Travel Agencies<br><br>2, 3, 4 | Government<br>National · State · Local<br>  ├ City<br>  ├ Ward<br>  └ Village<br>Law & Order<br>  Police ┬ Prevention<br>    └ Detection<br>  Justice<br>Defence<br>1, 3, 4 | Education & Basic Research<br>Health: Human, Veterinary<br>  Paramedical<br>  Testing & Diagnosis<br>  Hospital Care<br>Libraries, Museums, Galleries<br>Weather Services<br>Personal Artisan Services<br><br>3, 4, 5, 7 |

*Note*: The numerals refer to the applications listed in Figure 4.

The extensive use of technology creates new jobs by extending the scope of economic activity to cover hitherto unexplored domains. In contrast to this, the intensive use of technology almost always results in loss of jobs (if not of existing jobs, at least of potential jobs). Productivity increasing through the intensive use of technology enables the expansion of ongoing economic activity without simultaneously having to create additional jobs to cope with this expanded activity. Although in the short term this kind of intensive use of a technology creates social apprehension and consequent resistance to the technology on the part of the working population, historically, in the long term, increased productivity in an economy has always contributed to the general betterment of society. Thus, from the long-term viewpoint, for industrial development and modernisation, both the extensive and intensive uses of technology should be considered positive and desirable features of an economy.

As already noted, information technology, embodied in the form of computers and, more recently, microprocessors, is the technology underpinning the service sector. In an extremely interesting paper Thompson (1978) has pointed out that at present most of the applications of information technology in the service sector are intensive in nature. This technology is being introduced to increase productivity and consequently results in decreasing labour growth. This is perhaps most immediately evident in the office automation area. And, as already emphasised, most of the current apprehensions about the microprocessor technology in the West directly derives from its intensive usage. Thompson emphasises the need to explore the feasibility of deploying information technology in an extensive manner. The necessity for attempting this becomes especially relevant for the developing countries in the context of creating the hundreds of thousands of additional jobs that are needed to meet the demands of an ever-growing working population. We shall return to this topic a little later and discuss it in more detail.

In Figure 4 I have listed the major software application areas. Their relevance to the various service sectors are indicated by the corresponding numerals in Figure 3. From these two figures it will be seen how comprehensive is the applicability of information technology to the service sector.

**Figure 4. Some Major Software Application Areas**

1. Planning & Decision Making
2. Transaction Processing
3. Information Storage & Retrieval
4. Text/word Processing
5. Design Calculations
6. Message Management
7. Real Time Control

## THE IMPORTANCE OF THE SERVICES SECTOR TO DEVELOPING COUNTRIES

THE PRIMARY AND SECONDARY production sectors which are the principal consumers of the services (shown in Figure 3) are shown expanded into their constituent sectors in Figure 5. From this figure it will be seen that every developing country is more or less industrialised. There are very few developing countries with no industries whatsoever. In countries like India, for example, the industrial base is quite comprehensive. Each one of the primary and secondary production sectors identified in Figure 5 is a constituent part of the industrial infrastructure of India. Why is India, then, not an industrially advanced country?

The answer to this would become obvious if one looks at Figure 6. Over a 50 year period during which time this comprehensive industrial base has been built up in India, there has virtually been no structural shift of the labour force from the traditional occupational categories. The socio-economic significance of this fact is as follows: although on the surface much industrial development has taken place in India, the Indian society as a whole has not been involved in this industrialisation process. Industrialisation in India has not been the outcome of a total transformation of the socio-economic structure of the society (as it was the case in the West) but has merely been grafted on as an appendage, more or less, to a society that continues to function in its traditional mode to a large extent. One result of this has been that the many occupational roles that accompany industrialisation, which should have been created and made available to individuals, have not been created and are unavailable to increase the diversity of employment opportunities in the country.

Apart from their employment creation aspects, these missing occupational categories are precisely the ones in the service sector which contribute to the efficiency and productivity of the primary and secondary production sectors. A very large proportion of the industries shown in Figure 5, which are operational in India, function, for the most part, without many of the service inputs shown in Figure 3. The absence of the right kinds of service inputs into the primary and

**Figure 5. Primary and Secondary Production Sector Industries**

| Level 0 | Level 1 | Level 2 | Level 3 | Level 4 |
|---|---|---|---|---|
| | FISHERY | FOOD, BEVERAGES, TOBACCO | CHEMICALS, PETROLEUM & COAL PRODUCTS<br>Basic chemicals<br>Fertilisers<br>Paints, Inks, Explosives<br>Pesticides, Soaps,<br>Pharmaceuticals,<br>Cosmetics<br>Liquid Fuels,<br>Lubricants | VEHICLE MANUFACTURE<br>Road<br>Rail<br>Sea<br>Waterways<br>Air |
| | AGRICULTURE<br>Food Crops<br>Cash Crops<br>Livestock<br>meat<br>poultry<br>dairy | TEXTILES, CLOTHING FOOTWEAR | | |
| ENERGY<br><br>Wood<br>Biomass<br>Wind<br>Oil<br>Electricity<br>Thermal<br>Hydro<br>Nuclear<br>Fusion<br>Solar | | WOOD, WOOD PRODUCTS | | |
| | FORESTRY | PAPER, PAPER PRODUCTS | WOOD & METAL CRAFT WORK | ELECTRICAL GOODS<br>Components<br>Equipment |
| | | | PRINTING & PUBLISHING | |
| | MINING | IRON & STEEL, NONFERROUS METAL | FABRICATED METAL PRODUCTS<br>Metal Containers,<br>Tools, Nuts & Bolts<br>Spring & Wire products | ELECTRONIC GOODS<br>Components<br>Equipment |
| | QUARRYING | NON METALLIC MINERAL PRODUCTS<br>Glass, Glass Products,<br>Bricks, Ceramics<br>Cement, Concrete<br>Plaster, Stoneware | RUBBER & PLASTIC PRODUCTS | ELECTRONIC SYSTEMS<br>Communications<br>Computers |
| | | | PRODUCTION TECHNOLOGY GOODS<br>Machine Tools<br>—Manual<br>—NC<br>CAM<br>ROBOTICS | |

**Figure 6. India: Labour Force and Its Distribution**

| Year | Total Labour Force Millions | % Distribution across sectors | | |
|---|---|---|---|---|
| | | Agriculture | Mining & Manufacturing | Others |
| 1921 | 120 | 73.1 | 9 | 17.9 |
| 1931 | 124 | 72 | 8.7 | 19.3 |
| 1941 | 121 | 74 | 9.2 | 16.8 |
| 1951 | 140 | 72.8 | 9.3 | 17.9 |
| 1961 | 188.7 | 73 | 10.4 | 16.6 |
| 1971 | 226.9 | 73.8 | 9.8 | 16.4 |
| USA | | | | |
| 1976 | 96.9 | 4 | 23 | 73 |

secondary production activities shows up in India and other developing countries not only in the form of nonexistent occupational categories, but also in the form of low levels of information consciousness in the entire spectrum of socio-economic activities.

Whether one measures information consciousness in terms of printed matter production, circulation and consumption (dailies, periodicals, books), or communication channel availability and usage (telephones, telex), or the deployment of mass telecommunications media (radio, television), one finds that in developing countries like India the level of information consciousness is less by factors of several tens to a few hundreds compared to the industrialised countries of the West. We have already seen that information technology is of importance to upgrade the service sector and, thus, improve productivity in the primary and secondary sectors. But apart from this fact it is worth discussing in some detail why information technology is an intrinsically relevant technology to the developing countries to improve the quality of life in these countries.

## THE RELEVANCE OF INFORMATION TECHNOLOGY TO DEVELOPING COUNTRIES

Two of the major stumbling blocks to achieving rapid socio-economic development in countries like India are massive illiteracy and an extremely poorly developed communications infrastructure to support information transactions among the people at large. The implications of these two impediments, as they affect India, are forcefully brought out in Figures 7 and 8. It is seen from Figure 8 that the rural population which forms 80 per cent of the total population of India lives in economically unviable groups. Over 90 per cent of the 575,000 odd villages have each a population of less than 2000 people. Except for a few southern states, in the greater part of India the individual villages are also geographically too far apart for easy physical communication (by walking, bullock-carts, etc.).

In the absence of viable communications and information technologies to bridge the geographical distances between the communities, and to counter illiteracy, many hundreds of millions of people are kept out of participation in the ongoing cultural and political life of the country. Aside from actual illiteracy, for want of access to functional information channels, all but a minute percentage of the country's population lead a life of disguised illiteracy: although classified as literate, they are unable to play an effective role in the socio-economic and political affairs of the country.

Provision of education to eradicate illiteracy, healthcare to improve the standard of living, and community services to upgrade the quality of life, to the geographically dispersed small communities of people in countries like India pose a tremendous challenge. It is clear that the mere magnitude of the problem (the number of people to be serviced and the distances to be covered) makes it infeasible to tackle these problems through conventional means. One cannot hope, in the foreseeable future, with the resources available, to provide con-

**Figure 7. India: Manpower Profile (1978)**

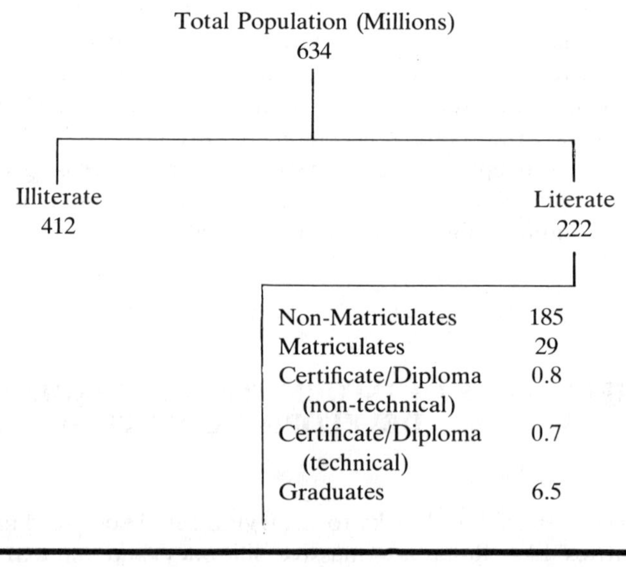

**Figure 8. India: Population Distribution**

Rural-Urban (per cent)

| | 1901 | 1921 | 1931 | 1941 | 1951 | 1961 | 1971 |
|---|---|---|---|---|---|---|---|
| Rural | 90 | 88.8 | 88 | 86.1 | 82.7 | 82 | 80.1 |
| Urban | 10 | 11.2 | 12 | 13.9 | 17.3 | 18 | 19.9 |

**Distribution of Villages according to population**

1971 Census: Total Villages: 575721

| Population | <500 | 500–999 | 1000–1999 | 2000–4999 | 5000–9999 | ⩾10,000 |
|---|---|---|---|---|---|---|
| Percentage of Villages | 55.3 | 23.1 | 14.2 | 6.3 | 0.9 | 0.2 |

1901 Census: Total Villages & Towns: 730753

| | | | | | | |
|---|---|---|---|---|---|---|
| Percentage | 79.3 | 13.2 | 5.5 | 1.7 | 0.2 | 0.1 |

ventional schools, diagnostic clinics, hospitals and so forth to service the inhabitants of each village. Apart from the problem of constructing structures to house such service centres, the generation of qualified manpower to manage them cannot be coped with. Radically innovative, unconventional solutions would have to be found and tried.

Information technology—as the technology underpinning the service sector—could be made to assist the developing countries in accomplishing this task. Preprogrammed, prepackaged service equipment which could be operated by para-professional personnel—if developed and deployed in sufficient numbers—could offer a viable solution. Microprocessor technology could be deployed to meet this need and could function as an appropriate technology to service the cause of development in the community services sector. Such a use of microprocessors would be an example of extensive use of information technology.

The use of sophisticated technology to shrink vast geographical distances, and also to bridge the wide gulf between the urban and rural life styles, is of immediate and critical importance to developing countries. Computer-based information technology has a vital role to play in solving these problems: through the extensive use of microprocessor technology as indicated above, through the creation of public utility data-transmission networks for community, governmental and industrial uses, through the extensive use of computerised printing technology to assist small-group and mass-level publications of all kinds, and through the use of computerised message management techniques to promote wider interpersonal and group-level communication.

I would conclude this paper by re-emphasising my central thesis. The service sector constitutes the critical backbone of a developing society. Information technology is the technology that underpins activities in the service sector. The development and the appropriate deployment of information technology is, thus, essential to extend the scope and improve the standards of functionally available services in a society. In this sense, information technology is an appropriate technology for socio-economic development in the third world countries.

## References

Bell, D. (1973) *The Coming of Post-Industrial Society.* Basic Books, N.Y.

Parsons, T. (1968) *Professions* in *International Encyclopaedia of the Social Sciences.* Vol. 12. MacMillan.

Thompson, G.B. (1978) *On the relation between information technology and socio-economic systems.* Phil. Trans. R. Soc. London A289, 207–212.

# Technology Perspective of Microelectronics in the Coming Decade and its Implications for Developing Countries

***Sang Joon Hahn***
*Institute of Electronics Technology,*
*Korea*

## INTRODUCTION

I QUOTE FROM Dr. Adam Osborne's statement: "In 1960, President John F. Kennedy declared that within the decade an American would walk on the moon. One did, but history may well be more impressed by unanticipated consequences of President Kennedy's declaration. Unwittingly, he triggered an industrial revolution." This revolution, as we all know, has been based on microelectronics.

The remarkable developments or breakthroughs made during the Space Age can be summarized as follows:

(1) Miniaturisation of the circuit as in integrated circuits (IC)
(2) Analogue/digital converter
(3) Sensor development
(4) Memory device, the cost of which will be decreased by VLSI development.

The above evolution has been made for qualitatively or quantitatively improving, for reducing the power consumption of, for speeding, or for reducing the costs of devices or systems. For example, in the analogue/digital area, scientists understand colour in terms of a number; i.e., wave length of light.

The economic impact of microelectronics for the advanced as well as developing countries is becoming widespread. In the advanced countries such as the United States, Japan, and Europe, there is enough capital and technology to push the state-of-art, and the social and economic structures of these countries are readily accommodating sophisticated applications. It is also anticipated that the increased affluence of microelectronics-based industry will inevitably spill over to developing countries.

The effects of microelectronic developments for the advanced countries and for the economy of the rest of the world are too serious for the policy maker to neglect. The development of microelectronics is providing an infinite range of applications for automating the process and operation of the machine and the advanced countries are substituting labour with automotive machines. Since a microelectronics system can control a process, material movement, shaping, cutting, mixing, assembly, quality inspection, testing, etc., there is theoretically no need for any labour except the start-up instruction. As a matter of fact, the relative numbers employed in the advanced countries is sharply decreasing in manufacturing and in agriculture. In the United States, for example, industrial manpower decreased from 61 per cent in 1950 to 37 per cent in 1980. In West Germany, the industrial change is moderate but the agricultural ratio is down to 5 per cent in 1980 from 22 per cent in 1950. In other words, microelectronics has not only improved modern life but has also induced a transfer of manpower from production to other areas. In order to absorb the shift in employment, the service and information sectors are in a state of growth.

In future, microelectronics will automate most domestic appliances, and make TV and telephone function as terminals of the available public services. There is hardly anything left, the accuracy and efficiency of which cannot be improved by using a semiconductor device. The impact on society, including socio-economics, is endless and future life quality will mainly depend upon microelectronics. The microprocessor can be produced and programmed for such a cheap price that it can be used everywhere. In other words, the digital signal computer and analogue signal from the real world will be processed in a microprocessor to get the desired function. The analogue signal is generally converted to digital form for data processing and the D/A converter will change back to the analogue signal. The future microprocessor will carry out A/D, data processing and D/A functions on a single chip, which means one microprocessor unit can act as a local brain to control, process or direct complex activities. The next interface required is either local to local or local to central, which can be achieved by telecommunications. The gradual merging of data processing and communications will upgrade the efficiency and life quality of future society. Microelectronics, it is feared, will adversely affect developing economies. Some economists even claim that microprocessor-based automation will widen the economic and technical gaps between advanced countries and DCs. Before making such a hasty judgement, one must look into another function of the microprocessor—it can enable an unskilled person to do a sophisticated task which otherwise could only be performed by a highly skilled person.

In spite of such optimistic anticipation of the achievement of industrial civilisation, there are, however, many people suffering from 'absolute' poverty on the globe at present. The number is claimed to reach eight hundred million people. Many people are still suffering from malnutrition, disease, lack of safe drinking water and illiteracy.

Since the end of the World War II, a strategy for industrialisation has dominated the policies of developing countries causing problems such as urbanisation and a critical shortage of infrastructure. In contrast to urbanisation, the strategy calls for rural development, labour intensive production, low scale production, and decentralisation of industries. This concept coincides with so-called intermediate, soft, alternative, or appropriate technology.

Recently research centres in developing countries have started pouring out appropriate technologies. Furthermore, the emerging technology in the field of microelectronics is expected to promote the use of local resources in place of expensive imports and to remedy the trend of urbanisation and to make the rural area more viable.

In spite of the excellent ideas and efforts, the realisation of "the third wave" in developing countries is vague and inconclusive. Therefore, the most essential task for developing countries will be the setting up of national policies for, and effective implementation of meeting the basic requirements of, infrastructure building such as the following.

## COMMUNICATIONS

In some developing countries with a low population density like Cameroon, it is difficult to install an all-cable communications system for economic and geographic reasons. But some technological trials may overcome such difficulty. For example, the Subscriber Radio System invented by Farinon in the United States is often a more economic method of extending telephone services to scattered population centres. And the system can be installed at the place where geographical obstructions preclude the installation of cable.

The advantages of radio over cable and open wire have only now become economical for normal agricultural and other rural areas, with the development of Subscriber Radio. The exchange area can be confined to a typical area where the subscribers in an agricultural region are served by a combination of Subscriber Radio and buried cable. This has considerable economic advantages over all-cable plant. Studies of actual areas similar to this example revealed that capital cost savings of approximately 20 per cent are realized over all-cable plant, together with similar savings in maintenance and depreciation costs.

Subscriber Radio (SR) is a point-to-multipoint microwave system consisting of a central station and several out-stations. It provides transmission quality equal to that of cable plant and will serve subscribers within a 40 Km radius of the central telephone exchange. The central station is located at the telephone central office and can accommodate 90 to 100 lines.

The SR system seems to be more economical through a little modification. The elimination of the cables around the outstations may give a more feasible communication system in developing countries with a low density of population. It is well-known that the cost of a communications system is a function of the number and the length of cables. Some experts in the communications field recommend that it is better to reduce the consumption of cable required for the system as much as possible.

As an initial programme in the communications area, the implementation of a pilot programme in line with the preceding scheme is recommended.

## EDUCATION

ALTHOUGH TELEVISION has been with us for about forty years, it has not been able to play a significant educational role in developing countries. However micro-electronics has been changing the role of television. Television actually substitutes for a teacher where no teacher is available. Students are given the chance of learning at home.

The development of computer-based instructional systems has moved into a demonstration and dissemination phase. The educational promise of CAI (Computer-Aided Instruction) lies in its ability to individualise and personalise the instructional processes and to simulate experiences not readily available. CAI lessons can serve as text, test and tutor, while compelling students to be active participants in their own learning. Students work at their own pace while their CAI lesson monitors their progress and commonly prevents them from continuing to more advanced instruction unless mastery is demonstrated.

The most basic equipment used with CAI includes a computer which stores and transmits educational material and information by means of a specialised computer language. The computer is less often seen by students and teachers than the familiar learning stations. The learning station appears as a television or teleprinter which displays instructions and graphic information and has a keyset attached to it.

In the near future, we will see a rich array of educational systems. National distributed computing networks will make it possible for universities to specialise in areas of interest and to co-operatively share resources and programs without concern for equipment or location. It will be possible to launch a communications satellite, totally devoted to science and education, and eliminate distance as a physical and economic barrier to the access and use of CAI programs.

The most significant trend is the positive change in the public attitude towards computers in education. The ever-widening acceptance and use of computers by scientists, engineers, and businessmen, and the broad public enthusiasm for programmable calculators and computer-based games is evidence of a transformation in social values which will have a profound impact upon education in the future. By 1990, the cost of computer-assisted instruction will be so cheap and its applications so broad that it will be viewed as an educational necessity.

This world-wide development will do much to contribute to advancing the level of CAI acceptance. Currently, CAI activity in the United States has focussed on basic skill development and has had its greatest impact on elementary and secondary students. In the future, CAI utilisation will be extended to those areas where teaching is currently difficult, hard to visualize, and almost impossible to understand with current instructional systems. This means that the major utilisation of CAI will be centred at the preschool level and the higher end of professional development.

## ENERGY

THE OIL CRISIS has caused considerable attention to be paid to primary sources of energy: solar, geothermal, wind, tides, ocean thermal, wave power, organic materials and waste. Power sources applicable to many rural areas of developing countries call for solar energy. The first half of this section leads up to the analysis of the generation of electrical power from solar energy, the last half deals with the energy savings due to the increase of energy efficiency.

### Generation of electrical power from solar energy

Solar energy can be converted into electrical energy. The photoelectric effect causes an electron to be emitted, thus generating a current when light strikes certain substances. A wide variety of schemes have been proposed, but the only one that has reached commercial development is the so-called photovoltaic effect. The heart of the photovoltaic power system is the solar cell. The most usual solar cell consists of a tiny chip of silicon. Single-crystal silicon of extraordinarily high purity is prepared and sliced into chips; one end of the chip is "doped" with a tiny amount of a trivalent impurity such as phosphorous. These different impurities set up a voltage gap across the 'junction' between the two portions. When a photon of visible or ultraviolet light strikes such a cell, it creates a pair of charge carriers (an electron and a "hole"), one of which drifts to the junction, and in the process of travelling across the junction creates electric current. Such a cell can create at most a large fraction of a volt and a small fraction of an ampere. By connecting a number of such cells in series, the voltage can be raised to a high level, and by connecting a number of groups of cells in parallel, the current can be raised to a higher level still. With the best presently available technology about 14 per cent of the incident solar energy can be converted into electricity, and there is hope that this efficiency can be doubled. Makers of cells are also struggling to bring down costs. Since nobody has yet found a simple and cheap way to get more sunshine onto a photovoltaic cell, large numbers of cells are needed for any energy-generating system. Costs will come down with volume production of single-crystal silicon cells or with new devices utilising polycrystalline silicon cells. Further reductions will be achieved by making solar cells of less highly purified silicon, by improving sawing methods and by the more efficient assembly of the modules.

### Energy conservation by higher energy efficiency

Energy conservation is vitally important to our energy future. Energy conservation is generally not limited by technology, but is limited by economic factors. And conservation here does not involve significant changes in the traditional growth of economic activities, changes in lifestyles, or major shifts away from energy-intensive activities, other than those that would result regardless of our overall economic assumptions.

(1) Energy savings in buildings can be achieved by using computers and covers such things as: energy consumption; hourly shift and daily operating information; control utilisation by shift; performance information (stack gas temperatures, steam pressures, etc.); electrical power generation, fuel usage, water usage, purchased power, etc. and air quality information.

(2) Energy conservation in transportation involves measuring improvements in the efficiency of transportation equipment; switching from efficient to more efficient modes; and switching away for overwhelming dependence on oil. Energy saving is achieved through electronic engine control and fuel management systems. Engine control systems use many integrated circuits. The main function of the system is to provide a precisely-controlled air/fuel ratio to the catalytic converter. However, a microprocessor in the system also allows control over the following functions: electronic spark timing; idling speed; choke; exhaust gas recirculation; gear-changing; and exhaust stroke.

## MICROPROCESSOR APPLICATIONS TO AGRO-INDUSTRY—FOOD PROCESSING

THE GREEN REVOLUTION is being accelerated by the emerging technologies based on microelectronics. Among promising sectors of agro-industry, food preservation and processing seem to be the most important and urgent subjects to be considered.

The microprocessor has not been intensively applied to food processing. Most microprocessors are currently used in such commercial OEM and computer-oriented products as traffic control systems, blood analysis, data entry terminals, point of sale terminals, cheque processors and automatic typesetting. One of the historical reasons for this was the high cost of microprocessors. But recently the costs of CPUs including hardware like memory, I/O devices, clocks and others have been considerably reduced.

In addition to the lower cost, the advantages of a microprocessor system over the manual or semi-automatic system make it possible for the microprocessor to be applied in food processing. The flexible programming obtainable with microprocessors may give a detailed level of control. One area of the food processing industry, yoghurt processing, depends on manual control. This manual process consists of many in-out valves, temperature, mixing and level controls.

The old manual method often has problems with quality and process controls. The microprocessor appears to warrant replacing manually operated process control systems with an automated distributed process controller in order to improve the quality and productivity. The process as originally designed is:

(1) Tanks for raw materials like milk and additives
(2) Flow control valves which control inputs and outputs
(3) Mixing with level switches and mixer on/off
(4) Temperature control with superheated steam in/out and cooling water in/out.

The process will be converted in the following stages by using a microprocessor:

(1) Automatic process input and output with programmed timing
(2) Temperature sensing and control through automatic input and output of steam and cold water
(3) Mixer control via main control system
(4) Operating software and debugging the hardware
(5) Microprocessor for the application is 8 bit and maximum memory capacity is 9 K of ROM and RAM.

Electronic monitoring systems have also been developed to provide readings on soil-moisture. More sophisticated use of computers is to forecast harvest dates. It measures daily temperature at which there is no growth for each crop.

Agriculture is also benefitting from electronics in glasshouse culture. Fruit and vegetables, among others, are being grown in greenhouses with a microcomputer controlling water supply and temperature and conserving energy. In fact, the entire concept of agriculture is changing in accordance with this new digital age.

## CONCLUSION

I WOULD LIKE TO QUOTE again from Dr. Adam Osborne's statements: "We are already five years into a new industrial revolution, the impact of which rivals the first industrial revolution. Of all the jobs in the industrial world today, perhaps half will be eliminated during the next 25 years ............ But, without adequate planning we could be heading for a time of anguish and chaos. The misuse of computers, for example, is already an urgent problem which must be addressed immediately."

In developing the potential for microelectronics applications in developing countries, it is important to consider the past performance of the national economy, its current strengths and weaknesses, and the goals and targets set forth by the planners for the future. And also the ability to regulate the inflow of microelectronics technology depends upon the level of development of the country concerned and local technological capacities in developing countries. Many of the policies towards microelectronics technology inflow must be pursued at considerable efforts for international action in order to exchange information and strengthen each developing countries' knowledge of alternatives and practices.

Microelectronics technology has been developed in the developed countries while most of the current R & D activity in developing countries is not directed to the development of microelectronics technology. International assistance may be needed to help provide resources of both finance and manpower to create institutions to promote the inflow and application of this technology.

## SECTION 2

# International Transfer of Technology

In the whole field of technology, but especially in that of informatics, the developed countries are the main repositories of knowledge and expertise. This puts developing countries at a disadvantage in that while they are anxious to gain access to informatics technology, they may lack sufficient knowledge to enable them to evaluate the best and most suitable hardware, software and systems for their particular circumstances. Their negotiating skills regarding imported technologies may also be deficient. Another acute problem affecting developing countries is that, having acquired the necessary computing facilities, they may lack the educated personnel to make best use of the technology. This Section describes the activities of national and international organisations in assisting the developing countries to cope with the problems enumerated above and to promote the transfer of informatics technology. Such transfer, however, is not a simple process. It involves a wide spectrum, e.g. Governments, the higher education and industrial sectors etc, and can have very marked social, economic and political effects.

# The Role of the United Nations Development Programme in Promoting Informatics

*Erling Dessau*
*United Nations Development Programme (UNDP)*

I WOULD LIKE TO FOCUS attention on three issues: first, development and informatics, second, the UN role in promoting informatics in support of development and third, a particular aspect of UN efforts and particularly the UNDP on setting up a development information network. The distinguished philosopher from Denmark, Kierkegaard, has said that man shines his light forwards but understands, usually, only backwards. It is a question of whether we can break this vicious circle and provide tools to try to understand forwards also. The word development, it has often been said, is a delightful concept that nobody can really oppose. The root of the word development means unfolding. It is the realisation of the potential of the human race.

## DEVELOPMENT PROCESS

THE STARTING POINT for my brief discourse is the Willy Brandt report on the Programme for Survival, in which he discussed the various ingredients of the development process, particularly the question of promoting self-reliance, transfer of know-how, upgrading and increasing educational facilities, improving training, facilitating the planning and decision-making process and finally improving the quality of life, particularly through the application of technical co-operation amongst the developing countries. The major difficult issue confronting us, as spelled out in the Brandt report, is the serious question of poverty. We know today that up to one billion people live in abject, absolute poverty, two to three billion people on this earth are still in the underdeveloped sector. There is high infant mortality, a shortage of food, malnutrition, the energy crunch, the need for new alternative energy sources, a lack of housing, rapid urbanisation, environmental and pollution problems, a lack of educational facilities and the related need for technology transfers, for appropriate technology, for industrialisation and for finding employment. Consequently, any information system must be supportive to our concern for and handling of such basic issues.

## THE UNITED NATIONS SYSTEM

THE UN SYSTEM has a very great responsibility in promoting development as understood in its broader sense. The UN system is acting as a catalyst in many ways. The UNDP which is, as you may know, the largest financial agency for technical co-operation, is, together with its many co-operating and specialised agents, such as our sponsor for this Conference—UNIDO—supporting a programme of many billions. I can just quote that for the current five-year period we are envisaging a programme of $6\frac{1}{2}$ billion US dollars for the various activities sponsored by the UNDP, and in addition we have the other UN agencies, including the International Fund for Agricultural Development, the Population Fund, UNICEF, the World Food Programme, and of course the World Bank. They are all heavily engaged in development activities either through technical co-operation or capital investment.

It is, of course, understood that out of this vast amount of effort, computers and informatics can naturally only play a limited role. However, in discharging its responsibilities for furthering the quality of life and development, for the majority of mankind, the UNDP and the specialised agencies are focussing their attention increasingly on the applications of informatics. It is realised that the role which the UN system can and must play is to educate people, transfer software, in a broader sense know-how, not just computer programmes but the whole spectrum, make data available, ensure ready and reliable information to help interpret information, create understanding and promote action and decision-making; in brief, modernise public life through informatics and improve the industrial process. And here I want to pay a special tribute to the UNIDO activities which have been already widely discussed and which will be discussed further.

## THE UNITED NATIONS DEVELOPMENT PROGRAMME

The UNDP and its agencies have been supporting the setting up of national informatics centres, promoting education and research, as I have already mentioned, also supporting various industrial projects in which computers play a strict role. Microprocessors in real time dedicated systems for management, control and planning of essential economic activities is one of the new areas in which we are involved, also computer assisted design and numerical control. The developing countries, as you may know, carry out only 5 per cent of the world's research and development activities. Their participation is often hampered by economic factors, and also because of the lack of information. We consider it crucial to support the marketing of technology, marketing both making the product available physically and learning how to apply it since even when know-how is available it cannot be easily put to use without assistance. Therefore appropriate information must be provided and appropriate information and technologies are the problem of the day.

## THE UNDP INFORMATION NETWORK

Let me continue now to specially mention the development information network, where UNDP has been preparing what they think is the response to some of the needs which I have just briefly outlined. Recognising that information is the very ignition key to the engine of multi-purpose co-operation among the developing countries in respect of trade, technical and industrial collaboration, economic development and joint research activities, it must be made available; also realising that about 90 per cent of all information flows in a North/South direction and only 10 per cent or so is South/South, which means that the developing world is to the greatest extent depending on the information coming from the industrialized and developed countries, this is the process which perhaps this development information network can break. The distortion of the 1980's image of human communication lies behind many of the problems that confront this planet and it is necessary and urgent to search for a new and more just, more universal, beneficial and more dynamic international economic system. This must be based upon the symbiotic relationships between communication and development. The development information network proposal is that a new flow of information with distinct characteristics can be created through a South/South system, based upon microcomputers and assisted by a satellite-based telecommunication network and providing a full horizontal exchange of mutually supportive development information between these developing countries.

The development information network is a project which is envisaged to last about four years and the idea is to set up something like sixty national bureaus in the developing countries. The idea is to generate a two-way flow of development information. The information flow will be of current economic and financial and trade data, technical and social development data, also in-depth analyses, statistics and facts. It should be timely and readily available. The information will cover such areas as current national and international economic trends, including financial or balance of payments situations, prices of raw materials and commodities, capital formation and investment and lending opportunities and other business opportunities. It will also provide information about science and technology related development. We hope that innovation somewhere out in the rural areas can be readily packed and brought into the centre and then transmitted to other centres in the developing world and thence sent out to the rural areas again, where all such discoveries are made.

The question I would like to propose, in summing up, and perhaps for the workshop discussion, is to ask is this kind of development information network an answer to real needs? Is such an approach at all feasible? Is it relevant and beneficial? Do we, through such an effort promote the small, the medium or the large industries? Who will monitor the information, edit it and transmit it? Our project, which is being promoted now, calls for four years of development activities for setting up the 60 national focus centres to the tune of some 30 million dollars. And of course the support has come in from many sources and we are now in the phase of in-depth soul-searching to see whether it responds, as I just mentioned, to the real needs. The question is, are we dreaming, are we naive or are we realistic? Will and can information lead us to a better life? As you may know Oscar Morgenstern once said: "Information in itself is silent: it is the use to which it is put that is important."

# Role of UNESCO in the International Transfer of Informatics Technology

*E.A. Owolabi*
*United Nations Educational, Scientific and Cultural Organization (UNESCO), Paris*

## INTRODUCTION

In the light of the growing importance of informatics and its applications in various fields, UNESCO's activities in informatics have been strengthened for the 1981/1983 triennium. This growth in the informatics programme is particularly geared towards providing necessary assistance to the developing countries to enable them to take advantage of present and future capabilities of informatics technology. To achieve this purpose, transfer and adaptation of appropriate technology in informatics is very much emphasized. the modalities by which UNESCO is contributing to the transfer and adaptation of informatics technology are described and a brief analysis of the classification of nations according to their development in informatics is presented.

## CLASSIFICATION OF NATIONS ACCORDING TO THEIR DEVELOPMENT IN INFORMATICS

The UN Economic and Social Council proposed (United Nations, 1971) that nations could be classified into four categories according to their level of development in informatics: initial, basic, operational, and advanced. To appreciate the variety of requirements for effective transfer and adaptation of informatics technology, this summary presents an analysis of the classification based on the current situation.

Countries in the ***advanced stage*** of development in informatics continue to achieve accelerated progress in the technology. Assisted by discoveries in microelectronics which led to the realisation of microprocessors, they have been able to expand the application of informatics beyond mere data processing to include activities on the production line, including robotics and process control. They have been able to advance the technology of microelectronics into very large scale integration (VLSI), which made it possible to put up to 100,000 transistors on a chip by 1980. The forecast was that, by the end of the 1980's, it would be possible to put up to 1,000,000 transistors on a chip. The effect of this miniaturization of products, including computers, is leading to availability, in large numbers, of microcomputers, which are cheaper considering their capability. In these advanced countries, home computers have been introduced into the consumer market demonstrating the extent of the impact of informatics. Furthermore, informatics technology applications include space probes, remote sensing, data base and data communication networks, data processing as well as secretarial (e.g. word processing) activities. There exists a complete range of high quality training and education in informatics which is well complemented by the activities of professional societies including national and international conferences. The national economy of these countries is boosted by informatics activities.

Countries at the ***operational level*** of development in informatics will be able to adapt the technology provided there is appropriate transfer from the advanced countries. The rate of progress in technology is very rapid and countries at this level need to accelerate their effort if they want to catch up with those at the advanced stage. At this operational stage, however, government is no longer dependent on the advice of manufacturer's salesmen because an extensive understanding of informatics technology exists in the public and private sectors. Component service can be obtained from specialists who operate local computer installations. In addition to business applications, computers are used in such fields as science, engineering and medicine where these activities involve design, development and production of software and some hardware manufacturing. Of course, the standard of informatics education in countries at this level is very high. However, informatics technology has not been having an appreciable incremental effect on the economy but its application is gaining momentum.

Countries at the ***basic level*** of development in informatics have to learn from those at the operational level as well as those at the advanced level in order to leap-frog and catch-up with developments. Transfer of technology to countries at the basic level from countries at the operational level may be appropriate in terms of the possible rate of adaptation but it is not sufficient in terms of the progressive advance in technology. Direct transfer from countries in the advanced stage of the technology to the basic coupled with those from the operational stage will be more advantageous for developmental purposes. In countries at the basic level of development in informatics, some nationals are already involved in informatics-related activities and some informatics training and education exists. Governments and private decision centres have begun to understand the role that informatics can play, and computers are used in basic government and business operations but hardware and software are mainly imported. Thus informatics activities in such countries constitute a drain on the economy.

The countries at the ***initial level*** of development in informatics are those where only a few nationals have contact with informatics. There is no computer installation operating in government and the main source of information relating to informatics are often manufacturers' salesmen. These countries need to be encouraged to learn from the first three so as to have an appropriate mix of experience for the purpose of transfer of appropriate technology. Many poor countries are in this category and one of the factors contributing to their inability to benefit from informatics technology is its cost.

## EDUCATION AND TRAINING OF PERSONNEL AS A MEANS OF TRANSFERRING TECHNOLOGY

UNESCO IS VERY CONCERNED with the education and training of personnel as a vehicle for appropriate transfer and adaptation of technology. Several recommendations of the SPIN Conference underlined the importance of education in informatics and stressed that UNESCO should strengthen its programme to enhance informatics education at different levels including post-university, university and pre-university, as well as public awareness education.

In response to this call, UNESCO, in its informatics programme for the 1981/1983 triennium, decided to increase its support for post-graduate training courses in informatics and to encourage the establishment of centres of excellence in the discipline. The following post-graduate training courses will be supported:

Computer Science, University of Lagos, Nigeria
Computer Management Studies, University of London, U.K., and
Mathematics and Informatics applied to Research, University of Bucharest, Rumania.

Other training courses which will be supported during the triennium include:

International training course in computer technology, Tokyo, Japan;
Computer applications in industry and management, University of Patras, Greece;
Informatics training course, Institut Africain d'Informatique, Libreville, Gabon;
Training course in computer science, China.

All the courses will be encouraged to admit candidates from many developing countries, and to provide practical and relevant experience to the candidates.

## RESEARCH AND APPLICATION OF INFORMATICS TO DEVELOPMENT

ALSO IN THE TRIENNIUM PROGRAMME, international joint research projects are planned in order to speed up informatics activities in developing countries. Several national research institutions will be encouraged to undertake joint research projects which should attract international support. Joint participation of national research institutions in both industrialized and developing countries will be promoted, thus creating channels for effective transfer of technology. Encouragement will be given for regular exchanges of R + D staff.

Projects involving informatics applications in support of development will be accorded priority, particularly in the countries least advanced in informatics. To assist in identifying such projects and to create a forum for exchange of experiences in applications areas, seminars and workshops have been or are being planned for different regions.

To encourage lateral transfer of informatics technology among countries in the same region, UNESCO has been promoting the establishment of regional networks of informatic centres. Initially each network will comprise existing informatic institutions which will co-operate in research and development, solution of common problems in hardware and software procurement and maintenance, exchange of personnel, and training. Activities were initiated in the 1979/1980 biennium by organising regional meetings of computer centre directors in Africa, Latin America, South and Central Asia, and South East Asia; the regional meeting for Arab States will be held in 1981.

At the request of Member States, advice has been and will continue to be provided towards the formulation and implementation of national strategies and policies in informatics. Great use will continue to be made of UNESCO's international competence and contacts to provide services of consultants and staff members to Member States.

## DEVELOPMENT OF INTERNATIONAL CO-OPERATION

TO HELP THE TRANSFER of technology especially in informatics, there should exist an avenue for international co-operation. Informatics being an intellectual activity developed from an amalgamation of disciplines and technologies it could be that its benefit should be shared by the international community in the spirit of the new international economic order. Interaction between informatics specialists from industrialized and developing countries is a rational means of exchanging ideas and experience.

UNESCO encourages cross-fertilisation of ideas by promoting and supporting activities of Non-Governmental Organizations (NGOs) in informatics. The International Federation for Information Processing (IFIP), founded under the auspices of UNESCO in 1959, is a leading NGO in informatics. At the IFIP Congress '80 held in Tokyo and Melbourne, UNESCO arranged the participation of about 30 informatics specialists from developing countries. IFIP with the support of UNESCO established the IFIP Committee on Informatics for Development (ICID) to encourage participation of informatics specialists from developing countries in IFIP activities and thus enhance international co-operation in the advanced technology. This committee organised seminars on microprocessors/microcomputers and computer networks in Ankara, Turkey; on data communications, in India; and on computers in developing nations, as part of activities in IFIP Congress '80. In collaboration with the Computer Association of Nigeria, ICID organised an international conference on applications of mini- and microcomputers in the University of Ibadan, in 1981. Informatics '81, an international symposium on Informatics for Development was organized by ICID in New Delhi, India. Other seminars and workshops supported by the same committee in 1981 and, of course, by UNESCO, included: Working Conference on Computer Applications in Food and Agricultural Engineering, in Cuba; and European Symposium on Data Communications/COMNET '81, in Budapest, Hungary.

In addition to utilizing the competence and resources of relevant international NGOs to benefit developing countries, UNESCO has co-operated in its informatics programme, with other organisations of the United Nations system as well as with intergovernmental organisations such as OECD and EEC. Collaboration with the Intergovernmental Bureau for Informatics (IBI) is well maintained.

## INFORMATION ON INFORMATICS

AT THE UNITED NATIONS CONFERENCE on Science and Technology for Development (UNCSTD) held in Vienna in 1979, the importance of access to technological information by developing countries was very much emphasised. Also at the SPIN Conference referred to earlier, dissemination of information on informatics was identified as one of the important means of transferring technological know-how. To reduce the communication gap between developed and developing countries, UNESCO has included in its informatics programme an important publishing programme.

The Organisation published a brochure called "Informatics: a vital factor in development" (UNESCO, 1980) meant for the general educated public; this describes UNESCO's activities in the field of informatics and its applications.

Also to be published is a document entitled: "Bibliography on recent literature on strategies and policies for informatics". In addition, a comparative study on strategies and policies for informatics is being prepared and a study will be undertaken to compile information on computer science information centres, educational and training establishments and courses offered, research and experimental development institutions, research opportunities and fellowships.

To have a proper framework for the dissemination of the information and to meet one of the aspirations of Member States concerning informatics expressed at the General Conference of the Organisation in Belgrade, Yugoslavia, September and October 1980, efforts are being made to set up a clearing house for information exchange. In addition to the dissemination of information on informatics generated by UNESCO, this clearing house will encourage Member States to utilize its resources for mutual information exchange on strategies and policies, research and development, and national legislation on, as well as application of, informatics.

## ACTION PROGRAMME IN MICROCOMPUTERS AND MINICOMPUTERS

In the implementation of UNESCO's action programme in micro- and minicomputers, a pilot project to establish microcomputer laboratories in developing countries is planned. This project will include purchase of equipment, transportation, installation and training of staff. Initially, the microcomputer facility in UNESCO headquarters will cater for training and evaluation purposes.

Support will be given to institutions in developing countries, especially teaching and research institutions for the establishment of minicomputer centres. In addition, advisory services will be provided on request for the establishment of minicomputer centres, e.g. assistance in definition of purpose, requirements, procurement, installation, management and staffing.

## CONCLUSION

The foregoing illustrates UNESCO's activities in the appropriate transfer and adaptation of the advanced technology of informatics to the developing countries. The programme reflects the important principle that, for effective transfer, the technology should be introduced in the context of indigenous capability. There has been increasing recognition that technology in general cannot merely be imported and applied effectively; in most instances it needs to be adapted to suit the new environment, different economic situations, governmental or community policies, and different social and cultural factors.

These principles contribute to the design of UNESCO's programme in informatics which strongly emphasises the development of indigenous capacities through education and training, workshops and seminars, development and implementation of national strategies and policies, interaction between informatics experts from developing and industrialized countries, practical activities including the application of micro- and minicomputers, and information on informatics.

UNESCO is ready to promote co-operation and assist in applying informatics to development.

## References

United Nations (1971) *The Application of Computer Technology for Development.* First Report of the Secretary-General. United Nations E.71.II.A.1, New York.

UNESCO (1980) *Informatics: A vital factor in development.* UNESCO, Paris.

# IBI's Role in the International Transfer of Informatics Technology

***L. Castelli Avolio***
*Intergovernmental Bureau for Informatics (IBI), Rome*

## INTRODUCTION

THE INFORMATICS TECHNOLOGY transfer process has specific characteristics which are additional to the general aspects of technology transfer problems. Current scientific and technical progress attributes a key role to technology, in its various fields, for the social and economic development of countries. Revolutionary technologies in the field of informatics have developed rapidly, taking a place of essential importance in the organisation, control and evolution of all aspects in the life of our society. The contribution of informatics technology, which requires no explanation, primarily covers key issues in economic development, such as public administration and management, information on economic and human resources and control of production.

The transfer of informatics technology is therefore subject to specific national policies and international co-operation in which technology is considered an instrument of independent economic development.

**National policies:** Formulation of national policies regulating technology transfer in developing countries is important for informatics technology, which is linked to the specific aim of independent development and of avoiding or reducing technological dependence in this sector.

**International actions:** International co-operation promoted by international specialised agencies is associated with the action of national policies in planning the installation of informatics equipment and services of foreign origin, and in developing negotiating capacity vis-a-vis the major producing companies in order to counteract technological dependence on these companies and to maintain adequate national technical and scientific progress.

The main objective of co-operation provided by the international agencies to member countries consists of promoting the capacity of the least developed countries to exercise their independent judgement when negotiating on technology transfer. Thus, in the field of informatics, they favour the free flow of information and knowledge with respect to the world market for informatics equipment, services and software by means of periodic publications, seminars and working groups, dealing with national policies and strategies and informatics methodologies and systems.

## IBI ACTIVITIES IN INFORMATICS TECHNOLOGY TRANSFER

IBI INTENDS to establish itself from a technical, scientific and political viewpoint as an 'institutional organisation' in the field of informatics technology transfer. Moreover, it intends to act as a promoter of regulations in this field, specifically in the sector of informatics applications. The policy of IBI as an intergovernmental body in the field of informatics mainly promoting its use in developing countries will be one of reconciling the needs of the manufacturers, managers and users.

This involves the responsibility of proposing policies necessary for the use of informatics systems. Such policies should also concern the legal and legislative sector, particularly for transmission networks, whether national or international.

**SPIN Conference and the New Informatics:** The SPIN Conference (Conference on Strategies and Policies for Informatics), held in Torremolinos (Spain) in August 1978, represented a milestone in the life of IBI, and conferred upon it responsibility for promoting the policies for the 'new' informatics, which consist of analysing and forecasting the impact of informatics on society through:

(a) the applications of microcomputers;
(b) the social, scientific and cultural impact of the compilation and management of data bases;
(c) above all, the proper distribution of information and computation capacity through the use of teleinformatics.

**Department of Co-operation:** To execute the new policies on science and technology dissemination and the use of informatics, the structure of IBI was changed. To promote the indigenous capacity of member countries, especially developing ones, the Department of Co-operation was created, and with the aim of transferring advanced technology applications to developing countries, some activity has already taken place. The activities of the Department of Co-operation in the field of the technological transfer are:

(a) assistance and co-operation missions for policy planning for member states;
(b) specialisation courses;
(c) flying seminars;
(d) pilot projects.

Pilot projects were considered the most efficient means for technology transfer. The most important ones, at present in operation, concern the following sectors:

**Judicial and Legislative Informatics:** This system is in action at the Electronic Documentation Centre of the Italian Supreme Court of Cassazione in Rome. It was set up to provide magistrates, attorneys and others engaged in the field of law, as well as, in the future, every citizen, with all information required for knowledge and application of the law. It is also, in co-operation with the Ministry of Justice, aimed at automating the services of the Supreme Court and of the other judicial offices. The system will also be used by the Argentine Government.

**Parliamentary Informatics:** This system is in action at Camera dei Deputati in Italy. Since April 1980 a number of meetings have been held, and studies and analyses carried out on the priorities for computerising the data on parliamentary activity and the possibility of adapting the information that already exists in the Argentinian Congress to be used once it has been recorded and encoded by the Chamber of Deputies' system.

To create the parliamentary data programme of the legal informatics project, a Commission in the IBI Projects Department, composed of professional staff and technicians from the Italian Chamber of Deputies, has been established.

**Arabisation of Informatics:** The project comprises writing input and output utility programmes in Arabic, making the necessary changes to the file systems for introducing these characters, making the necessary studies and changes to links with specific peripherals. The Arabisation process will be completed at the level of programme-users. This project is being set up for the Tunisian Government.

**Transcription of LEDA software:** A project to standardise data during the compilation of statistics, specially surveys and censuses to be presented in tabular form. This project is being set up for the Moroccan Statistics Authority.

**AIDS Data Base:** In relation to technology transfer, IBI felt that there should be a data base to provide an information storage and dissemination service for users in developing countries. Accordingly, the Automated Informatics Documentation System (AIDS) was created.

This system was conceived as a tool to help decision-making by national authorities responsible for informatics. They receive the information from the system and, in turn, feed information into it thereby enriching it with their own data. Thus, AIDS becomes, in a way, the common memory of its users. The AIDS Central Unit supplies its corresponding institution regularly with a complete copy of its documentation (e.g. information on informatics policies and legislation, national informatics plans, bases installed, computer industry, computer applications in the public sector, etc).

The functioning of the AIDS network is based on a convention which defines the framework within which practical modalities are adhered to.

The official linking to the EURONET—DIANE network opens up further development prospects for AIDS. IBI is examining the possibility of offering its installation as host, thus allowing on-line access to the AIDS base by users of the network. Such a feasibility study involves the examination of prospects for future connections with countries outside Europe—as already foreseen in the EURONET project for countries belonging to the Lomé Convention—and the possible extension of links to other areas with special attention to developing countries, possibly using existing transmission networks.

## IBI'S PROGRAMME FOR THE BIENNIUM 1981–1982

**Decentralisation and Regional Centres:** In regard to informatics, the fundamental feature of the 1981–1982 biennium is one of creating structures to enable the Organisation to acquire the potential for expanding activities appropriate to its constitution, without necessarily incurring an increase in administrative costs and personnel.

To this end regional centres will be set up, an important activity of each centre being that of offering logistic and administrative support to IBI activity in the region. This decentralisation will allow more efficient functioning of the Organisation, providing relevant technical assistance or training and adequate for the conditions of the region and its social economic environment.

**Informatics Development Fund:** The favourable conclusions of a study in 1979 led to a proposal for creating an *Informatics Development Fund.* This fund will provide loans and guarantees for obtaining international financing for equipment procurement, systems and software development, and for carrying out feasibility studies. As in the past, IBI will continue to advise, on each project while maintaining control of its execution.

**IBI's Institute for Informatics Development:** An IBI Institute for Informatics Development will be created in Italy, accommodating 30–40 people, offices, classrooms and research areas. It will offer, to senior officials and experts of Member Countries, high level, short-term, courses and seminars related to informatics and policy-making.

**Pilot Projects for biennium 1981–1982:** The valuable experience already gained by IBI with current pilot projects has led to the establishment of a general plan for pilot projects, in their constitutional, organisational and operational aspects.

Proposals for new projects for the next biennium will be agreed between countries concerned and IBI according to the interest of the countries and up to approved amounts. Among projects which may be examined and agreed upon, the following are at an advanced stage of definition:

**Pilot Teleprocessing System:** The object of this is to establish a pilot teleprocessing system allowing commencement of a training and experimental programme for technical personnel on a data transmission network based on the package switching technique, and it may be the basis for establishing a Switched Public Data Transmission Network.

**The Automated Preparation of Postal Despatches:** Handling bulky objects such as parcel post, printed matter and nonstandardised objects is an onerous task because of dimensions, weight, and quantity, making manual operation necessary for distribution and despatch.

**Cadastral Information System:** This system is a network of interconnected data banks, connection with and consultation of which can raise technical problems not yet resolved. Solutions could be facilitated by creating standardisation criteria.

**Classification of Territory through Image Processing:** This project is based on microcomputer with all the necessary peripherals for the operations of display processing and restitution of the image (video, plotter etc.) connected to a host for cartographic data base enquiry and other applications.

**Joint Action Programmes:** These are seen as lasting five years for the informatisation of key sectors of the national administration of the beneficiary country. Such programmes are carried out thanks to a threefold participation: a beneficiary Member Country (from among the least favoured countries), a supplying Member Country (from among the developed or developing countries), and IBI. Assistance will cover training, equipment and systems. Naturally, mechanisms for the control and follow up of the projects through their execution will be supplied. Financing of these Programmes will fall under the charges of the three parties.

# Role of UNIDO in the International Transfer of Technology

***G.S. Gouri***
*United Nations Industrial Development Organization (UNIDO), Vienna*

## INTRODUCTION

WE ARE IN THE MIDST of an information explosion. Challenged by this situation and at the same time faced with the increasing demand for acquiring that information in a usable way, substantial developments have taken place in hardware, primarily through computer facilities and the marriage of computers and telecommunications. This has further resulted in the development of systems and approaches to compile, process and retrieve information as easily and as inexpensively as possible. The current estimate of available data banks is well over 200, the majority in the United States.

Existing informatics systems facilitate access to a vast amount of information, but what is available is essentially information about information; i.e. addresses, abstracts, references, etc. and not the information itself. There is an assumption that the user will have facilities to obtain total information from printed material, books, articles, etc. The end user would very often like to refer to the totality of information of his special interest so as to contribute to the desired end result of his search. Therefore, the problem of availability of detailed information appears to be an important bottleneck. The problem, however, may be much simpler in regard to information retrieval from a book or data bank since it could be easily obtained from numerous data banks which are developing, assuming, of course, the end user makes very clear requests about the subject he is researching.

There is also the problem of imperfection of the user-system interface. From the viewpoint of the user, the data base is the work of document scanners, analysts, abstractors and indexers but it is possible that items may be missed or that the level of understanding of their content may not have been complete enough. In any case, the user will have to interrogate the data base using a retrieval language which is not quite as flexible as natural language. Some of these problems have demonstrated the need for intermediaries to act for the end user on the terminal; the intermediary having a professional knowledge of the structure of the relevant data base and information retrieval systems. The number of such intermediaries or the capacity of such intermediaries is also subject to question since the end users are not homogeneous and serious problems are created in fulfilling their requirements.

The main problem, however, is whether data exists which is needed by the end user, particularly in the field of industrial development in the developing countries. Recent UNIDO experience has shown that the usefulness of available data through the existing system is not directly relevant for decision-making purposes. The bulk of information that has been developed seems better suited to researchers, scientific personnel and R and D personnel than to Government officials who have to make decisions, or entrepreneurs who have to negotiate the contracts.

## INFORMATION REQUIRED FOR SELECTION AND USE OF TECHNOLOGIES

THE END USERS of the information system include policy and decision-makers in Governments, financial institutions, manufacturers and potential entrepreneurs, Chambers of Commerce, technical institutions as well as research and development organizations, etc. A chain of these end users interact at different stages to facilitate the transfer and absorption of technology, with implications for choice of technology and the negotiation of terms and conditions for the acquisition of such technology. This could also be expressed in another way; the various stages of a project development involve feasibility studies, detailed project reports, negotiation of contracts, construction, and operation of a factory. Information is required at different stages of the project cycle. In all these matters, the basic decision will revolve around a choice and use of technology. In order to make such a choice, it is necessary to have detailed information on the various technological options available, their proven commercial nature, special advantages to suit specific situations, prices of equipment, and follow-up matters over the life of that technology in production.

Information on all these aspects is very scattered and not readily available unless a special effort is made to gather, access and provide certain options with a specific request or a project in mind. Even the collection of information presents problems as technology comes in various forms, being often embodied in equipment but also in vital but costly technical documentation. Consequently, information considered critical for decision-making purposes would have to be specially compiled, reviewed and made available as information packages. No doubt, the existing science and technology information (STI) system provides some information but it is only the beginning of the type of exercise mentioned.

Since it is a specialised area of work, UNIDO has tried to provide critical information on the subject areas mentioned above through the Industrial and Technological Information Bank (INTIB). This institution was established as a part of UNIDO, first on a pilot basis and, as of 1980, as a constituent body of UNIDO. Its main function is to provide information to developing countries which permits the proper selection of technologies. Its other functions include:

(a) broadening the base of knowledge on available technologies and identifying alternative technologies, both quantitatively and qualitatively.

(b) providing details of alternative technologies in the various industrial sectors and the criteria for selecting from among them. It also provides guidelines for choosing technologies that would help developing countries in assessing the full implication of their choices;

(c) providing information and advice on the unpackaging of technologies. The ability of developing countries to separate the cost of know-how from that of hardware, engineering and other services depends on the capability of assessing each component separately;

(d) supplying analyses of the results of operating different technologies in various economic and social environments.

## INDUSTRIAL AND TECHNOLOGICAL INFORMATION BANK (INTIB)

THE WORK OF INTIB comprises three components which represent three distinct stages in the flow of technological information covering the 'up stream', 'downstream' and dissemination aspects:

(i) generation of technological information, including the establishment of a networking system for INTIB;

(ii) dissemination of information by INTIB; and

(iii) identification of, and establishment of, linkages with and among INTIB clients.

Since user needs finally determine the nature and content of information flow, it is appropriate to start with item (iii) above.

### Linkages with and among INTIB Clients

The objective is to develop linkages and communications with end-users of INTIB in developing countries, in particular through regional, sub-regional and national institutions so as to disseminate relevant information based on needs and to promote greater availability and utilization of such information in decision-making processes, namely planning and financing agencies; investment authorities; industrial development agencies; technology registries and regulation agencies; industrial and technological information centres; R and D centres; and enterprises.

The task of promoting INTB services to target end-users and interacting with them is done by systematically informing end-user institutions on the services available from INTIB and eliciting their information needs. In addition, INTIB services are discussed in important international meetings.

Given the importance of technology selection, it is useful to develop groups of developing country end-user institutions which can articulate their current information needs (as derived from the national development programmes and also exchange information and contribute to strengthening each other. Periodical meetings of heads of technology transfer registries led to the establishment of the Technological Information Exchange System (TIES) which has enhanced the field of technology acquisition by up-grading the type and quality of information and imparting dynamism to co-operation and information exchange. Likewise, end-user institutions such as development financing agencies, investment authorities, R and D centres, etc. could serve a useful purpose, *inter alia*, providing UNIDO with current information on sectors and areas in which they are likely to have substantial information needs. It is proposed to organize meetings of end-user groups in selected INTIB sectors and thus bring the services of INTIB to as close a contact with actual decision-making as possible.

**Networking with Sources of Information**

The objective of this "upstream" activity is to generate technological information relating to the pre-investment choice of technology in the 20 industrial sectors selected for operation of INTIB; and to establish and develop the INTIB network of suppliers of technological information and improve its flow. This feature of INTIB distinguishes it from 'data bank' or other storage-oriented activities and is of particular importance, in view of both limited resources and expanding sources of information.

The activity involves identifying top-level experts who have necessary information 'on their finger tips' and can advise INTIB and decision-makers in developing countries on technology selection and evaluation at the pre-investment stage. Agreements are being made with 50 to 70 international experts in various INTIB sectors, as the first phase in creating a data base of experts capable of answering specific inquiries on recent and forthcoming activities and developments in the technologies in question. Another activity has been to broaden the base of institutional correspondents to cover more INTIB sectors, ensuring good geographical distribution. Emphasis is being given to identifying correspondent 'centres of excellence' in the developing countries. The Directory of Information Systems and Services in Developing Countries would also provide a basis for sources of information from developing countries. In addition to two volumes of information on *Technologies from developing countries* already published,* a third volume is being produced.

**Dissemination of Information**

In regard to information dissemination by INTIB, inquiries relating to technology selection will be answered through the Industrial Inquiry Service but using the INTIB network (including UNIDO technical staff and providing processed information on technology selection). In addition, information dissemination by INTIB will include the preparation and dissemination of technological information profiles on the selected sectors allowing comparison of the principal industrial and technical data. Such profiles are being prepared in areas such as solar energy equipment, alcohol fuels, and pumps for agriculture. The first two respond to the emphasis on dissemination of information on energy-related technologies made by the New Delhi Declaration and Plan of Action. The pump study is a follow-up to a recommendation from a consultation meeting on agricultural machinery.

In addition, as substantial user demands are identified, information packages relating to technology choice in the relevant areas of demand are prepared using material available at UNIDO headquarters.

INTIB has sought to promote and maintain contacts with other information systems and services, in keeping with the role envisaged for it in the global network of technological information. Other UN information systems are being utilized to meet specific requirements for information. The work of preparing technical memoranda in co-operation with ILO is continuing. In co-operation with WIPO, users' guides to patent information in the iron and steel, fertilisers, agro-industries and agricultural machinery sectors have been prepared.

Through its networking activities already described, INTIB is geared to fulfil its expected role in a global industrial and technological information network. Its distinguishing feature when compared to other systems is that it operates as a service, and not as a bibliographic, documentary or referral system. Its mission is not to provide documentary or 'raw' information *per se*, but processed information of immediate relevance to technology selection. Such practical services have been welcomed by users in developing countries and INTIB will assume increasingly specialized functions to have the greatest possible contact with and relevance to practical decision-making.

* Development and Transfer of Technology, Series 7 and 7(II).

## STRENGTHENING THE CAPACITY OF DEVELOPING COUNTRIES TO NEGOTIATE THE ACQUISITION OF TECHNOLOGIES

In general information on technology contracting is hard to come by. Since this has usually been considered a matter for negotiation between two parties, information available to each party depends on their ability to marshal and use it effectively for negotiations. Unless better information systems pertaining to terms and conditions of acquisition of technology exist, the negotiating capacities of the developing countries will remain one-sided and weak. Information, in this connection, is also needed on legislative and administrative matters, including policies, norms and procedures followed by participating countries. To permit better choice of technology and to obtain better conditions, information is also needed on indigenous technologies, services and facilities. More specifically, basic information is needed on prices of know-how, engineering, technical services, etc.; royalty rates; methods of calculating running and fixed payments; prices and terms of delivery of raw materials, components, and intermediate products; scope of sales and manufacturing rights; duration of agreements; and analysis of information based on sectoral investigation as well as that received by monitoring selected transactions.

During the last decade, some developing countries have introduced laws and regulations to ensure a proper selection of technology according to their requirements, and more particularly, to obtain better terms and conditions. As a result, many countries have established national technology registries or similar offices which offer a rich data base. On the recommendation of Ministers of Industry who met in New Delhi in January 1977, UNIDO has established the Technological Information Exchange System (TIES), to facilitate information sharing and exchange on a reciprocal and confidential basis. This system has provided the participating developing countries with an opportunity of comparing respective experiences and to share such experience for negotiation of contracts. Participating developing countries have formed an informal senior level group which meets annually to review the entire area of technology information flow and other areas of mutual interest in regard to acquisition of technology.

The TIES system concentrates on five main areas and operates through UNIDO.

(1) Periodic reviews by participating countries of trends and features of foreign technology inflow and regulatory control of such inflow.
(2) Exchange of general or average data at industry level.
(3) Exchange of detailed data on individual contracts.
(4) Bilateral co-operation among Technology Transfer Registries of participating countries through visits, study tours and information on contracts being negotiated.
(5) Exchange of experiences by heads of Technology Transfer Registries.

With the co-operation of participating countries, UNIDO has developed the

TIES system to the point where it is a fully operational computer system with data on some 4,000 contracts. The system is operated under the auspices of INTIB.

According to their degree of participation countries provide data on:

(1) Summary of terms and conditions of contracts by industrial sector.

(2) Data on a contract by contract basis.

UNIDO processes the data and produces a series of computer reports for each participating country providing initial data. Information collected is commercial information relating to the acquisition and use of foreign technology and not confidential proprietary information.

Three follow-up meetings of TIES participants have discussed matters relating to development of the system and to explore areas of further co-operation. Countries participating in this system include: Peru, Ecuador, Mexico, Malaysia, Iraq, India, Republic of Korea, Argentina, Philippines, Spain, Algeria, Venezuela, Portugal, Egypt, Colombia, China, Guatamala and Nigeria.

To facilitate greater linkages between the participating agencies and to provide continuous information, UNIDO publishes a newsletter entitled TIES every two months.

The last meeting of the heads of Technology Transfer Registries held in Buenos Aires recommended the growth of the TIES system in terms of the generation of information and further analysis of data by industrial sectors. .The meeting recommended the following:

(a) Sectoral studies, with particular emphasis on identifying technology gaps in developing countries and harmful restrictive practices followed by the sellers of technology.

(b) Documentation on emerging new areas of technology transfer, e.g., microelectronics, bioengineering, non-conventional energy sources; and possible restrictive practices against developing countries in these areas;

(c) Interpretation of government policies on transfer of technology by recipients as well as transferers, to enable Governments to mould their policies in desired directions and to frame guidelines for e.g., purchase of technology, promoting indigenous technology, and commercialisation and export of local technologies. Inter-country comparative studies would be particularly useful;

(d) Strengthening the negotiating capabilities of countries through wider exchange of information, improved information regarding transnational corporations and other sellers of technology, extension of the monitoring role of registries, wide use of UNIDO Technology Advisory Services (TAS), and training through workshops, bilateral country visits by experts and other similar means.

There should be a constant interaction between those who contribute to information and those who use it. Only through this interaction can information content and quality be increasingly changed to suit the requirements of developing countries. The information function must be a dynamic one. In the 'upstream', information sources and access to them are expanding and new technological options are emerging in several fields, including energy. In the 'downstream', user needs vary not only with countries but with changing industrial structures and new industrial development plans and programmes. The vitality of an information system and the conditions for its success depends upon its receptivity and resilience to changing signals of availability, as well as requirements of technological information.

# EDP Training—A Developing Need

*B.N. Platts*
*UK National Computing Centre (NCC)*

## EDP TRAINING IN THE DEVELOPED COUNTRIES

The last five years have seen an unprecedented growth in investment in computers throughout the developing world. Precise figures are difficult to obtain but it is likely that this growth has been around 30 per cent per year, and one of the spin-offs from the large capital input has meant that the world-wide shortage of computer personnel has grown proportionately.

The UK National Economic Development Office reported in June of last year that the United Kingdom is suffering from a shortage of 25,000 skilled computer staff. The greatest need is for programmers, where 500 new staff are required every month until 1986 to cope with the demand. The problem is exacerbated by a 20 per cent per year turnover of staff, brought about by 'inappropriate manpower planning and management policies'.

In the West, companies are creating shortages for themselves by poor selection techniques based on misleading ideas about educational qualifications. Seen against the backcloth of growing unemployment, there has been a serious failure of the education sector to come to terms with the computer industry's needs.

It is a little early to be drawing tight comparisons between the UK and the developing countries—who have their own special problems—but the writing is on the wall. Criticisms of 'inappropriate manpower planning and management policies' could apply equally.

A successful computer department will only be possible if sufficient thought is given to its place in the organisational structure. Currently, computers are attracting some of the ablest minds in the developing nations but this situation will not remain if long-term career patterns cannot be established. The positioning of analysts, programmers and operations staff, within the authoritative hierarchy, will have a bearing on the long-term development of systems. Equally there must be clearly established roads into the department from non-related areas for staff with low levels of expertise if future staff shortages are to be avoided. The failure to provide effective and well regulated training development programmes is having serious consequences.

The problem is born, perhaps in part, of the colonial tradition of reward for long and patient service. Too often, training is awarded as a perquisite and this is especially true where a trip overseas is involved, as it often is. Much needs to be done to restore the central vital role of training in corporate development plans.

## POSITION IN THE DEVELOPING COUNTRIES

In the developing countries, the need for diverse technical skills is comparatively new and professional training in computer skills has not generally reached a high local standard. In many areas, people have not been in the industry long enough to recognise the type of training required, nor even for the industry as a whole to recognise its importance. Decisions about training have often been left with the hardware manufacturer, with the consequent emphasis on operational skills. Management skills, particularly in systems and methods, have often been neglected.

'Unbundling' has gone some way to alleviating this problem. Users are beginning to recognise that training is not free, and so they are becoming more concerned with the quality and relevance of the courses. Having visible costs has also meant that budgetry provision has had to be made and controls to monitor performance are appearing. The signs are that training may be moving away from being a personal responsibility—and this is encouraging.

There is a need for one person in every computer department to be responsible for personnel development, for the planning and co-ordination of training programmes. If the installation is not large enough to justify a full-time appointment then the function should be given to a suitable manager. The job of redefining attitudes to professional training can then begin.

The goal will be to provide skilled local manpower for the rapidly expanding number of computer installations in the developing world. For all practical purposes the training for this skill must be provided locally. This means that we must get away from the old ideas which treat training as a reward. It has a role in improving morale, but this should not be to the detriment of its primary purpose which is the supply of technical skills where and when they are needed. The ending of the era of apparently free training courses overseas at the manufacturers' expense will go some way towards ensuring that technical skills are acquired in response to need rather than seniority.

### Costs of EDP Training

The acquisition of training budgets will pose further thorny questions for EDP management. Long-term academic courses leading to certificates will be attractive to prospective students, but cannot fit the needs of the installation as well as would shorter, specialised courses. Taken in sequence over a period, a series of short courses will provide training in specific skills which can be reinforced by immediate application in the home environment. Undoubtedly, the installation will see quicker results following this policy but the cost of air-fares to Europe or to America will weigh disproportionately heavy for a short course and might not be seen to be cost-effective. The long-term, cost-effective solution will be to provide specialised training locally. With this goal in mind the UK National Computing Centre has developed a training strategy for developing nations.

## THE UK NATIONAL COMPUTING CENTRE

THE NCC WAS ESTABLISHED in 1966, with the primary objective of promoting an increased and more effective use of computers in every field of activity. It is a membership organisation, governed by a board which is representative of over 2,000 member organisations. It serves as a bridge between different parts of the computing community; between government and industry; between the public and private sectors; between the supplier of computing facilities and the user. It is a non-profit distributing organisation with revenue derived from members' subscriptions world-wide, government contracts for projects and sales of products and services.

In the seventies NCC extended its international involvement. Through a network of 25 international agencies its products and services are in widescale use in over 60 countries. The Centre has gained international influence in many spheres, participating in co-operative projects and discussions worldwide.

From its inception the NCC recognised the central role of education and training. Over the last fifteen years it has developed a comprehensive range of material for educating computer users, from school leavers to top level management, and for training the wide range of computer professionals in a broad spectrum of skills and disciplines. The Centre's expertise is recognised worldwide.

A good example of the way NCC tackled a large-scale training problem may be observed in the NCC "Threshold Scheme". In the mid 1970s the British Government asked the Centre to develop a scheme for reducing youth unemployment and increasing the supply of trained computer staff. The "Threshold Scheme" is the result.

The Scheme is funded by the Government, designed and administered by NCC and is implemented by Colleges of Further Education in collaboration with private sector employers.

Training lasts 42 weeks, of which 24 are spent, in two separate periods, in the data processing departments of host employers. In addition to providing valuable practical experience, these two periods give the young person and the employer an opportunity to evaluate each other. While there is no commitment on either side, many job offers do result from this exposure. Some 60 per cent of those who enter the Scheme find employment as computer programmers or computer operators, with a further 15 per cent accepting offers of clerical positions, some computer-related and some not.

## Qualifications Arising from the Threshold Scheme

Two qualifications are available to those who perform to an acceptable standard during the course and in the formal examinations: the NCC Threshold Certificate and the National Certificate in Computer Studies of the Business and Technician Education Councils. While all trainees are encouraged to try for these qualifications, there is nothing to stop someone leaving the course at any time, whether for employment or through a decision against working in data processing. The overall wastage rate from dissatisfaction on the part of either the trainee or the college is fairly constant at 10 per cent.

The unique feature of the Scheme, apart from its "double sandwich" format, is that entry does not depend on performance in school examinations. The selection procedure consists of a half-day of pencil and paper tests and an interview. The tests, monitored by continuing research, are all administered and assessed by NCC; interviewing is shared between NCC and the staff of the college. Final selection based on a combination of test and structured interview results, is done by NCC.

From 180 places in seven colleges in 1976/7, the Scheme has grown to 1400 places in 60 colleges spread throughout the UK in 1980/81.

Beneficial results, apart from the direct and intended ones—in particular the rescue of some hundreds of young people who, for one reason or another, have grossly under-achieved at school—include enabling many colleges to develop their computing facilities and establish themselves with local employers as a training resource and a source of competent and reliable employees.

## Recent Developments

Another recent development is the inception, in some colleges, of additional full-time courses to take the overflow of Threshold applicants (over and above the available Threshold places) whose test results qualify them to enter a National

Certificate course though their school results do not so qualify them—a quite unplanned multiplier effect.

The current cost of the Scheme is approximately £2,200 (Sterling) per entrant, and this includes all tuition administration costs and subsistence payments made to the students by the Government.

This enormously successful scheme is continuing to grow in the UK and is having a significant impact on staff shortages. The bonus for the developing world is that while the problems it faces are not identical to Britain's there are common needs and solutions can be shared. The lessons learned and the material developed in the Threshold Scheme are on offer for quick adaption by any nation represented here today.

## INTERNATIONAL ASPECTS OF NCC TRAINING ACTIVITIES

Looking to the 1980s and particularly to the needs of developing nations, the International Division of NCC has geared itself up to provide long-term cost-effective training on a large scale. Many of the courses which the Centre has been running in the UK have been 'packaged' and are available for purchase complete. There are essentially three main categories of training materials.

### Training Materials Available

The first is lecture support material. The conventional lecture situation is preserved but the lecturer is not entirely dependent upon his own experience—lecture guides and notes, visual aids, case study materials and student exercises and notes have been researched and developed by experts in various fields over a lengthy period. The result should be a top quality performance by a lecturer very heavily supported by his packaged material.

The second category involves still more support with structured lecture material being reinforced or replaced by video-taped programmes. These programmes have won international television awards and the packages will take care of those parts of the course which are the most difficult to teach. This means that the course leader need not be skilled in teaching procedures. Any experienced data processing professional should be able to run a small course successfully.

Finally, the third category involves a higher degree of support again. This time the lecturer is replaced altogether as the student picks his way through the self-instruction package. This is a development of the idea of the programmed text but incorporating a variety of media. The student will be provided with workbooks, audio or video instruction tapes, exercises and self-assessment tests.

So long as he has access to an experienced colleague, his instruction can proceed privately, and at a pace which suits his circumstances.

### Suitability of Various Training Materials

One can draw no hard and fast rules about which format is best suited to which application. Much is dependent upon the skills of the tutor and the level of knowledge and personal motivation of the students. For these reasons the packaged material needs to be flexible to allow a variety of approaches. Systems analysis training, for example, is a large and complex training task. In this case the National Computing Centre has produced 50 volumes of lecture support material (the basis of 180 hours of lectures) with visual aids and extensive practical case study work, and an optional 20 hours of video support material in the more technically difficult areas. This degree of support has been found by long experience to provide highly effective systems training courses for users ranging widely in their levels of teaching resources.

## ACHIEVING NATIONAL EDP TRAINING SELF-SUFFICIENCY

It may not be appropriate for a user or educational establishment simply to buy the material and begin running training courses. If this is thought to be the case then NCC has a structured pathway to the goal of self-sufficiency.

First, one or two key members of staff come to the UK to participate in a public course or series of courses in the conventional way. These people should then be well placed to assess the local training needs and any particular problems which might be encountered. Second, having decided upon the most appropriate programme they will set up an initial training course on their own territory for which NCC will provide the consultant tutors. The students on this phase will eventually be responsible for the running of the local courses. At this stage the course can be tailored to fit closely the needs of the particular installation.

The third and continuing phase of the training project is the acquisition of materials to form the basis of the local courses. Here again, NCC will act as consultants in the choice of the most appropriate material to suit the particular case. This three-phase strategy has been employed with success in a number of territories in the developing world. Clearly, it is ideally suited to any continuing training requirement, as has been proven in Cuba, Kuwait, Iraq and Iran.

## CONCLUSION

NCC IS NOT UNIQUE in offering training materials to developing nations, but it is one of the few large, hardware-independent, training authorities and it is well placed to develop and conduct long-term, co-ordinated training strategies.

Commentators agree that local skills training is thc kcy to success. Unless it is provided there will be a severe, perhaps tragic, waste of resources which must inevitably lead to a slowing of development. However, in the end it is up to the nation concerned, to face the challenge of providing sufficient well-trained personnel for its computer industry, and to take advantage of the help which is available. Here is a field where the record of the 'developed' economies is not too impressive. Developing nations should act now to avoid repeating mistakes.

# Case Studies: The Role of the Consultant

***H.J. Helms***
*Commission of the European Communities Joint Research Centre, Ispra*

## INTRODUCTION

IN MANY SITUATIONS where informatics is to be introduced to a new environment or major changes are to take place, the decision-maker calls on the services of a consultant. This occurs in administrative, commercial, industrial or scientific environments and consultants are used both in cases where informatics is well established as well as in cases where it is new. In principle, the role of the consultant for a developing country is not different from the role of the consultant elsewhere in the world. In a developing country the role of the consultant may, however, be particularly critical and the advice and proposals suggested may have a stronger impact than elsewhere. The margin between failure and success may be narrower than in other situations, and advice based on use of technologies and vested experiences from elsewhere may turn out to be irrelevant and lead to consequences far from those originally envisaged. These perspectives call for a more detailed analysis of the role of the consultant prior to any plans for introduction of informatics.

## THE CONSULTANT

IN PRINCIPLE the consultant is supposed, as a first step, to appreciate and analyse a particular problem to be solved, assess the boundary conditions, and synthesise the findings. The next step is to draw on knowledge of technologies (in a wide sense) available, compare with known experiences from similar problem situations and project this into the problem under discussion. Based on this, the consultant will suggest a solution, or a limited number of alternative solutions, present the advantages and disadvantages of the proposals, including likely benefits, resource demands, impact on the organisation and short term and long term operational constraints of the solutions.

The consultant is not the decision-maker and should not interchange roles with those who ultimately will have to take responsibility for the solution adopted. He should however provide for the decision-makers a clear strategy which may bc applied towards arriving at a final decision and indicate pertinent criteria for a selection from among the options available.

The success of the consultant is measured against the success of the ultimate solution decided upon. This in turn, as far as the consultant is concerned, will call for the following qualities: a neutral attitude towards technologies, a deep knowledge of technologies available, an ability to draw upon experiences from elsewhere without being bound by preconceived solutions, and a talent for human communication.

Whereas the first qualities outlined may be obvious requisites, it is often seen how much success depends in fact on the latter qualities. No solution proposing introduction of a new technology including informatics technology can be viable unless all boundary conditions are well appreciated. These may relate to physical conditions, organisational conditions, availability of trained personnel, possibilities for further training or retraining, and, particularly in developing countries, it may also include cultural factors.

The consultant is often thought to be the wise man whose wisdom may have a radical impact on the problems to be solved. Even the deepest wisdom is however, valid only if it can be transmitted to its receivers in a form they can appreciate. The success of the consultant is therefore much dependent on his ability to transfer information to environments and persons whose background will be different from his own. Nobody should expect his wisdom and advice to be accepted unless the content is clearly understood, and nobody should accept advice unless they clearly understand its assumptions and implications. This may be trivial, but the world is full of examples of failures in informatics and other technologies which have their roots in a communication barrier between a consultant and a decision-maker. The prime role of the consultant may be to overcome this communication barrier.

## DEVELOPING COUNTRIES

THERE EXISTS A SCHOOL of thought which argues that developing countries as a rule should oppose computers because of their implications for unemployment.

Several years ago the United Nations, however, recommended "Each developing country should formulate a broad national policy, consistent with its national goals, on the application of computer technology".

Several authors link the development of computerisation of a country with economic potential and progress, and there are convincing macro comparisons. In this way the United Nations recommendation can be seen in a larger perspective, and no doubt is interpreted as such by many countries and in several quarters.

The recommendation calls for a national policy, and indeed computerised solutions in several countries are imbedded clearly into a formalized overall policy. Formulation of policies may have benefitted also from advice given by consultants, and national policies often bear direct reference to the use of consultants. This can take several forms: the establishment of a national body serving as the central point for coordination of computing activities and having its own consultancy function, the provision of consultancy capacity on a long term or short term basis through international organisations, the provision of consultancies from consultancy firms and organisations, and the provision of consultancy from individual experts, quite often university professors or experts drawn on an *ad hoc* basis from an appropriate professional environment.

All of these sources have their particular merit and in practical terms many countries draw upon a mixture of the available channels for provision of advice. It will no doubt remain so in the future in developing countries, and other countries as well. In line with the United Nations recommendation a central national body may be of key importance for developing and implementing a national policy on informatics. It will serve a useful role at the overall executive and administrative level. Its role may develop substantially provided it is given also sufficient expertise to serve in a neutral way as the prime consultancy body. This may be one of the most effective ways to overcome the communication barrier discussed previously.

It may also be a diligent way of overcoming one of the mistakes often made which is the introduction of solutions which may work perfectly elsewhere into a new environment.

## SOME SOLUTIONS TO PROBLEMS ENCOUNTERED BY DEVELOPING COUNTRIES

THE TRADITIONAL CENTRAL service bureau type of operation works satisfactorily in many environments and, due to the lack of experienced computer professionals, it may be the only alternative to the purchase of expensive equipment with scattered applications for which it is difficult to maintain a professional organisation and staff.

Distributed processing is often suggested as an alternative, and indeed provides a satisfactory and effective solution, at least to more developed environments. Distributed computing requires only a limited initial investment, the risk is low and the running expenditure can be predicted rather precisely. It permits a gradual introduction to computer use growing as experience, confidence and training are built up.

It is however a technical solution which should not be confused with the mere distribution of computing centres as simply replicas of the central service bureau operations. Distributed computing as a technical solution depends on the availability of reliable, fast and cost effective telecommunications facilities. It has, therefore, been suggested that developing countries should first attend to the technology of telecommunications and only then to computer technology.

This is one example of the opportunities for developing countries to make a generation leap. Another opportunity may be found in promoting microprocessors at an early stage, instead of the traditional development pattern via large computers and microcomputers.

The imaginative consultant who clearly understands the environment which he serves may have a major impact in placing a developing country in the forefront of computing in a realistic way rather than suggest solutions which others are now about to leave.

The inspiration for these developments should, in an ideal situation, be indigenous and not only be imported by external advisors and experts. Hence the concluding lesson is to build up an internal consultancy capacity alongside other sources of knowledge and expertise.

## References

Chevion, D. (1978) *The role of a government computing centre in public administration*, in *Computer Applications in Management: The Asian Scene*. Asian Productivity Organization, Tokyo.

Kalman, R.E. (1979) *Appropriate Informatics Technology, Criteria for Selecting Appropriate Technologies under Different Cultural, Technical and Social Conditions*. IFAC Symposium, Bari, Italy, FAST Milan.

Price, H.C. (1979) *Microcomputers: the key to an easy and peaceful transition to 20th century expertise by the Third World*, IFAC Symposium, Bari, Italy, FAST Milan.

Smyth, W.F. and Jayasiri, N.W.N. (1980) *Effective computer use in developing countries* in *Information Processing '80*, North Holland Publishing Company.

United Nations (1973) *The Application of Computer Technology for Development*. Second Report of the Secretary General. United Nations Publications, E.73.II.A.12.

# Transferring Informatics Technology to Developing Countries: Report of Some Findings

***R.C. Barquin***
*IBM World Trade Asia Corporation,*
*Hong Kong*

## INTRODUCTION

THIS PAPER PRESENTS the results of research done on the transfer of computer technology to developing countries. The field work was done in Latin America in the early to middle part of the 1970s. While there is an obvious time lag in presenting these findings, important insights can still be obtained applicable to the generic process of transferring systems technologies. Some definitions are:

> By *technology* we mean the set of disciplines, methods, techniques and supporting instruments which make up the process by which a tangible or intangible product is elaborated.
>
> A *systems technology* is one having a large number of components which must interact in a systematic fashion to produce its characteristic output or service.
>
> *Technology Transfer* is the process by which a technology is transplanted or diffused from one location to another.

By studying in detail the computer industry in Latin America and the process by which the technology is transferred, we can draw several conclusions about the process of technology transfer in general and about the transfer of systems technologies in particular.

## REVIEW OF THE LITERATURE

TECHNOLOGY TRANSFER has been dealt with at considerable length in the technical literature. Several aspects of the transfer of technology process have been the subject of scholarly research. From the management point of view, the problem of how to transfer technology is of special interest, since it requires knowledge upon which to base decisions. Management research on the topic has been directed primarily at determining which are the factors involved in the transfer, and to quantify the interactions in the process. Thus, Myers and Marquis (1969) and Gruber and Marquis (1969) are probably the best initial sources for such research from the vantage point of management. But the problem of diffusion of innovations, which is at the heart of technology transfer, has been the subject of extensive study in the social science literature even prior to its formal treatment by management researchers. Many cases studies have described innumerable occurrences of the adoption of new tools, practices, or techniques. Rogers (1962) lists a large number of these studies and categorises them according to the research tradition into six distinct groups: anthropology, early sociology, rural sociology, education, industrial, and medical sociology.

## DIFFUSION OF INNOVATIONS AND SYSTEMS TECHNOLOGIES

THE AGGREGATE INVESTIGATION of these case studies offers good insight into the process of diffusion of innovations. But all of these cases deal with discrete situations. That is, they study the adoption of a single new hardware component, processing technique, or agricultural strain. Sutherland describes the introduction of high-draft spinning in the cotton processing industry; Coleman *et al.* (1966) deal with the adoption of gammanym (a medication) by physicians; Ryan and Gross (1943) research the acceptance of hybrid corn by Iowa farmers; Sharp (1952) studies the adoption of steel axes by the Yir Yorunt in Australia.

When we deal with a systems technology, however, we must consider simultaneously a large number of factors which are all innovations to the new user. In fact, we must be able to adopt each integral part of the system or forfeit productive use of the technology as a whole.

In addition, systems technologies create information networks which transcend organisational boundaries. These networks have certain inherent attributes which inject further elements of complexity into the process of the technology's diffusion. The aggregate of all these factors provides a strong argument for a separate treatment of the transfer of systems technologies in the literature.

Dealing with specific innovations, Rogers (*op. cit.* p. 18) establishes a framework for imbedding the process into a time structure. For this, he selects awareness (first hearing of the idea), trial (first use), and adoption (100 per cent use), as the three possible stages in the process of diffusion of an innovation.

Since systems technologies have to be adopted en masse—with all the necessary hardware, software and operational techniques, and in such a manner that actual production is possible—they do not lend themselves to such a structure. It might be purposeful to speak of awareness, trial, and adoption of a new programming language, or magnetic tape drive, but it is not very meaningful to apply this scheme to computer technology as a whole.

The principal cause for non-applicability lies in the need for analysing, not only the status of each factor within the systems technology, but also the interaction among them. Thus, it appeared more adequate to determine the penetration of the technology at any specific point in time, by identifying the number of existing installations, and determining the characteristics of said installations individually, and for the host country as whole. In this context penetration was seen as having two components, depth and breadth. By depth was understood the degree of acceptability achieved in the host environment as measured by the number of installations established over time relative to the size of the economy. By breadth of the penetration was meant the variety in operational formats which the technology adopts in the host environment.

## DESCRIPTION OF FINDINGS

Because systems technologies are in themselves complex, it was found that multiple decisions had to be made in the process of their introduction to a developing country. These decisions were made at distinct levels depending on the aspects of the technology which were under consideration. It was found that the decision maker needs to be well aware of all the possible alternatives open to him at each specific point of the process, or run a high risk or not making the best selection. In the transfer of systems technologies to developing countries, there appear to be at least two distinct levels of decision making, national and institutional.

At the national level we are dealing primarily with the problems of how the technology affects the recipient society, polity and economy. The decisions to be made here seem to be such as regulation or non-regulation of the industry, manufacture or importation of the hardware, official participation or private enterprise involvement, means of transfer, national costs, etc. At the institutional level we are dealing mainly with those aspects of the technology affecting the adopting institution. Decisions made at this level seem to be typified by such as:

choice of hardware and software for an installation, identification of application areas, selection and training of personnel, institutional costs, etc.

There are other levels of decision-making also involved, such as the multinational or non-institutional. Nonetheless, the national and the institutional are the two most relevant levels.

## TECHNOLOGICAL GATEKEEPER

A DECISIVE ELEMENT for the effective transfer of a technology appears to be the identification of technological gatekeepers. These are people who because of their intellectual capabilities, investigative zeal, and contacts with the centers of relevant technological development, become vital sources of technical information for the rest of the organisation.

The existence of brokers in the process of diffusion of innovations had long been established in the literature.* Gruber and Marquis (1969) later showed that organisations in developed countries counted on similar persons who functioned as internal consultants for technological information. Extending the concept to organisations in developing nations, the term "international technological gatekeepers" was introduced by Allen *et al.* (1979). We have found that in developing countries the information networks created by systems technologies become national in scope. These networks also feature technogical gatekeepers; and they fulfil the same consultant function, previously noted inside organizations, outside narrow institutional contexts. It was found that the identification of these people was an important prerequisite for the effective introduction of a new technology, or innovations within existing technologies.

In his research Allen devised a method of identifying the gatekeepers by examining the information networks about a technology which occur within an organisation. We utilised a different method, based on groups of interviews of technologists. The interviews were conducted in snowball-like fashion, with each interviewee suggesting a set of names which were the subjects for the next round of interviews. However, the names of potential interviewees repeated themselves. Independent chains of interviews were initiated by resorting to distinct initial sources. In most countries these were the manufacturers, the universities, the professional organisations, and the government entities which dealt with the problems of the technology. This process allowed us to identify the existence of a technological network for the diffusion of computer information in every one of

* Such brokers were, for example, the early adopters of hybrid corn in Ryan and Gross or some of the pioneering doctors who commenced to experiment with gammanyn and were sufficiently respected by their fellow physicians as to be followed, in Coleman *et al.*

the Latin American countries, and also yielded sets of gatekeepers for each one of these nations.

It was also found that the identification and characterisation of the gatekeepers by the decision makers in charge of managing the technology transfer process at a national level, or by those entrusted with establishing an installation at the institutional level, was very important. If the gatekeepers can be identified, then measures can be taken to let them function more effectively. Facilities to attend international conferences, keep up with the literature, maintain foreign technological contacts, and go on occasional work assignments, can then be encouraged for this key group of people.

## THE ROLE OF MANUFACTURERS

In the initial stages of the introduction of a systems technology to a developing country, basic components are likely to be tightly controlled by the hardware manufacturers. In our case study we found that the role of the computer manufacturers was crucial at that stage in Latin America. As we mentioned above, 26 per cent of all gatekeepers are manufacturer-affiliated. That does not represent the full impact of the manufacturer's role in the process for a number of reasons. The figures for sources of education showed that 78 per cent of all programmers and systems analysts in the region had studied with a manufacturer. Of that figure, approximately two thirds mentioned IBM as the source. In addition, although no specific numbers were determined, a large percentage of the respondents to the programmers and analysts survey studied with IBM manuals as texts. Moreover, there is also a relatively large set of DP managers and other high DP-related executives, or senior programmers and systems analysts throughout Latin America, who are former employees of a manufacturer. Hence, the figure for gatekeepers would show an even larger proportion with manufacturer's affiliation if it reflected prior employment with Borroughs, IBM, NCR, or any other of the principal manufacturers.

The possibility of manufacturers exercising a negative or restrictive influence on the transfer was considered a potential problem, but no such activity could be presently identified on the part of any major manufacturer. The host countries have available a number of measures to minimize that eventuality. For example, economic incentives could be created to ensure that a manufacturer releases and distributes information. Alternate sources of information can be obtained via the literature, or attendance at foreign trade shows and conferences; or by the encouragement of commercial competition among the manufacturers.

## THE ROLE OF UNIVERSITIES

Most important of all, however, is the role of the national universities. As centres of research and depositories of knowledge that are relatively free from business or other commercial influence they can act as a counterweight to bias in the transfer process on the part of any manufacturer.

Through active investigation, as well as via the channels of communication open for international interchange in the academic world, national universities can inject a measure of vigilance over the process. In addition, they can provide a much needed source of technical information on the systems technologies of relevance to the country; and become an ideal setting for pioneering new application areas and developing new concepts on how system technologies can benefit their nation.

## CHANNELS AND MECHANISMS

In addition to the international technological gatekeepers, which were prime individual actors, some channels and mechanisms of the transfer process were identified. Three types of organisational actors, or channels, were found: vendors, users, and others. The vendors were basically the multinational corporations which manufacture the computer hardware. The importance of their role was found to be crucial, as was discussed above. The users were found to be a diverse group, with the national governments and the large multinational corporations (non-computer manufacturers) being most in evidence.

It was found that governments played an important role in the transfer process as users of computer technology because in many cases they were the only organisations with sufficient financial resources to acquire the hardware. In other situations, only national governments had both the massive data processing needs and sufficient resources to take advantage of economies of scale in computation. Because of this, government installations often become leading centers for the diffusion of computer technology to the rest of a developing country.

The multinational corporations were also found to have a very important role in the transfer process as users of computer technology. These corporations are vital channels because they can bring to a developing country the resources and experience of the developed lands through their internal routes usually benefitting the host nation in many ways.

Among the organisational actors, or channels, designated as "others" it was found that professional organisations and user groups had a significant role to play in the process of transferring computer technology. These organisations were found to have, in general, a low membership. At the time, only about 16 per cent of all the programmers and systems analysts in the region belonged to a professional organisation. In addition, although the countries that had them were among the most active, many nations still did not have any professional computer organisations or user groups.

It was found that these organisations were important because they provided centralised sources of information and manpower for the technology, as well as being bridges for the transfer from the foreign sources to the national. They also play an important part in the establishment of standards and the quest for excellence in the professions associated with the technology. Moreover, it was found that a large number of the technological gatekeepers were affiliated with a professional organisation or user group; illustrating the vitality of such associations, and their importance as a source of technological gatekeepers.*

Five mechanisms for the transfer of systems technology were examined. These were: (i) education, (ii) technical literature, (iii) conferences, conventions, and trade shows, (iv) cooperative working programs, and (v) personnel migration. Of these, education was found to be the most important, since it was the only permanent and continuous mechanism existing in the region for the transfer of computer technology. The principal sources of education for programmers and systems analysts, in order of importance, were the manufacturers (48.2 per cent), followed by the universities (37.4 per cent), and in-house education (14.4 per cent). Private DP schools were found to be important only in the instruction of the very elementary functions, such as data entry, and basic machine operation.

Technical literature was not found to be effective for two reasons. First, there was a widespread unawareness of its existence. Second, almost all relevant technical journals were in English, whereas it was found that only 22 per cent of all DP professionals feel they had a good grasp of this language.

Conference and trade shows were found to be effective, but they were relatively infrequent in Latin America. In addition, attendance at the regional or international events was relatively poor.

Personnel migration was found to have important effects on the transfer of computer technology, but the occurrence of such migrations were not readily controlled or channeled. Some of the effects of Cuban and Chilean technicians being utilised in other subsidiaries within a vendor's network illustrate this.

* Although the figure obtained for technological gatekeepers affiliated with professional organisations was 11 per cent, the number should be much higher. That datum simply indicates the percentage of respondents which gave their full-time employment as being with a professional organisation or user group.

## SOCIAL, POLITICAL AND ECONOMIC EFFECTS

EVERY NEW TECHNOLOGY when introduced in any society has a number of social, political and economic effects. Systems technologies are especially important in this context, because they deal principally with information. Examined in the light of occurrences in the more advanced countries, the social, political and economic effects of computation were found to be minimal as of yet in Latin America. Elements of the same types of problems and fears which are confronting the North American public were detected in the region, but were not yet of major importance.

The problem of unemployment was also reviewed. Although there was no hard evidence as of yet, in the long-run the net effect is believed to be similar to that in the industrial nations, where computerisation has not brought about net unemployment, but a shift in skill levels. Nonetheless, it was noted that using the computer purely as a labor saving device in Latin America could have social costs in the short-run, because of the possible displacement of personnel who cannot find a DP related job within the same organisation.

The computer can be used as a tool for the consolidation of technocratic rule. At the same time, it is an instrument that no modern state can really ignore or do without. The dangers implicit in the use of mass data banks, and population registers to invade individual privacy and violate certain civil rights are real, but each country can and should implement its own safeguards to address the issue. Systems technologies may have considerable impact on host societies.

The case of mass communications is another example of a systems technology and several authors may be quoted as sources documenting its influence (Lang & Lang, 1953; Bauer & Bauer 1960; and Pool & Adler, 1960). Because the effects on the receiving societies are important, governments become increasingly aware of such technologies and attempt to monitor and control their activities and diffusion. This was clearly seen in the case of computation in Latin America. It was found that governments in the region were becoming very conscious of computer technology and that they were ever more inclined to implement measures to monitor and control it. The principal reasons for this awareness and concern were found to be four: (i) the cost of the technology as affecting the national balance of payments, (ii) dependence on foreign suppliers, (iii) that the computer is a source of power through its information handling capabilities, and (iv) that the computer is a status symbol. Because of these reasons measures have been taken in every country of the region to monitor some or all aspects of the data processing environment.

## CONCLUSION

In summary, it was found that there are strong arguments for approaching the transfer of systems technologies to developing countries in a different manner from the transfer of other technologies. Also, that Roger's model for the process of adoption of an innovation was not applicable to the transfer of systems technologies because of the large number of component factors and the complexity of their interaction. Nonetheless, it appears that a measure of penetration of a systems technology can be obtained by identifying numbers of installations relative to population or other economic magnitudes, as well as by characterizing the individual installations in a host country and obtaining some form of a national aggregate. It was also found that due to the multiple alternatives which exist at any one point in the transfer process, decision makers must be well identified with the variants in the technology in order to be able to arrive at the best possible solutions, both at the national and the institutional levels.

The role of technological gatekeepers was found to be crucial, and their identification and characterisation a key requisite for an effective transfer process. Likewise, it was found that the computer manufacturers were essential channels for the introduction of computer technology.

The universities were found to have potentially crucial roles to play in the transfer of computer technology, but their orientation toward scientific computing detracted from their mission. It was found that national governments and multinational corporations as users of computer technology also play a very important role in the transfer process. Similarly, professional organisations and user groups were found to be dynamic channels for the diffusion of technical information, and significant sources of technological gatekeepers.

Of the mechanisms for the transfer of computer technology, only education was found to be permanent and continuous. Conferences and trade shows, personnel migration, cooperative working programmes and the technical literature were found to vary in degrees of effectiveness, but were all much less important than education. The social and political impact of computer technology in Latin America was found to be minimal up to now. Certain potential dangers were identified, but none sufficiently great or urgent as to warrant major reactions. With respect to economic effects, it was found that the primary one was on the balance of payment of the host countries. This, together with other political perceptions of the potential role of computers, has caused governments to be concerned about the technology, and implement certain measures to monitor and control its diffusion.

This research should provide some insights toward the establishing of sound guidelines by which a systems technology may be effectively and efficiently transferred to a developing land.

## References

Allen, T.J., Piepmier, J.M. and Cooney, S. (1971) *The International Technological Gatekeeper.* Technol. Rev. March.

Bauer, R.A. and Bauer, A.H. (1960) *America, 'Mass Society' and 'Mass Media'.* J. Soc. Issues, **16**, 3–66.

Coleman, J., Katz, E. and Menzel, H. (1966) *Medical Innovation, a Diffusion Study.* Bobbs-Merrill Co., Inc., Indianapolis, IN.

Dexter, L.A. and White, D.M. (eds.) (1964) *People, Society and Mass Communications.* Free Press of Glencoe, Glencoe, IL.

Gruber, W. and Marquis, D. (1969) *The Human Factor in the Transfer of Technology.*

Lang, K. and Lang, G.E. (1953) *The Unique Perspective of Television and its Effects: A Pilot Study.* Amer. Sociological Rev. **18**, 3–12.

Myers, S. and Marquis, D. (1969) *Successful Industrial Innovations.* NSF 69-17, National Science Foundation, Washington, D.C.

Pool, I. de S. and Adler, B. (1960) *Educational Television, Is Anybody Watching?* Center for International Studies, M.I.T.

Rogers, E.M. (1962) *Diffusion of Innovations.* Free Press of Glencoe, Glencoe, IL.

Ryan, B. and Gross, N.C. (1943) *The Diffusion of hybrid Corn in two Iowa Communities,* Rural Sociol. **8**, 15–24.

Sharp, L. (1952) *Steel Axes for Stone Age Australians.* In Edward H. Spicer (ed.) *Human Problems in Technological Change.* Russell Sage Foundation, New York.

# Control Informatics—Its Role in Developing Countries

*Utpal K. Banerjee*
*Computronics, New Delhi, India*

## THE TECHNOLOGY GAP

IT IS WELL-KNOWN that more than 90 per cent of the world's technology is generated from the R & D efforts of the developed countries, leaving less than 10 per cent to the care of developing ones. This has led to an arrangement of technology transfer either on a rental basis or outright sale. In both cases, the transferee country in the developing region accepts the technology at a specific time and usually remains stuck with it. The transferrer country (or the multinational corporation) keeps the transferee updated on the technology, once it has been transferred.

Developing countries, therefore, are left with what are essentially 'frozen' technologies. Both financial resources and the political support are usually lacking to salvage the antiquated technologies, for three reasons. First, adequate funds are not available to back a political programme of technological innovation. Secondly, vested interests often denigrate indigenous efforts even when they are trying to compete with foreign suppliers, on the basis of strict schedules of production and delivery dates. Finally, the principle of the lowest tender is weighted by a variety of discriminatory duties on the import of components.

The way to antiquated technology in developing countries is often helped by some degree of obsession on self-reliance at the cost of competence. Such self-reliance is without the refreshing touch of indigenous research and development, and can reduce the best technology to so much junk (Thapar, 1980).

The above factors have facilitated the creation and the perpetuation of technology 'poor' countries. They hover on the periphery of transferred technology, looking always for sustenance to a centre of R & D-sustained technologies beyond their reach. The technology 'rich' countries and the multinationals ensure that the flow of technology is forever a one-way traffic.

## ADVENT OF INFORMATION GAP

THE UN ACAST WORKING GROUP (United Nations, 1973) on Computer Technology recognised the existence of four levels of computer usage: initial, basic, operational and advanced. An index of computer industry development potential (CIDP) has been suggested (Barquin *et al.*, 1976) based on number and size of computers, D.P. education, computer applications, utilisation by government, degree of technology in hands of national organisations, official policy towards computerisation, international assistance in computer technology, existence of professional DP groups and user organisations (government, universities, computer manufacturers, multinational enterprises and UN agencies).

Developing countries were placed according to their CIDP level as follows (Perelet, 1977):

(a) *Initial*: Afghanistan, Bangladesh, Bhutan, Botswana, Burma, Burundi, Cambodia, Cameroon, Central African Republic, Dahomey, Ethiopia, Haiti, Laos, Lesotho, Liberia, Malawi, Mali, Nepal, Niger, Rwanda, Senegal, Somalia, Southern Yemen, Togo, Tonga, Uganda, Upper Volta, Western Samoa, Yemen;
(b) *Initial to Basic*: Albania, Algeria, Bahamas, Barbados, Bolivia, Congo, Cyprus, Costa Rica, Dominican Republic, Ecuador, Egypt, El Salvador, Fiji, Gabon, Gambia, Ghana, Guatemala, Guinea, Guyana, Honduras, Indonesia, Iraq, Ivory Coast, Jamaica, Jordan, Kenya, Libya, Malagasy, Mauritania, Morocco, Nicaragua, Nigeria, Pakistan, Paraguay, Saudi Arabia, Sierra Leone, Sri Lanka, Sudan, Swaziland, Syria, Tanzania, Thailand, Trinidad, Tunisia, Zaire, Zambia;
(c) *Basic*: Chile, Colombia, Cuba, Iran, Lebanon, Malaysia, Panama, Peru, Philippines, Republic of Korea, Singapore, Turkey, Uruguay;
(d) *Basic to Operational*: Bulgaria, Greece, Hong Kong, Hungary, Puerto Rico, Rumania, Venezuela;
(e) *Operational*: Argentina, India, Mexico;
(f) *Operational to Advanced*: Brazil.

The 1978 International Conference of 75 nations on strategies and policies in informatics (SPIN) in Spain recognised that data flow was central to building a "New World Information Order", with access and control of information sources being critical to the eventual development of a nation's economy.

The 1980 World Conference in Rome of 60 nations on transborder data flow policies was a follow-up to SPIN, and presented overtones of the 'North-South debate'—the dominance of the industrialised nations in international information services and the fear that developing countries might become mere client states of the North American and European multinational corporations which controlled existing information sources. Kirchnir (1980) summarised the debate as follows:

(a) The developed countries, including the US, Canada, Australia and European nations, generally argued for free flow of information, suggesting that developing societies have more to gain from unrestricted access to advanced information resources than from their regulation;
(b) Many developing countries, on the other hand, including Argentina, claimed that national, political, economic, technological and cultural considerations warranted at least some control over information flowing into and out of their countries;
(c) Brazil, as a relatively advanced information economy and a prototype for Third World countries, presented the viewpoint that any country, which was not concerned with the control of the strategic information resources it used, ran the risk of becoming intolerably dependent, through telecommunications, on the interests of political and economic groups outside its borders. Brazil contended that the North American and European multinational corporations were usually not interested in the legitimate desires of people in developing countries.
(d) The dilemma could be solved, according to Brazil, by encouraging indigenous development of data bases and data bank technologies which, along with outside sources, could provide a balance of universal distribution of information.

The advent of an information gap between the 'information rich' and 'information poor' countries that emerges from the above resumé is too evident to be missed.

## RESOLUTION OF THE INFORMATION GAP

In the context of bad experiences stemming from the technology gap between developing countries and developed ones, one must be doubly careful to try to close the emergent information gap.

Kalman (1981) observed that the material and human resources of nations were not unlimited, and their optimal allocation to various sectors of the economy was one of the principal functions of governments. The allocation of resources to the national information environment could not be an exception and needed a broad national strategy. Brauer (1979) characterised developing countries by the following statistics relating to their populations;

(a) 20 per cent seriously undernourished;
(b) 30 per cent without safe water and health care;
(c) 40 per cent unemployed or underemployed; and
(d) 50 per cent over 15 years of age illiterate.

Sackman (1981) approached the problems of the information gap from international, organisational and individual viewpoints and developed recommendations at all three levels, emphasising some of the possible educational, political, managerial and technical remedies for a more equitable balance of information power.

Kalman (1978) defined informatics as the complex of scientific, technological and engineering disciplines which deals with the inherent properties, structure, laws and rules of creation, transformation, conservation, transmission and application of information phenomena in natural and constructed systems. Bernasconi (1978) viewed informatics simply as the exploitation of information resources, according to an overall strategy implemented by appropriate policies. Rateau (1981) viewed the trends of informatics during the 1980s as follows:

(a) Improvement in the cost/effectiveness ratio, estimated as $10^4$ between 1955 and 1975 (effectiveness being measured by the number of operations performed per second);
(b) Miniaturisation of hardware;
(c) Production of software more adapted to the user;
(d) Development of tele-informatics with parallel progress of computers and telecommunications.

Informatics, becoming cheaper, more effective and more accessible, should allow the information-poor countries to face their data management problems in the following manner:

(a) to improve economic and social development planning based on mathematical models and statistical data bases;
(b) to develop information systems, and automated documentation centres and information networks;
(c) to use computer-aided teaching for increasing the numbers receiving education;
(d) to implement large epidemiological studies in the field of health;
(e) to gain (through satellite-based services) a better knowledge of natural resources;
(f) to help and accelerate industrialisation by greater use of automated process controls;
(g) to use computers in the services sector for improving productivity.

The conditions for success in reducing the information gap would depend on national informatics policies, active regional co-operation and essential international collaboration, the compilation of computer science plans, the training of experts and the establishment of software libraries.

## STRATEGY ON INFORMATICS

EVEN IF THE ABOVE STEPS were adopted to resolve the information gap of developing countries, Rateau (*op. cit.*) visualised that the informatics industry was still not within the reach of all countries. Moreover, databanks currently being established were essentially situated in industrialised countries. Informatics, therefore, involved a risk of increasing, instead of reducing, the gap between the industrialised and developing countries. Again, there was an important risk concerning the cultural identity of developing countries, since informatics was the product of a rationalist and basically western culture.

It is desirable, therefore, to avoid the above pitfalls while introducing informatics in the Third World. The following strategy is recommended.

### On-Line Transmission—Developed to Developing Countries:

There are a number of application areas where data bases have already been developed and are kept regularly updated in First World countries. The Third World can, with advantage, use these data bases without "re-inventing the wheel". An example is the Space Documentation Service developed by European Space Agency which offers a computerised information retrieval system, using interactive on-line dialogue, through remote terminals connected to a central SDS computer (at Frascati, Rome) and is used for bibliographic citation retrieval. An international telecommunication network (ESANET) stretches over more than 10,000 kilometers to link terminals by means of high-speed, data-transmission lines to cover direct dialing to Brussels, Copenhagen, London, Rome and Stockholm. The national CYCLADES network in France is also connected to ESANET, and similarly TYMSHARE nodes can be dialled from Brussels, The Hague and Paris.

The choice available of major scientific and technical data bases through ESANET is as follows:

| Name/Subject Coverage | Established | Number of Citations |
|---|---|---|
| "STAR/IAA": aerospace, geophysics, electronics, earth resources, biotechnology, thermodynamics. | 1962 | 800,000 |
| "Chemical Abstracts Condensates": pure and applied chemistry. | 1969 | 2,000,000 |
| "Metals Abstracts": metallurgy and associated disciplines. | 1969 | 200,000 |

*continued*

*continuation*

| | | |
|---|---|---|
| "Engineering Index": civil, electrical, mechanical engineering, electronics. | 1970 | 470,000 |
| "Government Reports Announcements": scientific, technical and social disciplines. | 1969 | 200,000 |
| "ESA Electronic Components Data Bank". | 1970 | 16,200 |
| "Nuclear Science Abstracts": nuclear science and technology. | 1968 | 485,000 |
| "INSPEC": physics, electronics, computers. | 1971 | 680,000 |
| "World Aluminium Abstracts": aluminium and associated technology. | 1968 | 45,000 |
| "Environmental Science Index": environment sciences, pollution. | 1971 | 50,000 |
| "Science Citation Index": Physical, chemical, engineering sciences. | 1972 | 600,000 |
| Aerospace and related subjects. | 1972 | 700,000 |

The above data bases are searched through terminals providing the following services:

(a) Choice of file;
(b) Selection of keywords and the logical combination of keywords;
(c) Citation displays and retention of useful citations;
(d) Printing of useful citations.

Satellite-based communication is possible with ESANET from developing countries and was tested in 1978. In this experiment, where a remote terminal was placed at T.I.F.R., Bombay, the author carried out an extensive literature survey from 1969 onwards on the following topics:

(a) Bottom-blown oxygen steel-making;
(b) Electro-slug refining technology;
(c) Direct-reduction processes;
(d) Continuous-casting technology.

## On-line Transmission Among Developing Countries

The UN Conference on Technical Co-operation among Developing Countries held at Buenos Aires in 1978, laid down a plan of action for the "increase and improvement of communication among developing countries, leading to a greater awareness of available knowledge and experience as well as the creation of new knowledge in tackling problems of development".

One of the applications of that plan is a development information network (DIN), designed to link up developing countries via satellite channels which are capable of 24-hour two-way communications with the following aims:

(a) to provide a continuous two-way flow of up-to-date technical, economic, social, cultural and development information;

(b) to promote co-operation among developing countries by increasing their own direct exchange of information which they need for their national and collective self-reliance.

Using existing satellite channels, DIN will set up the following centres over the next three years:

**National Centres:** Sixty national centres will be responsible for disseminating incoming information to subscribing users in the country, and for collecting outgoing information. Working to common guidelines, national centres will select, codify and feed information to regional centres;

**Regional Centres:** Regional centres will be in Latin America and the Caribbean, the Arab region, South Central Asia, Southeast Asia and the Pacific, East and Southern Africa, and West Africa. Equipped with microprocessors and short-duration information storage capability, the regional centres will be the structural pillars of the network. They will be linked to inter-regional centres;

**Inter-regional Centres:** These will be the network's main co-ordinating units and will also provide back-up to regional centres in case of technical breakdowns.

DIN will disseminate information in Arabic, English, French and Spanish. The information will be relevant; factual; set in context; continuous; up-to-date; concise; timely and professional.

DIN will be a user-oriented system, flexible enough to accommodate the changing information needs of subscribers. It will cover the entire spectrum of economic and social development, creating a new flow of information on themes such as:

(a) current international economic/financial trends;
(b) national development plans and programmes;
(c) resource discoveries;
(d) raw material supplies;
(e) balance of payments situations;
(f) trade and business opportunities;
(g) regional/inter-regional financial institutions;
(h) foreign trade activities, including market movements, loading schedules, shipping, capital markets and currency fluctuations, tariff rates and preferential trade agreements between different countries;
(i) international negotiations;
(j) technological developments and research;
(k) international/regional/integration and co-operation developments;
(l) socio-cultural aspects of development, such as health, education, housing, co-operative movements and social organisations;
(m) applied research findings.

DIN will have national centres linked to a wide range of organisations such as trade unions, public enterprises, industry, applied research institutes, regional economic commissions, commodity producers, professional associations, international organisations, the media, universities, government decision-makers, etc.

### On-line Transmission Within a Developing Country

It is possible to realise, within a developing country, functional integration of information activities of libraries, scientific and technical services and archives. This would enable a more exhaustive flow of primary and secondary information.

An example is the CDS ISIS system developed by UNESCO in 1978 to be used as a typical information retrieval software package for automation of a national system of scientific, technical and economic information. This is a standard software package whose advantages include possibilities of exchanging experience between information centres, and minimizing the cost of supporting the software.

Along with a few developing countries, CDS ISIS has been extended to Reyad computers in India for sophisticated data processing, including development of functions to support the needs of management information systems. The package uses a comprehensive collection of programmes for data input, management and output, running in "batch" or "on-line" modes. During "on-line" operation, users can be guided at their terminals by the system which can build indexes from key-words, thesaurus terms or free text, and can specify sorting and printing instructions.

### Transfer of Technology from Developed to Developing Countries

In a number of applications, it is possible to consider indigenous development of data bases and data bank technologies which, along with outside sources, can create a balance of universal information distribution. It is important, however, that the socio-economic and cultural identity of a developing country should not militate against the structure of automated information brought from outside sources.

An example is adaptation of a Burroughs computer system by the State Bank of India, along with the packaged banking software. However, this package has not been found very suitable and because banking practice in India is being organised on a 'social responsibility' ethos quite different from the American philosophy, the package has remained largely unutilized. Other nationalised banks have approached the Government of India to allow similar computer systems because of ostensible advantages of a banking software, but the government has rightly not encouraged such imports.

A slightly different point was made by the Algerian National Informatics Authority (NIA) at the International Colloquium on Informatics and Society, Paris, 1979. This was that developing countries are often able to supply very important information, but are not able to process it. Then the risk arises that an exploitation cycle may be created by developed countries from the raw data given by the developing countries in the first place (Rateau, 1981). Further, modern informatics is a reflection of a certain way of thinking, and of a certain socio-economic organisation. Since such informatics is the product of a rationalist western culture, developing countries should carefully examine the risk of cultural dependence arising from its adoption.

## Large Data Bases, Within a Developing Country

Within a developing country, distributed data processing (DDP) has now become feasible with the declining cost of computer hardware and the availability of network software. It is thus possible to connect to a network several computer systems with large data bases distributed all over a country.

An example of the above was the "star" network with the DEC-1077 computer at T.I.F.R. Bombay as its hub and DEC/PDP computer systems at Ahmedabad, Bombay, Bangalore and Delhi at various nodes, tried out in 1980 with 4 terminals located in Bombay and connected (via modems and synchronous communication lines) to a total of 11 systems, including front-end communication processors. The computer network, termed DECNET, was used to highlight the data communications infrastructure presently available in India and to help the participating organisations in gaining experience and developing expertise in computer communications software and hardware.

Quite separately, there can be large independent data bases in centralised locations which can be used by big user organisations for their own purposes. Such data bases are planned by TISCO and TELCO, which rank among the largest private sector enterprises in India, at their three premises in Jamshedpur and Pune.

## Feed-Back Information System, Within a Developing Country

It is possible to apply computers and data communications to socio-economic phenomena for controlling environmental processes through a non-automatic monitoring and control system design. Such applications have to collect data from the environment manually or through sensors; processing must be done on digital computers in an on-line or off-line mode; and on-line or off-line control of environment must be achieved within varying response times. Effective control would depend on the ability to respond in a reasonable time, keeping in mind the large volume of data likely to be communicated, the complexity of the analysis and the needs of follow-up on the corrective response. Obviously, sensing data from the social system and monitoring control process must be made continuous to ensure that the control is effective and yet responsive to social changes. On the other hand, it would be desirable to detect the out-of-phase conditions of control where rules may have to be changed, in phase with the social change.

An example is the use of the above approach for setting up operating procedures for multipurpose reservoirs or a group of reservoirs inter-connected as a system for optimal operation of both flood-control and power generation. This has been tried out at Nagarjuna Sagar reservoir for a feed back control and monitoring system with the following objectives:

(a) To put data, collected manually through rain-gauges and flow-gauges located on the Krishna river, at various State-run and Centre-run measuring stations;

(b) To fill up the reservoir sufficiently early in the season for timely release of water for purposes of irrigation, power generation and flood-control;

(c) To provide 6 to 8 hours' advance warning for release of water from the reservoir, whenever needed;

(d) To release water economically during the drought-prone years (when the rainfall is less than 25 inches for a specified year), normally for 3 out of every 10 years.

Krishna river, over which the Nagarjuna Sagar reservoir is built, has an inter-State basin involving Maharshtra, Karnataka and Andhra Pradesh, with 12 catchment-areas. Obviously, regulated use of water for reaping a planned return from the entire Krishna basin (and later, perhaps, from the Tungabhadra tributary basin and Godavari basin with their dams and barrages in a similar manner) is of the greatest utility in India.

## ILLUSTRATIONS OF CONTROL INFORMATICS

THE BASIC SECTORS of national development include education, employment, health, food and housing. In a recent survey by the German Foundation for International Development (DAC), these sectors have been identified in that order. Use of computers can be visualised for 'control informatics' in all the above sectors in a developing country. The concept of control informatics is based on the feed-back information and control instructions within a reasonable response time, as explained earlier. The concept will remain valid for both on-line and off-line use of a computer, either in a local or a remote mode, as long as suitable monitoring is adhered to and changes are introduced in a meaningful way.

**Education:** Milner and Wildberger described computer-assisted instructions (CAI) to impart knowledge to students through a "drill and practice" approach. Hawkins described a "tutorial" method to permit several students interact simultaneously with the computer on one or more subjects. It seems that the introduction of CAI into the educational system of the Third World could provide a welcome relief to teachers and enhance the quality of education.

The author has been instrumental in initiating a project proposal for CAI in formal and non-formal education. The idea is to use a stand-alone microprocessor-supported device to be installed on a mobile basis in district or taluk centres. Individual programme-lessons can be taken up for either formal education (e.g., geometry, algebra, etc.) or non-formal education (e.g., pest-control, insecticides-spraying, etc.), using a lot of computer graphics. Records would also be kept of the amount of usage of specific programme-lessons and the manner of learning so that lessons can be evaluated later for modification. Using digitised communication during off-broadcasting hours, programme-lessons can be renewed on casettes.

**Employment:** Longe (1981) recommended employment data bases on computers, to be used for matching a job vacancy with qualifications of unemployed persons. The author supervised a computer-based information system on medical manpower for the Andhra Pradesh medical services. By keeping full information on educational qualifications, practical skills, experience in city/town/village areas and other matters, it has been possible to regulate the posting of medical personnel in different regions of the State with the help of the above data base and implement the "social responsibility" scheme for posting doctors to outlying areas. What is more, the system is capable of being monitored on a regular basis.

**Health:** McLachlan (1968) implemented a hospital, employment and administration system for non-emergency patients. Whitby and Lutz (1971) used computers to keep medical records on patients while Holoien (1977) applied computers to clinical laboratories and intensive care units. The author has supervised computer usage for building up a comprehensive health care system in Andhra Pradesh State at different levels of hospitals (from taluk hospitals to teaching hospitals). This system was based on the records of diseases, the remedies applied, the medicine used, the facilities utilised and results achieved. Again, the system is capable of being monitored for extending different types of medical facilities and planning of resources.

**Food:** Computer applications have been suggested for agricultural research and development programmes. Particularly, computer based systems can help in the discovery, development, and wide distribution of high-yield pest-resistance varieties of crop plants.

The author had experience in organizing a detailed storage and distribution system for grain involving different regions of the country. The plan included location of warehouses for grain, demand for food in different localities and in different seasons, availability of food from different regions under different climatic conditions, and distribution of food-stock to meet demands fully or partially. The system has remained capable of being regularly monitored and controlled.

**Housing:** Sterling (1977) developed computer applications for regional plans of house building and reporting on potential violations.

The author has experience in working with the urban development authority in the city of Hyderabad. The data on existing housing units, their location, type of construction, ownership, vacancy and rent payable, could be collected and stored on a computer. The data can be used to monitor vacancies and implement urban land ceiling regulations.

## CONCLUSION

THE ABOVE ILLUSTRATIONS are meant to expand the concept of "control informatics" based on feed-back information and control instructions within a reasonable response time. It is evident that it is perfectly feasible to develop and implement such control and monitoring systems in the Third World so that the benefits of information technology can be reaped.

## References


Barquin, R.C. and Whitney, K. (1976) *Model for Progress in Developing Countries.* Datamation, September, p. 190.

Bernasconi, F.A. (1978) SPIN Final Report, Intergovernmental Conference on Strategies and Policies for Information, IBI, Rome.

Brauer, D., Ed. (1979) *Development and Cooperation.* German Foundation for International Development (DSE), 6/1979.

Hawkins, C.A. *Computer Based Learning: Why and Where it is Alive and Well?* Comput. and Educ. **2**, 187–196.

Holoien, M.O. (1977) *Computers and their Societal Impact.* John Wiley, London.

Kalman, R.E. (1978) IFIP Information Bulletin, IFIP Secretariat, Geneva, December.

Kalman, R.E. (1981) *Eight Strategic Issues for Informatics* in Bennett, J. and Kalman, R.E. (Eds) *Computers in Developing Nations.* North Holland Publishing Co., Amsterdam.

Kirchner, J. (1980) *World Congress Mulls New Information Order* and *National Control of Information Systems Urged.* Computer World, **XIV**, 26.

Longe, O. (1981) *Computer Applications to Basic Needs of Developing Nations* in Bennett, J. and Kalman, R.E. (Eds) *Computing in Developing Nations.* North Holland Publishing Co., Amsterdam.

McLachlan, G. and Shegog, R.A. (1968) *Computers in the Service of Medicine.* Oxford University Press, London.

Milner, S.D. and Wildberger, A.M. *Determining Appropriate Uses of Computers in Education.* Comput. and Educ. **1**, 117–123.

Perelet, R. (1977) *Minicomputer Systems to Manage Industries.* UNIDO/IOD.91, Vienna, p. 6.

Rateau, O. (1981) *Informatics: Evolution and Use in the 1980s, with Particular Reference to Developing Countries* in Bennett, J. and Kalman, R.E. (Eds) *Computers in Developing Nations.* North Holland Publishing Co., Amsterdam.

Sackman, H. (1981) *The Information Rich versus The Information Poor* in Bennett, J. and Kalman, R.E. (Eds) *Computers in Developing Nations.* North Holland Publishing Co., Amsterdam.

Sterling, B. (1977) *A Comprehensive Data Processing and Management Information System for State-Wide Housing Code Enforcement in New Jersey.* Urban Systems 2.1, 45–61.

Thapar, R. (1980) *Pressure From Lobbies—An Unequal Fight on Low Wages.* The Statesman, 16 December, p. 5.

United Nations (1973) *The Application of Computer Technology for Development.* Second Report of the Secretary General, United Nations E.73.II.A.12, New York, pp. 7–8.

Whitby, L.G. and Lutz, W. (1971) *Principles and Practice of Medical Computing.* Churchill Livingstone.

## SECTION 3

# National Policies for the Development of the Informatics Industry

An issue of major interest to policy makers in developing countries is the promotion of indigenous production capacities in hardware and software in order to reduce dependence on suppliers in developed countries for essential informatics products and also to ensure that a full range of products is available, in particular software, which is suited to the particular requirements of developing countries. A number of diverse national experiences are treated, including India, Singapore and Mexico. A wide range of policy options and specific measures are examined. The preconditions for, as well as the constraints relating to, an indigenisation strategy are emphasised and the consensus emerging points to the desirability of adopting a highly selective approach.

# Self-Reliance in Informatics for Industrial Development

*P.P. Gupta*
*Computer Maintenance Corporation Limited, India*

## DEVELOPING COUNTRIES

WHILE THE THEME of this conference is Informatics and Industrial Development, the emphasis is on policies for information processing for developing countries. The World Bank in its World Development Report, 1980, has classified 38 countries (listed in Table 2) with per-capita income of between US$80—360 as low-income countries. It is important to note that 30 per cent of the world population lives in this group of countries. If we include China which is treated as belonging to a separate category although its per-capita GNP falls in this group, it covers over 50 per cent of the world population. There are also another 52 countries with per-capita incomes between $390 to $3500 classified as middle-income countries covering 20 per cent of the world population. It is the relevance of informatics to these very large groups of people that is being discussed at this conference. It is ironic that at many such events, they are hardly represented!

Industrial development in an isolated manner is not what the developing countries are looking for. It has been argued that growth in the GNP of a country does not necessarily lead to improvement in the quality of life which is the major objective of most developing countries. Table 1 indicates certain indices about their quality of life.

An examination of this table indicates very forcefully the areas in which improvement is required and tends to show that the normal applications of computers in the industrialised countries may not have much, if at all any, relevance to the problems of the developing countries. It is well known that computer applications in the industrialised countries have been directed towards

**Table 1. Indices of quality of Life**

| | LIC | CHINA | MIC | IC |
|---|---|---|---|---|
| 1. Countries in each group | 38 | 1 | 52 | 18 |
| 2. Population in millions | 1294 | 952 | 873 | 668 |
| 3. Percent of total world population | 31 | 22 | 21 | 16 |
| 4. Projected population for 2000 AD in millions | 2050 | 1251 | 1409 | 736 |
| 5. Percent of total world population, projected for 2000 AD | 34 | 21 | 23 | 12 |
| 6. GNP per-capita in US $ | 200 | 230 | 1250 | 8070 |
| 7. Spread of per-capita GNP in US $ | 90 to 360 | 230 | 390 to 3500 | 3470 to 12100 |
| 8. Production distribution percent of gross domestic product | | | | |
| a. Agriculture | 38 | – | 16 | 4 |
| b. Industry | 24 | – | 34 | 37 |
| c. Services | 38 | – | 50 | 59 |
| 9. Labour force distribution per cent | | | | |
| a. Agriculture | 72 | 62 | 45 | 6 |
| b. Industry | 11 | 25 | 23 | 39 |
| c. Services | 17 | 13 | 32 | 55 |
| 10. Energy consumption per-capita in kgs of coal | 161 | 805 | 903 | 7060 |
| 11. Adult literacy rate (1975) | 38 | – | 71 | 99 |
| 12. Number enrolled in higher education as percentage of population (1976) | 4 | – | 11 | 36 |
| 13. Life expectancy at birth in years | 50 | 70 | 61 | 74 |
| 14. Population per doctor (1977) | 9900 | – | 4310 | 630 |
| 15. Daily calorie supply percapita as percent of requirement (1977) | 91 | 105 | 108 | 131 |

All data are for 1978 except where otherwise indicated.
Abbreviations used: LIC —low-income countries,
MIC—middle-income countries,
IC—industrialised countries.

Source: World Development Report, 1980, World Bank, Washington D.C., USA.

clerical labour-saving data processing, an activity entirely irrelevant to the rest of the world. Much harm has been done by thoughtless copying and introduction of computers for similar applications in the developing world. Unemployment is one of the major problems of the developing countries and such applications may only contribute further to the unemployment problem—if not direct unemployment, certainly the growth of unemployment—and redeployment of staff causing personal problems at the individual level. Entirely new approaches and innovative applications may be and are required for information processing to be relevant to the developing countries. Applications in agriculture, rural economy, health and education are of primary interest. Computers need to be used for development-oriented planning.

### The Objective of Self-reliance

It is the theme of this paper that for the meaningful application of informatics for developing countries an important pre-requisite is self-reliance. India, since its independence in 1947, has adopted this as its philosophy and framed its policies to attain this objective. Quite often people cite the example of many countries where, with the help of foreign support and foreign business, particularly spectacular improvements in the standards and quality of life may have been achieved. While this may be so, the relevance of this approach to most of the developing countries needs to be examined critically. For a sovereign independent country, in the long run it is essential that it stands on its own feet, depends on its own resources and skills. This seems an uphill task in the beginning. It also appears quite often that the policy of self-reliance stems from the thought that the country does not have enough foreign exchange, its balance of payments is not satisfactory, and therefore it wants to reduce imports. I would like emphatically to say that this is not so and whether a country has enough foreign exchange or not, the importance should be on long-term development of skills and resources and the independence to use them to meet national objectives.

## THE COMPUTER INDUSTRY

I WILL NOW GO on to the theme of self-reliance in the computer industry. Let us look at various activities which contribute towards self-reliance—research and development, manufacture, systems engineering, software development, applications development, maintenance and training.

Manufacture obviously suggests itself as the very first activity in this direction. However, keeping in mind the economy of scale and the very rapid development of technology worldwide, this is something which needs to be examined very carefully. While efforts must be made to achieve self-reliance in manufacture, it would be unwise to try and manufacture everything. Selective imports, particularly of sub-assemblies, units, etc., could be undertaken. Fortunately there is a large number of manufacturers in many countries who offer high quality and low cost components and equipment. Thus without getting too dependent on one country or one manufacturer, imports of optimum price-performance equipment could be undertaken.

By and large, manufacture in developing countries is limited to rather small computer systems based on microprocessors. Unless very carefully used, such systems would handle only very simple, routine data-processing applications. In India about 400 such systems have been supplied, which unfortunately are used largely for simple accounting purposes. This emphasises the need for technology to be used in an appropriate manner to ensure continued growth of employment in a country like India.

There are application areas which contribute directly to national development and deserve considerable attention e.g. process control applications, railway wagon information, power systems control, flood warning systems, etc. Computers are an integral part of such dedicated systems. Today most of these systems are supplied on a turnkey basis by a limited number of manufacturers from the industrialised countries. While the hardware costs are going down, the cost of manpower and software is increasing. A total system of the type mentioned above may cost anything from US$1 million upwards. However, the actual cost of hardware may be only 40 per cent of this cost. There is therefore no reason why the rest of the activities, e.g., systems integration, interfacing, installation, commissioning, maintenance, software development, training, etc., should not be done locally on a national basis. The options open to a country like India are:

(i) to import such systems in totality,
(ii) manufacture the hardware and other associated equipment, or
(iii) start doing systems engineering activity as explained above.

It is obvious that the last option is the most economical and beneficial. It retains the advantage of getting the latest and the most reliable hardware and yet depends for a large portion of the activity on one's own resources and skills. This activity would of course depend upon the availability of technical skills in the country. India, for instance, has skills to perform many of these activities itself and efforts in this direction have been going on in our country for quite some time. Similar activities are being developed in many other countries.

These activities also need an organisation which would be national in character, independent in approach and with the capability of rendering total support. The Government of India, keeping this in mind, set up CMC (Computer Maintenance Corporation Limited) in 1976 to provide one-point total support to computer users. This activity is now being carried out in a most meaningful manner by CMC which works closely with the Central Electrical Authority and the Meteorological Bureau, etc., to introduce computers for control and monitoring purposes. The same idea can be applied to large-scale general purpose computer systems where systems engineering activity can be carried out by integrating the best hardware available, imported if necessary, and developing software internally to reduce the total cost.

Another major objective is to detach the Indian computer user from the foreign manufacturer and to insulate him from normal marketing pressures. According to current policy in India, if a computer needs to be imported, careful evaluation of the requirement is carried out by the Department of Electronics of the Government of India. As pointed out earlier, the system is mainly used for applications which lead to growth rather than those which perform ordinary routine data processing jobs and mechanise standard clerical procedures. The support services, including hardware maintenance for imported systems, are provided by CMC. These services include site consultancy, site preparation, installation and commissioning of systems, software support, systems advice, user training, applications development, etc. It is interesting to note that while in 1975, computer equipment in India was mainly supplied by two multinational companies (which had marketing and support organisations) now there are about twenty manufacturers whose equipment is supported locally. It is doubtful whether such a diversity would have taken place without the emphasis on self-reliance in this important field of support and maintenance. Another example of this is the take-over of maintenance of IBM equipment in India in 1978, when IBM decided to close operations in India.

Enhancements to already installed systems, by integrating other manufacturer's supplied peripherals, are also being carried out.

Another advantage of having a national organisation is the pooling of spares. Even if more than one manufacturer's equipment is brought into the country, all spares can be maintained in a pool. Efforts for indigenisation of spares are being made in our country to reduce imports and develop self-reliance.

It is only when the basic skills for integration, installation, commissioning, maintenance, software support and software development are available from one organisation that the systems engineering activity referred to above becomes

**Table 2. List of low-income countries**

| | |
|---|---|
| Kampuchea Dem. | Sierra Leone |
| Bangladesh | Zaïre |
| Lao PDR | Niger |
| Bhutan | Benin |
| Ethiopia | Pakistan |
| Mali | Tanzania |
| Nepal | Afghanistan |
| Somalia | Central African Rep. |
| Burundi | Madagascar |
| Chad | Haiti |
| Mozambique | Mauritania |
| Burma | Lesotho |
| Upper Volta | Uganda |
| Vietnam | Angola |
| India | Sudan |
| Malawi | Togo |
| Rwanda | Kenya |
| Sri Lanka | Senegal |
| Guinea | Indonesia |

feasible in an economic manner. The above type of activity obviously presupposes a high level of skill availability in the country. Research and development activities, education and training in computer science and technology at academic institutions are a prerequisite for the development of such skills.

It is obvious that unless a conscious decision is taken to develop self-reliance and consequently develop the infrastructure and support services, a country would go on depending upon imported equipment and services.

It is the marketing pressures of manufacturers, political pressures and sometimes the availability of surplus capital which may make a country continue to depend on imported equipment and imported support services and thwart meaningful development in the country. In the short-term it is possible that many countries may not have the requisite skills. If, then, we have to depend upon outside support, it is perhaps better that we depend upon each other. Let a real transfer of technology take place under the TCDC (Technical Co-operation between Developing Countries) concept, and not just an exchange of goods and services for finance leading to a continued dependence on imports and restricting the growth of a nation's economy.

# The Informatics Industry in Mexico and its Future Prospects

*Carlos Enriquez Arcaraz*
*Ministry of Programming and Budgeting, Mexico City*

## BACKGROUND

A DEFINITION OF CURRENT PROBLEMS in Mexico's informatics industry involves an analysis of the conditions under which this new technology developed in this country and of the factors that have influenced its behaviour in the course of time.

Electronic data processing technology was introduced to our country in the mid-fifties. In the early sixties, after a few isolated acquisitions, the use of computers began to grow consistently and international manufacturers of these products took advantage of this growth to establish offices in this country. These facilities were solely for marketing, because in terms of actual demand the market was as yet too small to set up manufacturing or assembling operations for such an expensive and sophisticated technology.

However, there was a steadily rising trend in the use of this technology, as the installed computer capacity in the country increased from 100 units in 1966 to more than 1000 in 1978, a fact which caused imports to increase in the same ratio. Expenditure in this area amounted to 20 million dollars in the late sixties; this figure had doubled by 1970 and had grown five fold by 1978.

This dizzy growth of computer imports, as well as the country's need for informatics technology, is largely an outcome of Mexico's industrial growth.

However, the use of this technology gave rise to a number of difficulties due to the absence of an appropriate human, technical, industrial, financial and legal infrastructure. If all these things had been present, the potential benefits of the new technology would have been more efficiently capitalised, but their absence made it necessary to improvise mechanisms to regulate the import of these products and to rationalise their use, neglecting the creation of measures that would promote this new technology to meet national requirements and reduce Mexico's dependency on other countries. Multinational corporations took advantage of this lack of foresight to gain entry into this country and created an oligopolistic market.

It is estimated that the present universe of computer goods and services suppliers is made up of some 6,000 organisations, whose total revenue in 1978 was of the order of 38,000 million dollars. Eighty per cent of this amount (30,742 million dollars) was accounted for by only seven companies, known as the "Mainframers". These companies dominate the Mexican market, as they supply the largest portion of the informatics infrastructure presently in use in the country; an infrastructure which has been managed in keeping with interests that are not totally consistent with those of the nation.

The need to collect, process, store and retrieve an immense amount of data in a logical and orderly way and with maximum speed and minimum effort brought about the proliferation of electronic computers, and because of this need, computers became solidly entrenched in the country's organisational structure, creating a direct dependence on this tool, as it helped organisations to grow to such an extent that their management and control with the manual systems used formerly became virtually impossible.

However, because of this tool's sophisticated technology and high cost of manufacture—in addition to other characteristics—it cannot be manufactured entirely within the country at competitive prices. In spite of the financial benefits that such manufacture would generate, the technological complexity of the components makes domestic manufacture impossible, thus further strengthening our dependency on other countries through the international organisations active in the Mexican market.

This multinational presence in the growth of informatics has influenced in the past the adoption and development of informatics technology in Mexico, since almost all suppliers of computer goods and services in the country are part of multinational corporations or depend on such organisations.

## MEXICAN MARKET STRUCTURE

According to data being compiled by the General Bureau of Informatics Policy of the Ministry of Programming and Budgeting, there are some 350 companies in the country that may be considered as direct or indirect suppliers of the national market; of which only 92 can be considered important because of their total sales volume, which in 1978 amounted to 265 million dollars.

Of these 92 companies, 37 sell hardware, 28 sell supplies and 57 provide services. The difference in the total is because some companies sell both hardware and service at the same time.

A comparative analysis of the revenue percentage, worldwide, of the seven majors (80 per cent) compared to their market share in Mexico that same year (61 per cent) seems to show that their position here is not as strong as in other parts of the world.

However, an itemised breakdown of their revenue in Mexico shows that they had 71 per cent of the hardware market and only 14 per cent in supplies. Their market share for services was 83 per cent, so we must conclude that in all items where sales were dependent on technology, these companies' market share is very similar to what they have in the rest of the world.

As to other percentages, both in hardware and services, local firms compete with organisations which, although not considered part of the seven majors, do have foreign capital and a predominant market share, so that the purely Mexican companies cater only for marginal market requirements.

## DIFFICULTIES FACING INDIGENOUS SUPPLIERS

THESE COMPANIES ARE in an unfair competitive position vis-á-vis international firms, because they must import foreign technology by purchasing patents, manufacturing licences and specialised machinery at free market prices, since their volume is not enough for them to be considered internationally important clients (the country's consumption of informatics goods and services represents less than 1 per cent of the world market) and therefore they cannot demand special prices.

Moreover, they have no access to low cost financing or soft international loans. Their operations are conducted virtually on a cash basis (90 days at most) and thus they cannot be flexible in their marketing practices or sell on credit because of their inadequate liquidity, nor can they afford to wait a long time for buying decisions.

The supposed comparative advantage of cheap labour is cancelled out by the high relative price of components and other structural factors.

Furthermore, the educational system in the past had taken no measures to provide this sort of academic training, and the result is a shortage of personnel with formal training and the proliferation of technicians trained by vendors, who do so in such a way as to bind them to their own particular product.

In addition, most users who acquired hardware in the past did so on a lease basis because high prices made purchasing prohibitive, and the marketing policies of vendors were designed around the lease of a complete package (hardware, software, training, maintenance and service) claiming that it was better for users because of convenience and technical obsolescence. The result was that the suppliers of goods and services had a captive market, and to further their commercial interests they made it impossible for users to go to alternative vendors. This has been a great obstacle to the penetration of local firms in this market.

Other problems that frequently arise are an incorrect interpretation of tariff classifications and the time consuming procedures to secure importation permits and administrative red tape in case of claims arising from import problems.

With the purpose of helping a local informatics industry, in 1978 manufacturers were asked to submit manufacturing plans using a larger amount of domestic inputs to qualify for tax incentives.

Although many companies submitted their manufacturing programmes with domestic inputs, only eight can be considered of importance in the local market, and of these, five are totally Mexican and independent.

Of the other three, one is closely linked by its source of technology to a multinational corporation, under the OEM system (Original Equipment Manufacturer); and the other two are multinational organisations with manufacturing facilities in Mexico.

Their major products on the market are modems, microcomputers, teleprinters and magnetic disk holders. However, the ratio of domestic inputs is not more than 33 per cent in any case.

The local market is restricted and insufficient for them to have mass production with economies of scale. In spite of the Mexican market's natural limitation, the variety of computer models on the market is much too extensive. Of some 235 multipurpose computer models in the world market in 1979, more than 140 were sold in Mexico.

The total number of multipurpose computer models (235) in a large market, like that of the US, for example, with more than 60,000 computers, means that there are groups of a certain kind of hardware with sufficiently homogeneous characteristics to create specialised services or industrial activities warranted by the economies of scale.

An entire industry of specialised firms has been created around some computer lines, and even around some models, that build compatible peripheral equipment and design educational programmes and application packages, independently of the original manufacturers.

In Mexico, however, the users of most of the 140 computer models form very small groups which in no way justify the organisation of a special training program unless developed by the vendor at the home office.

The highly fragmented market is one of the main causes for the slow growth of a national informatics industry, even in fields where industrialisation is possible, such as software development, application packages and terminal construction.

Efforts to organise users of different suppliers and even of different models to improve service and act together for the development and exchange of software have, basically, succeeded in improving service.

The supply of minicomputers is even more fractional, not only because of the variety of models made by the same manufacturers, but also because of the increasing number of vendors.

The runaway increase in numbers of vendors hurts consumers used to certain rules of the industry, like maintenance service guarantees, technical support, training and others. Vendors of the latest generation of hardware do not supply these guarantees and frequently leave users without service or spare parts.

This damages both users and the incipient national microelectronics industry, because of the negative impact on the image of the mini and microcomputer industry.

Competition in the foreign market, particularly in LAFTA (Latin American Free Trade Association) was restricted by the Supplementary Agreement to Article 15 of the Montevideo Treaty, entered into in 1962.

By this agreement, production and exportation of "machinery for information processing" was assigned to Brazil, from where computers are exported duty free to all Latin American countries. Mexico, as a party to this agreement, is therefore excluded from any participation in this market.

## ACTIONS TAKEN

In view of all these problems, the Mexican government has undertaken a number of specific actions designed to promote and encourage the growth of a national informatics industry. The preparation of a National Plan for Informatics Resource Development is currently under way. This plan shall cover all informatics technology and its dissemination and application, including such things as:

Improvement of national productivity and efficiency by means of a rational application of technology;

Creation of growing awareness of the implications of using this kind of technology;

A reconsideration of the legal framework that now governs informatics, since the manufacture and use of this technology will raise new issues that must be covered by the legal system;

Co-ordination of research and development efforts of the various universities, research institutions and corporations in the country;

Use by the government of its massive purchasing power and requirements to implement strategies and actions for the benefit of domestic industry, and even its possible direct participation in financial risks.

In short, the plan is to establish the basic premises and guidelines for a policy regarding the development of informatics resources, so that an appropriate infrastructure can be laid down that will allow computers to be fully and profitably utilised throughout the country.

At the same time as the National Plan for Development of Informatics Resources is being prepared, several other actions have been initiated to organise users of the various makes and even of the various models, with the purpose of improving service and exchanging and developing software, through the Informatics Technical Consulting Committees of the Federal, State and Municipal Governments.

'Typical contracts' have been developed by users and the Federal Administration's major vendors, and an agency has been created to regulate federal acquisitions based on technical, economic and financial feasibility studies.

In order to regulate the establishment of foreign corporations, a number of policy guidelines have been proposed to deal with foreign investment in the area of electronic data processing.

There are several financial agencies with the basic objective of promoting the growth of small and medium industry, such as The Guarantee and Promotion Fund for Small and Medium Industry, the National Project and Study Fund, the National Industrial Promotion Fund and the Trust Funds for Industrial and Commercial Parks and Centres. Small and medium business, such as the informatics industry, can go to these agencies to secure loans for pre-investment studies, as well as short and long term loans for risk capital or installation of facilities.

Technical assistance has been provided by specialised institutions, such as the National Productivity Center, the Fund for Industry Information and Documentation, the National Science and Technology Council and the Industrial Financing Technical Support Institute. To perform all these functions an action tool has been designed called "industrial extensionism"—an innovation in this field—implemented by specialised and experienced agents who directly advise businessmen and industrialists, supplying a complete package of service and support, so as to gradually solve the various problems faced by industry.

A study is presently under way to analyse the current characteristics of the national informatics industry and of research being conducted in this field, so as to identify the major problems in the way of its development. This study shall be the base for the promotion of strategy guidelines and policies to orient this industry towards a development pattern in keeping with the country's requirements.

These measures have been adopted by the Federal Government in view of the need to promote various sectors of vital importance for the nation's development. Electronic data processing is considered to be one of these because of the many and complicated links that connect its technology to the different strategic areas and sectors of the country.

# The Computer Knowledge Industry—A Look at the Economic Rationale of a New Phenomenon from the East

***Robert Iau***
*Singapore Computer Society, Singapore*

## SOME FACTS ABOUT SINGAPORE

THE REPUBLIC OF SINGAPORE lies just north of the equator at the southern tip of the Malay Peninsula. It stands on the ancient trade routes between east and west and continues even today to be a vital link in the movement of people and goods from all parts of the world. This strategic location enables Singapore to establish close trade ties between all countries in the Asian region. To the north, Singapore serves Malaysia, Thailand, Philippines, Korea, Japan and all other countries in between, to the south, Indonesia, Australia and New Zealand, to the west the vast markets of the Indian sub-continent and Europe and to the east, the United States of America.

Singapore is an island city state with an area of 650 sq. kilometers. The population of 2.5 million makes it one of the most densely populated areas in the world. Singapore is multi-racial, multi-lingual and multi-cultural. All Singaporeans intermingle in a harmonious pot-pourri of languages, sounds and colours. Singapore has four official languages comprising English, Mandarin, Malay and Tamil. All citizens of the country are treated equally and no discrimination of any sort exists.

Literacy among Singaporeans is almost 100 per cent. The English language is the language of technology, science, government and business. Most of the population is bilingual with the English language as the common language. During the past 10 years Singapore has emerged as a centre of finance for the whole of the South East Asia region. There are all together 100 banks operating in Singapore either as full banks or offshore banks. By and large there is no import duty levied on capital and essential consumer goods and there is no foreign currency exchange control. (See Table 1.)

As a country Singapore is known as the cleanest city in Asia with a reputation for strictly enforcing the preservation of the high quality of the environment. Its people are highly disciplined and motivated. All male citizens are required to undergo $2\frac{1}{2}$ years of compulsory fulltime military service.

**Table 1. Banks in Singapore (by country of incorporation)**

Number of Banks: 99
Number of countries represented: 23

| Country of Incorporation | Number of banks |
|---|---|
| United States | 22 |
| Singapore | 13 |
| Japan | 11 |
| United Kingdom | 8 |
| Malaysia | 5 |
| Canada | 5 |
| India | 5 |
| West Germany | 4 |
| France | 4 |
| Switzerland | 3 |
| Hongkong | 3 |
| China | 2 |
| Australia | 2 |
| Netherlands | 2 |
| South Korea | 2 |
| Brazil | 1 |
| Philippines | 1 |
| Thailand | 1 |
| Indonesia | 1 |
| Italy | 1 |
| Pakistan | 1 |
| Spain | 1 |
| Taiwan | 1 |

**Growth of Commercial Banks in Singapore**

| | **1970** | **1975** | **1980** | **end-Jan 1981** |
|---|---|---|---|---|
| Full Licence | 37 | 37 | 37 | 37 |
| Restricted | — | 12 | 13 | 13 |
| Offshore | — | 21 | 47 | 49 |
| **Total** | **37** | **70** | **97** | **99** |

**Table 2. Unemployment in Singapore 1960–1978**

| Year | Number ('000) | Percentage of Labour Force |
|---|---|---|
| 1960 | 67.6 | 13.5 |
| 1965 | 70.6 | 12.3 |
| 1970 | 75.8 | 10.4 |
| 1974 | 34.0 | 4.0 |
| 1978 | 35.7 | 3.6 |

The picture painted of Singapore did not come about because of some fortunate circumstance. Some 15 years ago the picture was extremely grim to say the least. When Singapore became independent in 1965, it faced the frightful combination of high unemployment, the complete absence of a viable industrial base and total absence of natural resources. Furthermore the social and governmental system left behind by a colonial government made the rebuilding of a nation, together with its attributes, a matter of vital and urgent necessity regardless of how benign the past colonial government could have been. By a fortunate combination of geopolitical circumstances over the past 15 years and the pragmatic and sometimes even painful decisions which have to be taken to establish a long term economic base, Singapore was able to navigate the troubled and turbulent years of the '70s with reasonable success. (See Table 2 and Table 3.)

By 1973 Singapore achieved practically full employment. In the years that followed it became necessary to import guest workers to man the more labour-intensive industries. By 1979 about 10 per cent of the total work force in Singapore were immigrants. The long term implications of this trend and its impact upon the social fabric of Singapore were not lost on the planners. It became clear that the continuation of labour-intensive but low productivity industries could seriously affect the prospects of economic viability in the 21st century.

## RESTRUCTURING THE ECONOMY

In 1979 steps were taken to restructure the economy to aim specifically at industries with a high level of technology. Many of these target industries such as avionics require extensive inputs in terms of manpower training and capital investment. In this connection, encouragement is given in the form of attractive financial packages. Wages have been allowed to drift upwards as an incentive to workers to upgrade themselves. Re-training of staff by employers for higher levels of work

**Table 3. Singapore Indigenous Gross National Product At Current Market Prices ($Million)**

| | 1967 | 1968 | 1969 | 1970 | 1971 | 1972 | 1973 | 1974 | 1975 | 1976 | 1977 |
|---|---|---|---|---|---|---|---|---|---|---|---|
| 1 Total Gross Domestic Product | 3,746 | 4,315 | 5,020 | 5,805 | 6,823 | 8,156 | 10,205 | 12,543 | 13,373 | 14,615 | 16,091 |
| 2 Share of Resident Foreigners and Resident Foreign Companies in GDP | 384 | 624 | 800 | 1,072 | 1,234 | 1,516 | 2,090 | 2,990 | 3,059 | 3,344 | 3,636 |
| 3 Indigenous GDP (1) − (2) | 3,362 | 3,691 | 4,220 | 4,733 | 5,589 | 6,640 | 8,115 | 9,553 | 10,314 | 11,271 | 12,455 |
| 4 Net Factor Receipts of Singaporeans from Rest of the World | 236 | 231 | 252 | 256 | 237 | 244 | 294 | 413 | 481 | 442 | 524 |
| 5 Indigenous GNP (3) + (4) | 3,598 | 3,922 | 4,472 | 4,989 | 5,826 | 6,884 | 8,409 | 9,966 | 10,795 | 11,713 | 12,979 |
| 6 Number of Singaporeans ('000) | 1,978 | 2,012 | 2,043 | 2,075 | 2,110 | 2,147 | 2,185 | 2,219 | 2,250 | 2,278 | 2,308 |
| 7 Per Capita Indigenous GNP (S$) (5) ÷ (6) | 1,819 | 1,949 | 2,189 | 2,404 | 2,761 | 3,206 | 3,849 | 4,491 | 4,798 | 5,142 | 5,623 |

are subsidised by Government. Tax-holidays and soft loans for approved industries are also offered.

These measures resulted in the phasing out of a number of industries which were unable or unwilling to respond to the change in industrial climate. However it is not correct to say that the economic planners of Singapore are a callous or unsympathetic lot to entrepreneurs who have invested substantially in Singapore. The opportunities for the transformation of industries exist and generous assistance is always offered by the Government to help to upgrade the level of technology in any industry. It must be realised that, as a country moves away from the more basic forms of industry because of the increasing competence of its people and the need to look beyond the role of providing cheap labour for unproductive industries, the entrepreneurs must respond in like manner in order to stay in business. All developing countries have this aspiration and there is nothing adverse in such countries aiming to take their rightful places among the ranks of developed countries.

Asia is a vast landmass housing one third of humanity. It is largely underdeveloped and in many instances, technology is almost completely unknown.

As a notable exception to countries in Asia, Japan has been flexing its industrial muscles over the last two decades. Its entry into the automobile and consumer electronics markets and its aggressive marketing policies together with high productivity of labour aiming at zero-defect products have completely mesmerised other developed countries. There is evidence now to show that Japan is planning to play a larger role. The larger corporations are beginning to acquire interests in production facilities in the West and are extending their manufacturing bases beyond Japan. The smaller corporations in Japan with similar interests are also planning mergers and acquisitions to secure sufficient clout to improve on their efficiency and market positions. It is clear that Japan will attempt to preserve and improve upon its dominant market position by transforming her narrow production base at home to a world-wide network and also to take advantage of any indigenous research and development capability especially in the advanced Western countries in order to maintain its long-term leading-edge position.

In computer manufacture and informatics, Japan occupies top position in Asia. Its burgeoning informatics industry has overtaken many developed countries and today is second only to the United States. Research and development into VLSI technology and the unheralded but significant progress in robotics have resulted in a strong base to dominate informatics in the 21st century. Japan's past performance in penetrating established markets gives every reason to think that its efforts in the informatics market, which is still early in its product life cycle, will prove to be equally if not more spectacularly successful. (See Table 4.)

Asia represents the biggest area of potential growth in the world. The next 20 years will hopefully see a rationalisation of the political and economic pot-pourri of Indo-China and the Peoples Republic of China. The increasing stability of the South East Asia region through consolidation of ASEAN will obviously strengthen the purchasing power of the region and enhance its position as a market for advanced technology. It is not unrealistic to expect that this market could exert

**Table 4. Japan: Import and export of Computer Systems 1970–1978 (in yen million)**

| | IMPORT | | | | | | EXPORT | | | | | |
|---|---|---|---|---|---|---|---|---|---|---|---|---|
| Year | Main Units | Percentage Change | Peripherals | Percentage Change | Total | Percentage Change | Main Units | Percentage Change | Peripherals | Percentage Change | Total | Percentage Change |
| 1970 | 22,718 | — | 51,551 | — | 74,269 | — | 3,974 | — | 1,114 | — | 5,088 | — |
| 1971 | 21,659 | −5 | 48,873 | −5 | 70,532 | −5 | 5,587 | +40 | 2,382 | +114 | 7,969 | +57 |
| 1972 | 23,437 | +8 | 38,221 | −22 | 61,658 | −13 | 8,216 | +47 | 2,644 | +11 | 10,860 | +36 |
| 1973 | 32,469 | +38 | 49,564 | +30 | 82,033 | +33 | 6,062 | −26 | 5,728 | +117 | 11,790 | +9 |
| 1974 | 48,609 | +49 | 65,436 | +32 | 114,045 | +39 | 10,233 | +69 | 7,790 | +36 | 18,023 | +53 |
| 1975 | 45,504 | −7 | 51,870 | −21 | 97,374 | −15 | 11,089 | +8 | 9,628 | +24 | 20,717 | +15 |
| 1976 | 49,493 | +8 | 46,174 | −11 | 95,667 | −2 | 11,396 | +3 | 27,296 | +184 | 38,692 | +87 |
| 1977 | 54,491 | +10 | 54,755 | +18 | 109,246 | +14 | 17,866 | +57 | 22,135 | −19 | 40,001 | +3 |
| 1978 | 36,346 | −33 | 45,199 | −17 | 81,545 | −25 | 21,834 | +22 | 47,251 | +113 | 69,085 | +73 |
| Overall Percentage Change 1970/1978 | +60 | | −13 | | +10 | | 449 | | 4,141 | | 1,258 | |

considerable pressure on the direction which informatics will take in the region. Informatics technology applicable in the developed countries would no longer be foisted upon countries in South East Asia without regard for the differences in cultural, sociological and economic backgrounds. As with all products the market must ultimately decide on what it wants. In Asia this is no different and the informatics industry must accept the fact that the Asian environment will require Asian solutions to Asian problems.

Singapore is a traditional free port and almost all manner of legitimate businesses and enterprises are allowed to operate with practically no Government intervention. The system of education with its emphasis on the hard sciences and mathematics facilitates a natural shift to almost all areas of technology.

The first computer was installed in Singapore in 1963. As of this moment, there are over 400 computer installations of various sizes but not including microcomputer systems. It is anticipated that the number of installations will grow to a thousand by the end of 1983. This works out to about 0.04 per cent saturation over 2.5 million people in Singapore. Applications are varied and range from international network systems for airline reservations to banking systems, process control systems in the three oil refineries in Singapore and the normal business and management applications.

## DEVELOPMENT OF A SOFTWARE INDUSTRY

In planning the emergence of a software industry in Singapore over the next 10 years, the Singapore Government paid particular attention to the supply and quality of manpower. In this regard the education curricula were reviewed and new training institutions established for the express purpose of training software personnel.

To begin with, computer studies as an examination subject will be offered to all A-level students with effect from this year. Over 200 mini and micro processors will be installed in all secondary and pre-university schools within the next 18 months. Teachers have been undergoing training since mid-80 to ensure proper staffing in the schools. The National University of Singapore has revamped its syllabus to include a large element of computing in its curriculum in addition to pure computer science. This will ensure that graduates would be suitably trained to be productive almost immediately. Teaching of computer usage is also included in other disciplines such as engineering, science, business administration, commerce, economics and so forth.

A training institute known as The Japan Singapore Institute of Systems Technology was to be set up by the end of 1981. This institute, a joint venture between the Japanese and Singapore Governments, will train software personnel for both large and the minicomputer systems. It will also offer training courses for non-computer professionals such as engineers, managers and upgrading courses for existing computer personnel in the market. A second institute known as the Institute for Systems Studies will also be set up. The initial objective of this institute is to concentrate its training to basic entry level software personnel but the institute will work closely in future with the National University of Singapore and computing professionals on software at the leading edge of the technology. Computing studies have recently been introduced into the two technical colleges in Singapore. Another two technical colleges are expected to be established by mid-1980s and they will have computing included in their curriculum.

These steps have been taken to ensure a steady supply of high quality manpower in Singapore. The Singapore Computer Society has been asked to assist in ensuring professional competence and standards in the industry. Contacts have been made with the Institute for Certification of Computer Professionals in the United States, the British Computer Society and the Ministry of International Trade and Industry of Japan to explore the possibility of reciprocity in qualifications and membership. The development plan calls for the production of 10,000 highly trained personnel by mid-1980s.

On the business side of the coin, various incentives in terms of financial support and tax-holidays will be introduced by the Singapore Government. Purchased computer equipment can be written off against corporate tax over 3 years. Organisations who send their EDP staff for upgrading courses either in Singapore or overseas could be subsidised by the Government up to a maximum extent of 70 per cent of the total training cost. Guidelines are being prepared to allow tax relief on income from software developed in Singapore and exported overseas. The Government will also liberalise the issue of professional and employment passes for computer experts who wish to set up offices or work in Singapore.

You may well ask whether Singapore has gone overboard and created an overkill situation to develop an informatics and software industry. You might even feel that this mad rush to train a large manpower base is inconsistent with the nature of the constantly changing technology. You might also say that without first identifying the markets for different software, it would be wasteful of resources to embark on such an ambitious training programme.

These and many other questions have been asked. For a small country such as Singapore, having only recently overcome the frightful attrition of high unemployment and low skills content of its workforce, and having to contend with the stark realities of zero resource backing, the thrust towards the informatics industry must be one of the few options available to ensure the continued growth and prosperity of the country. Looking at the informatics world map, it is clear that the massive potential of Japan must be recognised and accepted as a force which will exert substantial influence in the shaping of informatics and the direction it will take. For a developing country like Singapore, it is impossible to compete with the industrial giants of the developed nations with their established product and marketing networks. Joint ventures with multinationals could at best place Singapore interests in the position of less than preferred shareholders.

Informatics as a market product is still very early in its life cycle and as the world moves towards increasing automation in repetitive chores to maximise productivity of manpower and diminishing resources, the life cycle of informatics as a genus is open-ended. Singapore's commitment to the informatics industry in future is no more than trying to secure for itself a leading-edge position in a growing market. The circumstances of education, language and economics all appear to be auspicious. Perhaps in the arena of computers and informatics small may indeed prove to be beautiful.

# Appropriate Informatics for Developing Countries

***P.D. Jain***
*Computer Maintenance Coroporation Limited*
*India*

## THE PROBLEMS OF DEVELOPING COUNTRIES

THE DEVELOPING COUNTRIES have common problems—problems of hunger, poverty, elimination of illiteracy, and the identification and management of resources. The role of informatics in these areas cannot be ignored. However, we cannot also ignore the fact that the role of informatics will vary significantly from one country to the other. The technologies developed in the industrialized countries are meant for mass production and large scale units. These are normally capital intensive with a heavy bias on labour saving devices. Further, these technologies assume the existence of certain infrastructures and stability of industrial activity. This unfortunately is not the situation in the developing countries. The extremely wide and unpredictable fluctuations in the supply of raw materials and other inputs put heavy strain on the small scale industries sector and have defied the use of informatics in any meaningful way.

There is therefore a need to differentiate between the model for development for industrialized nations and the one which would be of relevance to developing countries.

The biggest single factor of production in these countries is labour. Any program or model therefore suggested for developing countries has to ensure that the existing level of unemployment is first brought down to a reasonable level. Only after this primary task is completed can one visualize programs for a qualitative improvement of the work force.

According to the International Labour Office (ILO), despite projections of increased economic production, poverty will worsen in Asia and by 1987 more than 800 million people will be affected. General unemployment and lack of jobs are at the root of Asian poverty. The concentration of poverty is in the villages where the majority of the impoverished Asian masses live. India, for example, is a country of 600,000 villages. Approximately eighty per cent of the population lives in these villages.

## NEED TO CONCENTRATE ON RURAL SECTOR

A COMPARISON between various developing and developed countries' data on labour force distribution brings out very clearly the need to concentrate on the rural sector. The table below based on a World Bank study shows the distribution of labour for some selected countries:

**Labour Force Percentages**

| Country | Percentage of Working Population (15–64 yrs) | | Agriculture | Industry | Services | Av. Annual Growth of Labour Force |
|---|---|---|---|---|---|---|
| | 1960 | 1978 | 1978 | 1978 | 1978 | 1980–2000 |
| India | 57 | 56 | 74 | 11 | 15 | 2.0 |
| Brazil | 54 | 55 | 41 | 22 | 37 | 2.9 |
| UK | 65 | 64 | 2 | 43 | 55 | 0.4 |
| Japan | 64 | 68 | 13 | 39 | 48 | 0.7 |
| FDR | 68 | 65 | 4 | 48 | 48 | – |
| USA | 60 | 65 | 2 | 33 | 65 | 0.9 |
| Iran | 51 | 51 | 40 | 33 | 27 | 2.9 |
| China | 56 | 61 | 62 | 25 | 13 | 1.4 |
| USSR | 63 | 65 | 17 | 47 | 36 | 0.7 |

The concern of thinkers and planners should be the development of the rural sector first to ensure decentralization of economic growth and utilization of manpower. The role of agriculture in the national economy can never be overemphasized. It has been abundantly clear in the case of India and other developing countries and even some developed countries. A very important observation that one can make here is that any drop in the purchasing power of the rural sector in developing countries can wipe out all the efforts for industrial development. A corollary of this perhaps is that if development of the rural sector is taken care of, industrial development will be taken care of automatically.

Since the days of the industrial revolution, the emphasis has been on the growth of industries in urban areas. This has resulted in continuous pressure on the small scale and cottage industries, eliminated artisans and rendered millions of people jobless. There is enough evidence to show that in industrialised countries the percentage of the population engaged in industrial activity is going down although production is increasing.

Because of higher economic growth rates achieved in the developed nations, by more efficient production and by siphoning capital from the Third World, and extremely low annual growth rates of the labour force, the effects of a reduction of labour utilization are not felt so severely. But, in the developing countries, this causes critical problems of unemployment, concentration of investment in a few geographical pockets and large scale migration of rural population to urban centres. Some of these men and women get employment or perhaps under-employment. Others who do not get employment continue to stay in the urban centres in the fond hope of getting employment some day.

The socio-economic implications of such a population migration are multifarious. The strain on the already deficient infrastructure of the urban centres and the passenger and goods transportation system, the increasing crime rate in the cities, the impact of the alienation of the breadwinner from the family, the emotional strain a person from a rural area goes through when he has to close down his loom or his match factory due to unfair competition from the organised industrial sector—if all these factors could be converted into monetary values, I reckon these would add up to a figure which would be higher than the industrial output from the organised sector.

These are certain realities of life which have to be understood and examined before models of informatics developed in one country can be transplanted into another. According to India's Revised Draft Plan 1978–83, out of a total labour force of 273 million, 4.4 million are unemployed, 26.49 million are in recorded employment and 241.98 million are in informal activity. A very high percentage of this number is made up of a landless labour force. This segment of the labour force survives on a bare subsistence wage. Not only is employment seasonal, but it depends on the vagaries of nature. The second source of earning, the cottage industries, is denied to him.

## THE MODEL OF INFORMATICS FOR DEVELOPING COUNTRIES

WHAT IS THE RELEVANCE OF INFORMATICS, as being discussed here, to him? Will informatics continue to be accessible only to the elite and thus sustain its elitist status?

Having said this, I would like to examine the factors which are contributing to a lower rate of industrial development in many of the developing countries. For this analysis, it is necessary to divide the developing countries in two broad groups—the rich developing countries and the poor developing countries.

In the case of the poor developing countries, apart from the disruption in the supply of basic industrial inputs like petroleum, coal, electrical power, the output of industries can be increased only if the market can be broadened or the purchasing power of the people increased. Broadening of the market outside the boundary of the country is becoming more and more difficult due to barriers raised by importing countries. Thus one comes back to the problem of raising the purchasing power of the rural masses.

One of the common denominators for the developing countries is an abundant supply of human resources. The model of informatics for developing countries and their industrial development, therefore, can be christened the model for human resource utilization.

The world of computers and associated informatics is a very rapidly advancing one and a very glamorous one. The Third World blindly follows the developed nations and lives in a dream world of electronic mail, lightpens and laser beam devices. Many of the software packages used for information processing in the Third World countries may either not be relevant or may only serve the interests of the organised sector. These pockets of information processing have become islands which are cut off from reality and from the majority of the population. Their interests may not coincide with the interests of the people and the country but are adopted to suit the needs of their organisations.

International organisations, academics, administrators and others interested in the welfare of developing countries have to undertake the responsibility of weaning away developing countries from blindly following the industrialized countries. The emphasis of informatics should be on the exploitation of human and natural resources, and on the rural sector—both agricultural and cottage industries. An improvement in these sectors would generate sufficient savings and purchasing power to give a powerful impetus to industrial development.

To conclude I would like to quote from E.F. Schumacher's *Small is Beautiful*: "The new thinking that is required for aid and development will be different from the old because it will take poverty seriously. It will not go on mechanically saying: 'What is good for the rich must be good for the poor'. It will take care of the people—from a practical point of view. Why care for people? Because people are the primary and ultimate source of any wealth whatsoever. If they are left out, if they are pushed around by self-styled experts and high-handed planners, then nothing can ever yield real fruit."

# The Development of an Informatics Industry Sector in Ireland

***D.P. Hanna***
*Industrial Development Authority, Ireland*

## INTRODUCTION

THE EXPLOSIVE GROWTH of informatics and the movement towards the information society in this decade and the next has come about as a result of technological advances in a number of related areas—

1. Computer hardware, both mainframe and peripherals
2. Computer software, particularly on-line systems developments and database management systems
3. Advances in data transmission using telecommunications systems.

Through its industrial job creating agency, the Industrial Development Authority, the Irish Government has succeeded in establishing considerable activity and expertise in the country in both the hardware manufacturing area, and increasingly, in the export software area, and in conjunction with the Department of Posts and Telegraphs, in the manufacture of telecommunications equipment. Some comments on each of these areas may be of interest to the Conference.

## COMPUTER HARDWARE AND PERIPHERALS

THIS INDUSTRIAL SECTOR was initially selected due to its growth (up to 25 per cent per annum in Europe), the high value/volume ratio of its products, the easy transportability of the products, and its profitability.

The initial strategy was to encourage as many of the world's leading producers of these products to come to Ireland to service the European marketplace. They were attracted by:

1. A well educated workforce
2. Stable, and welcoming political structure
3. Readily available buildings on attractive sites
4. Cash grants towards the cost of land, buildings, equipment, and worker training
5. Generous taxation incentives.

The companies which responded to this invitation include, to date:

| Company | Products |
|---|---|
| Digital Equipment | Computers, Peripherals |
| Computer Automation | Computers, Peripherals |
| Apple Computers | Computers, Peripherals |
| Nixdorf | Computers, Peripherals |
| Prime Computers | Computers, Peripherals |
| Wang Laboratories | Computers, Peripherals |
| Perkin Elmer | Computers, Peripherals |
| Amdahl | Computers |
| Centronics | Computer Printers |
| Dataproducts | Computer Printers |
| Documation | Computer Printers |

Ireland is rapidly improving the relevant infrastructure (manufacturers of components, links with research institutes, universities etc.) to enable these companies to integrate their Irish operations, and thus enable them to increase the quality of their employment, their capability to become more innovative in Ireland, and to become, in time, a training ground for Irish entrepreneurs.

## COMPUTER SOFTWARE

In many ways, the production of computer software is an ideal employment creating activity for a country such as Ireland. The most important 'raw material'. is a supply of well educated young people, which Ireland has in plenty. The industry does not have a high demand for capital investment either directly or for physical infrastructure (with two important exceptions—telecommunications and educational facilities), and is expected to grow extremely rapidly in the foreseeable future.

For this reason, a new package of financial incentives is currently being designed to enable the Irish Industrial Development Authority to encourage both Irish owned and overseas owned software operations (both independent software operations, and the software departments of computer and instrumentation companies) to expand their businesses in Ireland.

## TELECOMMUNICATIONS

A KEY ELEMENT in the development of this industry is a reliable and efficient internal and external national telecommunications capability with full data transmission capability. The Irish Government has recently committed in excess of £600 million for the installation of digital switching systems, and other sophisticated telecommunications equipment to modernise our systems. Work has already started, and is proceeding rapidly.

## CONCLUSION

IRELAND has approximately 85 companies in its electronics industry. These have an output of £500–£600 million, and an employment of 15,000. This is expected to double by the mid-1980s. This industrial development has been 'forced' by the Irish Government which has created conditions which enable the industry to flourish here. The key ingredients to enable Ireland to take its place in the information society are already in place and are being actively developed. The whole process described above has taken place in less than ten years. Developments prior to 1970 were surveyed in Foster (1971). It is hoped that these experiences will prove of interest to representatives of small nations which are not yet highly industrialised.

### References

Foster, F.G. (1971) *Computers in Ireland.* Economic and Social Research Institute, Dublin.

# Some Measures for Promoting Indigenous Informatics Technology

***Robert E. Kalman***
*Hungarian Academy of Sciences*
*Budapest*

## INTRODUCTION

INFORMATICS, as interpreted here, deals with a wide spectrum of scientific, technological and engineering disciplines (Kalman, 1978). Informatics technology is the technological sector of informatics which includes both hardware and software for data processing, telecommunications and mass communication (Kalman, 1981). The Intergovernmental Conference on Strategies and Policies for Informatics (SPIN) attempted to identify a further segment, the appropriate informatics technology (Kalman, 1979). The informatics technology which is appropriate for one nation, might be rather unsuitable for others, due to the social, cultural and economic dissimilarities of countries. The social, economic and cultural differences between nations are well-illustrated in Figures 1 to 4 and discussed in detail elsewhere (World Bank, 1980; Kalman, 1982). The figures show the average life expectancy at birth (Figure 1), the number enrolled in primary, secondly and higher education, as a percentage of age groups (Figure 2), the percentage of the labour force in agriculture, industry and services (Figure 3) and the percentage of distribution of gross national product (GNP) in agriculture, industry and services (Figure 4) in relation to GNP per person.

These social, cultural and economic dissimilarities are also reflected in countries' level of development in informatics (Kalman, 1974). Thus, each country requires its own general macro-informatics strategy and to elaborate its corresponding sectoral policies. This paper concentrates on measures related to data processing only. We believe that an appropriate selection of these measures leads

**Figure 1.**

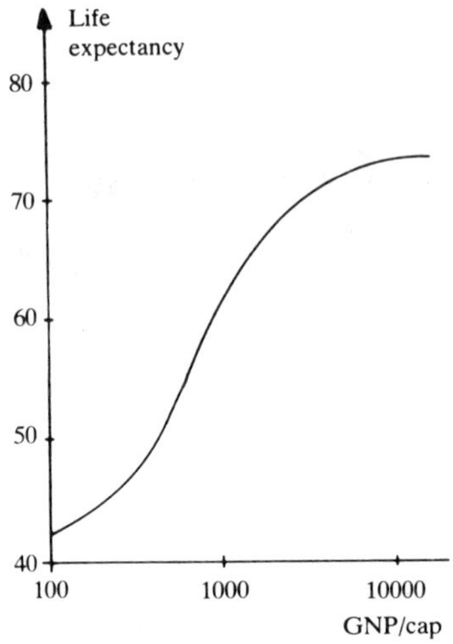


**Figure 2.**

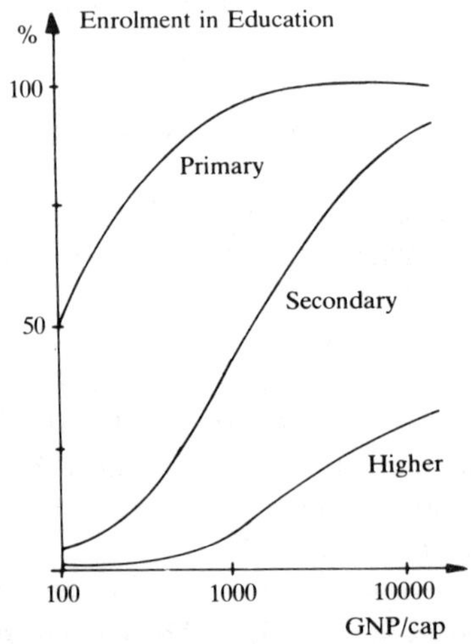


**Figure 3.**

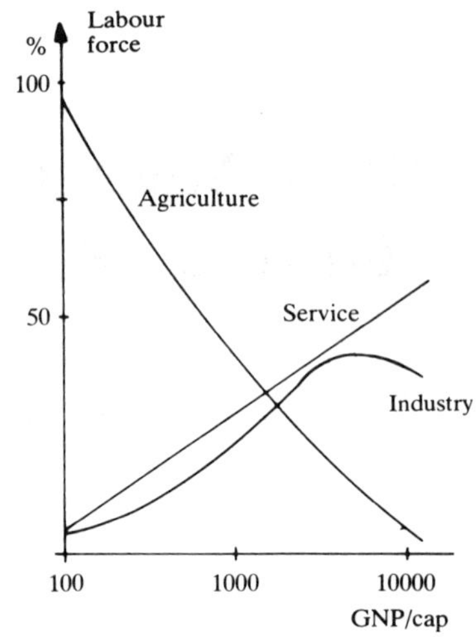


**Figure 4.**

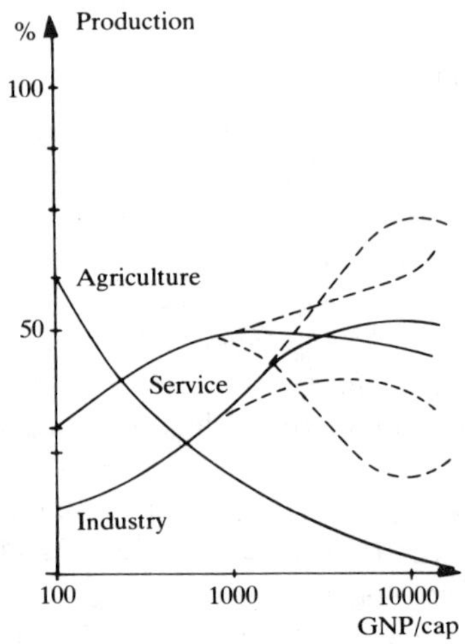

to the indigenous development of an important segment of domestic informatics technology

## PROMOTION OF INDIGENOUS HARDWARE INDUSTRY

THE HARDWARE AND SOFTWARE related to data and information processing, telecommunications and mass communications are in the realm of informatics technology. The research, development and production of hardware requires sufficient (Figure 2) and well-educated (Figure 3) manpower, as well as an advanced industrial production structure (Figure 4) which are typical of high GNP-per-person industrialized countries. The most important measures which can be beneficial for the development of a domestic hardware industry can be summarized as follows:

**Government financing of VLSI development:** The design and production of new generations of data processing systems require, among others, the development of advanced, very large scale integrated (VLSI) circuits. The experience of industrialized countries shows that special programmes should be set upand financed by governments for research and development of VLSI's. In Japan, for example, the government has covered 50 per cent of the research and development expenses of the five national computer manufacturers who were responsible for the introduction of VLSI circuits of new generations of data processing systems.

**Government financing of hardware development:** Direct financial assistance should be given by governments to the hardware industry to cover a certain percentage of any expenses related to a new generation of data processing systems, peripheral equipment and terminal devices. In Japan again, up to 50 per cent of the research and development costs were met by the government in 1976. In countries, however, which have no indigenous computer manufacturing industry and only the multinational firms have local plants for one or other line of their systems, the governments should aim at balancing their imports of goods with exports of locally manufactured equipment.

**Bank loans for restructuring industry:** In order to increase the competitive power of the domestic data processing industry, so as to be capable of meeting the challenge of foreign manufacturers in the local market, national development banks should grant loans for manufacturers to restructure their production.

**Bank loans for computer rentals:** In many countries a considerable proportion of computer sales is in the form of rentals. In order to reduce the financial burden on local manufacturers, the national banks should grant loans to supply the required rental funds. In Japan again the Japan Development Bank regularly grants loans to the Japanese Electronic Computer Company for this particular purpose.

**Special tax measures:** In hardware manufacturing countries the user who purchases domestic systems to improve his information processing capabilities is often encouraged by the government through special depreciations (up to 20 per cent in some places) against taxation for the initial period and a fixed asset tax at a reduced rate (4/5–2/3 of the ordinary rate) for another period of time.

**Government purchases:** Governments are usually the largest single users of domestically produced data processing products in industrialized countries and could be an important support for the domestic hardware industry. In the United States, for example, almost all computers used by the Federal Government and the state governments are of domestic make. In France, the central government offices, as well as the overseas territories were obliged by law to use French machines. Although this policy did not go into actual effect, the British National Economic Development Committee in its report recommended that all public enterprises in Britain be required to purchase domestically produced systems.

## PROMOTION OF INDIGENOUS SOFTWARE INDUSTRY

In spite of all these measures, the establishment of an independent, national hardware industry is an extremely difficult task, even for major industrialized countries like France or England and probably an impossible one for smaller countries of limited industrial potential like Belgium or Hungary (Figure 4). Certain segments of domestic software requirements, however, might be met locally if the country's educational system provides the required, skilled manpower (Figure 2). It should not be assumed, however, that setting up a software industry is a much easier task than the establishment of its hardware counterpart. The following measures are beneficial for the development of a domestic software industry.

**Government financing of software development:** To promote an indigeneous software industry direct government supports should be given to computer manufacturers, service bureaux, software houses, universities and users with the aim of helping them to obtain computer hardware, land, educational and training facilities etc. The Japanese government, for instance, is financing the development of high level languages, the automation of software production, on-line data processing technology and the promotion of informatics in the community.

**Bank loans for software industry:** To complement direct government support bank loans should be granted for service bureaux, training and educational institutions and users for software development and training of informatics personnel.

**Special tax measures:** In view of the importance of updating software, the local manufacturers should be allowed to set aside a certain percentage of the tax on their total software sales to provide reserves to cover expenses in making modifications to software already sold.

**Compilation of program register:** In order to avoid duplication of effort in software development and to promote the distribution of available programs among various users, program registers should be compiled. These program registers should be easily accessible for service bureaux, software houses and users, as well as potential users.

**Protection of software:** For the efficient operation of service bureaux and software houses in relation to the distribution of software, it is essential that strict measures are taken by governments to protect the copyrights of the program owner. Although the basic question of "what is to be protected against what" is studied in a number of industrialised countries, as well as in an international forum by the World Intellectual Property Organization, as far as we know, no important result has been achieved yet. In Canada, for instance, an attempt has been made for revision of the existing patent and copyright acts. It has been suggested, at the same time, that neither "mathematical methods or algorithms" nor "programs" nor "a general or multipurpose computing or data processing apparatus adopted to execute a given computer program as an instance of an ability to execute a variety of computer programs"—which is in fact the software itself—should be patentable.

## PROMOTION OF APPLICATIONS

The development and production of informatics technology is not a goal in itself but a tool to satisfy basic human needs through increasing the country's gross national product. In some industrialised countries the manufacturing and export of hardware and software products contribute directly to the national income, in other places they contribute indirectly to the GNP by supporting the country's agricultural and industrial activities (Figure 4). Consequently not all applications are equally important in all countries at a given point in time. The increasing demand in both developing and industrialised countries for a qualitative improvement in the standard of living in general and for food production and distribution, water resource management, primary health care and other government services are creating serious social problems. In order to be able to keep up with these increasing demands some governments may consider food production, or environment management or health care as of the highest priority.

**Identification of priority applications:** The developing countries usually place higher priorities on "traditional" activities related to government planning and control operations, such as food production, water resource management, energy production and distribution. The priorities in industrialised countries are rather different. A government report in Canada suggests the development of appropriate payments systems, and such communication interface standards which safeguard the rights of individuals, users and suppliers of the payments systems. The Japanese Government, with the help of specialists, has identified other priority applications such as pattern recognition, automobile traffic control, medical information processing, community video and international trade information processing.

**Applications in small enterprises:** Large and multinational enterprises can afford to purchase, and can extensively use, informatics technology. The economic situation of small enterprises, however, does not always allow them to benefit from data processing technology. Some governments, therefore, have initiated projects to improve the position of small enterprises. These projects usually include the development of locally made, standard management data processing systems for both production control and administration, special training for managers and data processing advisors in small enterprises, direct financial assistance for purchase of computers, terminals and software products and finally, the establishment of centres which make information services more readily available to small enterprises.

**Applications in central and local government officies:** Data processing technology has become an indispensable tool in government offices and consequently there has been a steady increase in both the number of personnel employed and in the expense incurred in data processing operations. It is required therefore, to initiate surveys to obtain a better grasp of the state of data processing operations in government, to encourage offices to evaluate their present activities, to promote studies, to consider effectiveness and efficiency in operations and to prevent the loss, damage, divulgence and theft of important data at government offices.

## IMPROVING THE HUMAN ENVIRONMENT

THE PROVISION of qualified manpower is at least as important for the development of domestic informatics technology as the allocation of financial resources. The experiences of capital-surplus oil exporting countries show that purchasing and installing informatics technology is often easier than assuring their operation and maintenance because qualified nationals are just not available and their education and training is a rather time-consuming task.

**Testing informatics specialists:** The education and training of specialists undertaken by educational institutions and vocational training schools play a central role in the advancement of informatics technology. The quality of preparation, however, is often very mixed. Qualification tests are needed for operators, programmers, system engineers and other data processing personnel to improve their technical skill. In order to improve the skill of data processing specialists, some governments allow those studying for the qualification tests to deduct their training expenses from tax.

**Computer-aided learning:** The extensive use of computers in university research is often the result of early support given by governments to universities for the acquisition of data processing facilities and for the introduction of computer science and other related courses. Computer-aided learning (CAL) is an attempt to make better use of available facilities and, on many occasions, is a joint effort by universities and industry. The educators develop the course materials, technologists in industry contribute to hardware and software and the government research agencies finance the projects. Some initial results have been reported: well over a thousand students have taken the CAL cardiology course during its seven years of operation at a Canadian university but, as the costs involved are still rather high, it is unlikely that the use of CAL will spread rapidly either in developing or industrialized countries.

**Promotion to the general public:** The extensive use of informatics technology in all sectors of the national economy requires, in addition to the improvement of technical skill of specialists, a growing interest and understanding by the general public, too. With this in mind information weeks on informatics would be organized; during this period special radio and television programmes would be broadcast, well written articles published in the newspapers and exhibitions held.

## CONCLUSION

Information is often considered as the third basic resource, in addition to matter and energy. Although data processing is just a segment of the whole information-related technology, its impact is growing both in industrialized and developing countries. In countries which manufacture and export hardware and software products, data processing contributes directly to the nations' GNP, in other countries it can indirectly support agricultural and industrial production. It is in the interest of all governments to understand what informatics technology, in general, and data processing in particular, can do for the country and to make well-conceived measures to promote its appropriate national development.

## References

Kalman, R.E. (1974) *Towards Quantitative Measurement of Development in Informatics* in Frielink, A.B. (Ed.), *Economics of Informatics.* IBI-ICC, Rome.

Kalman, R.E. (1978) *An Introduction to Informatics.* Impact of Sciences on Society, **29**, 3, UNESCO, Paris.

Kalman, R.E. (1979) *Appropriate Informatics Technology* in *Criteria for Selecting Appropriate Technologies under Different Cultural, Technical and Social Conditions.* FAST, Milan.

Kalman, R.E. (1980) *Emergence of Informatics and the Changing Employment Structure* in Berting, J., Mills, S.C., Wintersberger, H. (Eds.) *The Socio-economic Impact of Microelectronics.* Pergamon Press, Oxford.

Kalman, R.E. (1981), *Eight Strategic Issues for Informatics* in Bennett, J., Kalman, R.E. (Eds.) *Computers in Developing Nations.* North Holland Publishing Co., Amsterdam.

Kalman, R.E. (1982) *Comparative Study on Strategies and Policies for Informatics.* (manuscript), UNESCO, Paris.

World Bank (1980) *World Development Report.* Washington D.C.

SECTION 4

# National Infrastructures for the Utilisation of Informatics: Strategies for the Development of Institutional, Human and Physical Resources

This section is concerned with the infrastructural conditions necessary to support the harnessing of informatics for optimum national development. Emphasis is, in particular, placed on policies in the crucial area of education and training and this aspect is examined from a number of different perspectives and at a number of distinct levels. Other important infrastructural aspects and issues covered include communications, research and development, and institutional arrangements for providing information on industrial organisation.

# Basic Criteria for a National Infrastructure Specific to Developing Countries to Facilitate the Use of Informatics

*A. Mufti*
*Centre d'Etudes et Recherches Scientifiques, Syria*

## GENERAL ASPECTS OF THE PROBLEM

THE INFORMATISATION of society and informatics in the service of industrial development is now integrated into socio-economic thinking, having as a goal the improvement of human well-being by means of the 'man-machine' ergonomic system. This leads quite naturally to the acknowledgement of the general usefulness and international significance of informatics. Therefore an appropriate policy and infrastructure should be implemented to secure for society the socio-ergonomic promotion that informatics deserves.

The preparation of a national infrastructure for informatics is conditioned by the individual circumstances of each country, i.e. the requirements, the goals and the available means, which in fact constitute the specific criteria of a national infrastructure.

If these criteria are of a kind common to all countries, they are considered to be critical factors whose values specify the informatic potential of each country. Thus the industrialised countries, notably those with an indigenous informatics industry, enjoy some overwhelming advantages in achieving a policy of informatisation of society resulting from the concurrence of all the requisite means to this end, of which the principal ones are;

(1) The mastering of the informatics technology (hardware and software);

(2) The establishment of manufacturing capacity and the resulting employment;
(3) The creation of equipment for their own needs and conditions, for local application in the industrialised countries;
(4) Local availability of technical support for maintenance, repair and assistance in the application of informatic systems;
(5) Security and continuity of functioning;
(6) Theoretical and practical training at all levels;
(7) The linguistic advantages.

This group of advantageous factors are conducive to the formulation and implementation of a sound and far reaching informatic infrastructure.

By contrast, the developing countries, only partially enjoying these advantages, are not well placed to keep up with informatic developments, and even less so in the informatisation of their societies. Consequently, all informatic policies, insufficiently infrastructured, would be vulnerable.

The institution of a national infrastructure, adapted to the conditions of developing countries, would consist in the first place of the means to overcome the obstacles and constraints inherent in the circumstances of these countries.

In this paper an attempt will be made to set out the constraints and the basic criteria, specific to developing countries, with a view to elaborating an informatic infrastructure. It is not the intention of this short account to draw up an exhaustive list of the criteria, but rather to state precisely the most crucial factors of an informatic infrastructure. The basic criteria, forming a set of complementary and interdependent factors, are of a heterogeneous nature: socio-economic, technical, practical and strategic.

## THE CRITERIA FOR SOCIO-ECONOMIC ADAPTATION

A SOCIETY that has little or no experience of using the techniques of informatics must profit from the experience of others so as to avoid their mistakes, the effects of which could have very grave consequences for a society which has not attained technical maturity.

A national infrastructure adapted to the requirements of a society must take into consideration the socio-economic criteria set out below.

### The Social Criteria

The principle social criteria are:

(1) *The Ergonomic Aspect*: the ergonomic system consisting of man and machine, introducing a new methodology of work based on up-to-date

techniques, can only be assimilated by a society little acquainted with the use of this technique by dint of careful preparation by the most appropriate means. There should be a scrupulously planned and proportioned educational programme in terms of the normative circumstances of each country and particularly those which define man in society.

(2) *The Aspects of Training and Linguistics*: a national plan for education in informatics must be established by introducing informatics as a fundamental subject in primary, secondary and university studies. A plan of education of this scope can only be achieved in the developing countries with the concurrence of national and international organisations, and by setting up appropriate establishments.

It should be possible to use the language of the individual country throughout all data processing. It is difficult to accept that, in a scheme of informatisation of society, the language of the computer manufacturer could be imposed.

The most rational solution for developing countries would be to collaborate with the manufacturers at the initial stage of the production of informatics equipment adapted to their own languages. This could open the way to a large industry to produce linguistic interfaces.

(3) *The Systems Approach to Organisations*: against the numerous advantages of informatisation there is one major draw-back resulting from the systems approach to organisations. The flow and processing of information are restructured and the old structure normally disappears in the process of informatisation. The functioning of informatic systems depends on the technical team and on the informatic equipment. The failure of either carries the risk of paralysis of the informatised environment, since it would be practically impossible to put the old structure back into use. Hence the irreversibility of this phenomenon. For a developing country, having regard to the fragility of its technical potential, this phenomenon renders it vulnerable if back-up measures are not incorporated into the original systems design. Such measures, depending on the socio-economic circumstances and the informatic potential of each country, should be the object of in-depth study and constant revision within the framework of the national infrastructure programme.

## The Economic Criteria

On the economic plane several criteria specific to developing countries should be considered, such as the following.

(1) The concept of the "consumer society" as it is practised in the industrialised countries cannot be regarded as a factor of socio-economic evolution in the developing countries because of economic, financial and above all technical reasons: the degree of technical ability and rate of evolution in these countries remains constantly out of phase with the rate

of technical and industrial innovation. The introduction of an informatic system to a more evolved system rarely runs smoothly and is only achieved with difficulty in developing societies.

(2) Likewise the criterion of "quality and reliability" of equipment is a basic element for equally economic reasons, but above all for the security and continuity of functioning; indeed this last aspect has a preponderant influence in decision making regarding the informatisation of society.

## The Criteria of Technical Adaptation

The concentration of the informatics industry in the industrialised countries leads quite naturally to the conception and production of systems adapted to the needs and conditions of these countries, and it is obvious that the requirements of developing countries are not catered for (or are ignored) by the manufacturer, from which follow the constraints of adaptation at all levels: hardware, software and application.

(1) *At the Hardware Level*: hardware appropriate to the needs of developing countries must be capable of accepting the language of each country, both at the level of programming as well as at the level of input and output of information.

The social and administrative structure, in most of these countries, is characterised by a centralised decision-making system and decentralised data, data capture being the most important function. Moreover, in most cases, it is a question of data processing with characteristics such as:

introduction of data of little importance;

execution of very diversified and variable data processing, most frequently of low volume;

data capture at scattered locations.

The problem exists of defining the informatic configuration adapted to these characteristics, to optimise the triple factors: performance, profitability and price.

(2) *At the Software Level*: in the developing countries it is the tertiary sector which is the most important and the development of the active population benefits this sector.

Appropriate software must cover all the needs of this sector, the price of which should not include a subsidy to that of the other two sectors.

(3) *At the Level of Application*: it is at this level that the problem of adaptation is the most critical: because the informatic hardware is designed only to function in an applications environment, climatic and infrastructural, determined by the circumstances of the producer countries. It would not long maintain its efficiency when forced to work in appreciably different conditions. This is a case frequently met with in developing countries, for example:

*Climatic environment*

the safety margin of the manufacturers as well as the so-called normal values for temperature and humidity show up serious discrepancies in relation to actual values;

the temperature and humidity gradients are clearly higher than those of the manufacturers;

the seasons are practically reversed according to the geographic situation of the countries.

*Applications environment*

errors and anomalies appear in the data being collected;

there is a loss of information in transmission;

tests of the data can be ineffective.

These are due either to maladjustment of the systems (hardware and software) or to the physical circumstances of the infrastructural supports (e.g., the instability of the electrical supply, the poor quality transmission lines). It should be added that solutions in the form of air-conditioners and electrical stabilisers have proved to be even more clumsy than the hardware both at the installation and in application. Clearly, this state of affairs appreciably affects the profitability of informatics and consequently must be taken into consideration as a specific criterion of an informatics infrastructure.

**The Problem of Maintenance**

The effectiveness of the manufacturer's guarantee varies according to the geographic situation of the user: those closest to the manufacture having the greatest advantage.

In the developing countries where the manufacturers only provide basic representation (and this depending on their profit margin) maintenance remains unsatisfactory and constitutes a serious handicap to the use and development of informatics. This problem, which is a regional one, can only be solved within the framework of a multinational or regional infrastructure, and would consist of establishing a regional centre with common interests, having sound maintenance as its objective. This would have the support of: a training school for technicians in all branches of maintenance and a logistic network for carrying out repairs, providing supplies and spare-parts. This centre could be established through the co-operation of specialised organisations (IBI, UNIDO, ITM, etc.) and the collaboration of the manufacturers.

## CONCLUSION

THE INFRASTRUCTURAL preparation for the informatisation of a developing society, being the pillar of the establishment and development of informatics, must be considered from the point of view of apprenticeship and of ergonomic and socio-economic adaptations with a view to providing the technical potential necessary to master the informatics in order to facilitate its use with confidence in all areas of application.

# National Strategies for the Development of a Telecommunications Infrastructure in Ireland

*T.J. Byrnes*
*Irish Telecommunications Board, Dublin*

## REMOVING THE TIME/SPACE CONSTRAINT

MANKIND HAS DEVELOPED his social structure in response to several needs: the need to eat, to survive in a hostile environment and to continue as a species.

These needs and others less urgent have been met by the formation of social groups. As these social groups required some means of communication, sounds or symbols came to be used which transmit a thought in the mind in a way that was understood by another mind or minds.

In simple agrarian, pastoral or hunting societies these communication methods were simple and related to very fundamental needs and were specific to relatively small groups—families, tribes, etc.

As mankind's numbers increased and he mastered his environment, he began to produce goods and services beyond those necessary to meet his basic need for survival. Trade relationships developed and with them the demand for more extensive physical and verbal/pictorial communications links.

Until the development of the moving press within the last 500 years the extent of communications was severely limited to the capacity for manual reproduction of these symbols, words or pictures and the physical movement of them by the limited means of transport (foot, horse, sea) available.

The development of the moveable press and its refinement after 1445 dramatically increased the availability of a media for communication and significantly reduced its cost. The first real prospect for world-wide literacy became evident.

The distribution of the new printed works was still severely limited in time and space by its dependence on the physical movement of man or beast or vehicles which depended on their strength for land transport and the power of the wind for movement by sea.

If we look at the map of the world we will find the large cities have primarily developed in places where physical distribution or exchange took place in conjunction with transport by water. It was faster than any other method dependent on the power of men or beasts. I will mention only a few:

in Europe: London, Paris, Hamburg, Lisbon, Barcelona;

in the Americas: New York, Buenos Aires, Rio De Janeiro;

in Asia: Bombay and Tokyo;

in Ireland: Dublin, Cork, Belfast.

Water power was also used in the early mechanisation of manual methods and some early "industrial" centres developed along the banks of powerful water courses. Wind was also used as an auxiliary source especially to pump the useful water to where it was needed.

The development of the steam engine around 1700 and its evolution to the petrol powered internal combustion engine around 1885 provided a dramatic alteration in the time and space constraint on human development.

It was now possible to develop areas away from water courses and indeed it became necessary to do so because the energy sources required for the new "movers" were seldom to be found near the existing cities and towns. We can see this in cities such as Birmingham and Hannover which are close to large coal producing areas.

The accelerated development of man's physical capability to move anywhere in an ever decreasing time took a quantum leap at the beginning of this century when at last in the early years of this century he left the earth and took to the air.

Simultaneously with this breakthrough in physical movement we had the development of electrical "communication devices". First in code with the morse telegraph, and then in the spoken and written word with the telephone and telegram in the second half of the last century. Communicating time became nearly instantaneous and the space constraint was substantially reduced.

The continued development in this century of systems, devices and methods of communication by electrical energy reached its ultimate with the capture of impulses on recording media—impulses which stored, transmitted and reproduced symbols, voices, images, first on discs and then on film and tape. The need to have those communicating linked in time is gone; the need to have those communicating linked in space is gone; and with the use of satellites for the distribution world-wide of communication, the solar cycle of day and night is no longer relevant as information can be simultaneously broadcasted and stored world-wide.

## RESPONSE OF MANKIND TO CHANGE

THE RESPONSE OF MANKIND to change has always been slow. When the evolution of man's communication system was measured in generations this slow response time was acceptable. However in this century the speed of adaptation of man's social systems to change is inadequate and he has continued to act as if most of this change has never taken place.

Thus the traditional growth of already overcrowded large urban centres, most often based on those which developed because of their proximity to water, has continued. Across the world, people born in rural areas continue to gravitate to these centres in search of employment in industry, commerce or services. As the move from simple industry to complex industry takes place, there develops a need for complex commercial and service support which follows the industrial development. Government naturally grows on top of this base.

This development is particularly noticeable here in Ireland, with the former colonial capital Dublin still having a disproportionate amount of the population and services as well as Government workers.

It is not my purpose to comment on the human and social problems of large urban complexes. Suffice to say that they are quite obvious to any observer and in my opinion a development to be stopped, reversed and avoided whenever possible.

Common sense and considerable empirical evidence tell us that people are less frustrated, more helpful to each other and generally happier in smaller communities.

The purpose of this conference is to allow the developing countries to consider how the new information and communication technology can help them to develop both human and physical resources.

I am convinced that the information/communication revolution which is building on a foundation of sand—in the form of a silicon chip—will not collapse but offers real potential for increasing the quality of life for all the world's people.

The development of inexpensive, energy efficient information and communication systems offers the one hope to avoid the inhuman social conditions which exist in the large urban areas throughout the developed world. If you begin now to plan for the development of your countries on the basis of the fact that this technology combined with the development of air transport has removed the Space/Time Barrier to physical and human communication you can, I am convinced, look forward to a bright and humane future.

## DEVELOPMENTS IN IRELAND

As an island located on the western fringe of the European Continent, away from the major sea trading routes and centres, and not permitted to develop a strong sea presence under a national flag, this country remained primarily pastoral until the end of the 1950's. The surplus population moved to the industrial, commercial and administrative centres of the English speaking world in search of employment. For all those previous generations of Irishmen and women this was a sad fact of life. But for the present generation the fact that this country did not partake in the industrial revolution is a distinct advantage.

We are not burdened with the need to adapt old industry to new developments. We are not burdened with an infrastructure based on the traditional physical communication method for thoughts and goods and services. We have the potential to rapidly adopt the new technology in the development of our nation and of our nation's place in world trade.

As a result of a government decision at the end of the 1960's there is a positive policy to site industry away from the metropolitan area of Dublin. There has been a move of government departments to other parts of the country and this policy is being continued.

In 1972 the Government, in a preview of regional policy, announced as the basis of its planning for the next 20 years, the following strategy, and I quote:

> "In the government's view an overall regional strategy should not merely seek the attainment of required national growth rates but should also provide for the maximum spread of development, through all regions, giving an increased and wider range of economic and social opportunities and so minimising population dislocation through internal migration."

Accordingly, the strategy which the Government envisages should be pursued over the next 20 years or so, is as follows:

(1) Dublin development to be such as to accommodate the natural increase of its existing population,

(2) expansion in and around Cork City, the Limerick/Shannon/Ennis area, and Waterford, Galway, Dundalk, Drogheda, Sligo and Athlone.

(3) development of counties or large towns of strategic importance in each region, including the relatively large expansion of towns in areas remote from existing major towns.

(4) continuation of special measures for the development of the Gaeltacht.*

This policy of decentralisation of industry which has been entrusted to the Industrial Development Authority (IDA) for implementation has been made possible by the concentration of this body on industry which has a good future. It is indeed fortunate for us that one of these industries—the electronics industry—is

* Designated regions, mainly on the western seaboard, where Irish, rather than English, is the predominant language.

the primary contributor to, and benefactor from, the information/communication revolution. Thus we have in Ireland not only the opportunity to develop our industrial base in a decentralised way but also to benefit from the world-wide explosion in demand for the products produced by these companies.

In the 9 years from 1972 to the present, employment in this industry has doubled and output measured in current prices has grown from £38m to £500m (Punts), to a point where Ireland now supplies 10 per cent of Europe's office equipment products and 2 per cent of total demand for electronic products.

In the last two years the present government has taken steps to improve the infrastructure which is fundamental to the success of this programme. It has decided to invest £800 million in the telecommunications system in order to bring it up to the standard achieved in the EEC by 1984. It has taken the decision to base this development on digital switching and transmission which will permit the integration of voice, data, text and pictures for transmission through the system. It has decided to build a satellite earth station in the south of the country.

It has in recent months taken a decision to develop regional airports. This is important because many of the goods we produce have a high value added or light weight and suffer from the rough handling experienced by surface transport. Given our physical location on the edge of our market, we need to develop the air transport of the products of our new industry to the maximum extent. We are, of course, still a large exporter of beef and beer and minerals and therefore the commitment in the investment plan to invest in roads is also most important. I believe these decisions will begin to bear fruit in the latter years of this decade and if the policy of decentralisation is maintained there is a real prospect that the growth of Dublin will be halted, perhaps even reversed by the end of the century.

I believe that the developments taking place here in Ireland are far reaching in their consequence for the social development of this country. I think that those responsible for planning or administering the development of other countries can learn from Ireland. If you study what we have been doing and what we are planning to do you may find similar problems and a hint of some solutions.

# Notes on Computer Education and Training in Developing Countries

*Ibrahim S. El Miligi*
*University of Garyounis, Libya*

## IMPORTANCE OF THE HUMAN ELEMENT

THE PERFORMANCE of any system depends mainly on two sets of factors. The first set is the initial inception and design. If the system is not properly conceived and designed it will lack the necessary efficiency, reliability and coherence. Consequentlv the different parts of the system will not function properly and ultimately it will not achieve the goals for which it is commissioned.

The second set of factors includes the environment in which the system is operating whether that environment is organisational procedures or people interacting with the system in one form or another. However, among all the factors affecting the performance of a system the human element remains the least controllable. A necessary way of securing a proper interaction between the human element and the system is to ensure that proper education and training is given to all personnel interacting with or affected by the system. Failure to provide the necessary education and training will usually render the human element the weak link in the system.

The realisation that the human element could be the weak link led to the recognition of the education and training of computer personnel as a key factor in the success of computer applications.

In data processing systems we can at least identify three areas where people can directly affect the system. The first area is the area of the professional computer personnel, i.e. systems analysts, programmers, operators, etc. The effect of computer professionals on the system may, of course, differ from one category of personnel to another, i.e., a programmer may have a different effcct from an operator. The second area is thc area of the users of the system outputs. The end result of any computer-based system will ultimately be represented in a report of some form for the use of the person responsible for the application or for the use of the person who monitors the performance of the system itself. The needs of these users affect the system in its initial design brief and subsequent modifications. The third area is the area of input to the system. It is almost traditional to think of computer-based systems as being operated by data preparation staff, systems analysts, programmers, etc. The starting point of systems organised in that way is technically the input from the data preparation room. What happens before that is usually catered for by standard procedures or forms specified at the design stage. The interface between the computer-based system with its technical starting point of the data processing room and its input source may be a control clerk(s). Yet the predata preparation stage is a critical stage where difficulties could easily arise in developing countries given the substantial 'leap' from manual systems to computer-based systems.

The definition of areas where the human element affects the system is very important and has a direct bearing on the type of education and training necessary for each category of personnel. But before going into more detailed analysis the requirements of developing countries should be clearly stated.

It should be emphasized that developed countries have, through their economic, social and technical development, developed a suitable environment and trained personnel. Coping with new computer systems has not been difficult and benefits can be efficiently realised. In developing countries, however, the situation is different. Developing countries may import the most sophisticated non-human parts of computer systems from developed countries, but they cannot provide sophisticated human links in terms of local professional expertise. In certain cases they can acquire the services of expatriates to design and implement the systems, but subsequent running of the systems and use of information must be done by local personnel. The provision of well trained local people is both expensive and time consuming. Yet developing countries have no other alternative but to develop their own professional expertise if the introduction of computers is not to be slowed down or their capacities wasted.

Effective results in the field of computer personnel training and education, however, can only be achieved through the drawing up of adequate and realistic national policies. It is not enough to 'encourage' computer courses or even subsidise them. There must be a genuine government commitment to the provision of computer professionals and the training and education of people to use and cope with computers.

## EDUCATION AND TRAINING

ATTEMPTS TO ENCOURAGE developing countries to initiate national plans for the education and training of computer personnel are numerous. Analysis of some of the literature (United Nations, 1973; Pipe and Veenhuis, 1975) on computer education and training reveals that there are two areas of confusion. There is a confusion between education and training. The two terms are sometimes used synonymously although basically they are different and indicate two different types of learning.

Education, as defined by the Oxford Dictionary, is the 'development of a person's character or mental powers'. Training is defined as bringing a person to a 'desired state or standard of efficiency, etc. by instruction and practice'. The first impression one gets from the two definitions is that education is a more general process that aims at developing a person's mental powers to cope with unknown situations. Training on the other hand has a specific aim—to bring a person to a 'desired' state or 'standard' of efficiency. The desired standard of efficiency relates to how well a person carries out the specific job for which he is trained.

Training is usually required for persons whose tasks are well defined and do not vary in nature or content in the short run. Data preparation staff, programmers and computer operators, for example, do not require a high level of education but a varying degree of intensive training, whilst systems analysts and computer managers require a high level of education, supported perhaps with training in the use of certain techniques.

Training is easily catered for by manufacturers, institutions or various training agencies. What makes training easy to cater for is that it is specific and with a defined time span. Education on the other hand is rather difficult to cater for. It is not specific and has a longer time span. Moreover, before any person is educated in any discipline or science that discipline should be well defined.

## DEFINITION OF COMPUTER SCIENCE

THE EVOLUTION of new sciences has always been faced with problems of definition, especially boundary definitions, and computer science is no exception. Even now it is very difficult to find a definition of computer science that draws the unanimous consent of all computer scientists. In fact this is the second area of confusion.

The main reason why computer science is not yet well defined is that people associated with the early development of computers have come from other disciplines and as such they hold different points of view.

Recent discussions on education and training have even dropped the term 'computer science' and used the term 'informatics'. Yet even 'informatics' may not be the final term. The consequence of this confusion, or to be more precise, lack of proper and defined terminology, makes most of the discussion very general and theoretical. The problem of education and training has been approached in the context of the general problem of informatics (Hebenstreit, 1975) and thus practical details of how to solve the problem of education are not available.

This lack of definition of computer science or informatics makes it very difficult for education to be catered for by the same organisations that cater for training. Therefore, governments of developing countries must take increasing responsibility for providing the necessary education and training at all levels and should not leave it to non-governmental agencies.

## NATIONAL PLANS

As we have mentioned earlier it is not possible to adequately provide for education and training unless there is a definite national plan stating the requirements in terms of quantity of personnel and quality of education and training, and specifying the institutions that should cater for both education and training. National plans, however, need very determined efforts since they tend to be rather elusive and hard to achieve (Miligi, 1979).

One of the major efforts to encourage developing countries to draw up national plans for computer technology application is the United Nations second report of the Secretary-General (United Nations, *op. cit.*). The report outlined several recommendations for the enhancement of computer education and training. These recommendations, when subjected to a critical analysis, also revealed a certain lack of distinction between education and training as defined earlier. Such lack of distinction may perpetuate in whatever policies are based on the recommendations and the result will not be as desirable as the recommendations are aiming at.

The UN report also neglects the training and support of a very important category of professionals. Training of Organisations and Methods (O & M) experts is extremely important in developing countries since most of the prospective computer application areas lack formality, order, and clear definition of jobs and responsibilities (Miligi, 1978).

## AIMS OF COMPUTER EDUCATION AND TRAINING POLICIES

The aims of any computer education and training policy may be summarized briefly as follows:

(a) To produce computer professionals at all levels to man existing and future computer centres.
(b) To conduct concurrent courses of computer appreciation and orientation

for those who will subsequently use computers, e.g., decision makers, planners, middle management, etc., and to undertake the training of para-computer personnel (O. & M. experts).

(c) To conduct mass appreciation courses to create, in the general public, an awareness of computers, their power and shortcomings.

(d) To maintain a high standard in the profession. Computer professionals must be kept up-to-date with recent developments in the computer field. This involves a continuous transfer of computer technology from developed countries to developing countries. It would be highly undesirable and indeed impractical if such a transfer is not monitored and evaluated to see its relevance to the country involved so that only the useful parts are adopted. This monitoring and evaluation of development in the computer field can only be done by a research unit. Such a research unit can also help in initiating research in areas where problems are encountered and as such triggers an autonomous local research activity.

The type of education and training to be given to each category may differ. Even within one category training or education may differ in content, duration and intensity. Application professionals, for example, may receive a varying degree of training. It is here that the distinction between education and training helps in moulding the type of training and/or education to be given to each category or sub-category. Data preparation staff, computer operators, and programmers do not need a high level of education. Instead an intensive training course over a defined period of time is sufficient. Systems analysts and computer managers on the other hand require a high degree of education in addition to a certain degree of training. Computer scientists working in research units clearly need a high degree of education and also a high level of training so that they are producing realistic and relevant solutions and results.

Computer users need training in understanding the computer and how to utilise it while being aware of its shortcomings and limitations.

The training of the public is clearly not practical. What is needed here is a mass education programme. However, computer education in schools, especially secondary, technical and commercial, should be given priority. Use of the mass media, pamphlets, magazines and newspapers could be helpful in disseminating computer knowledge.

It is imperative that developing countries should realize that training by suppliers and other agencies is clearly limited and caters only for a small fraction of the overall needs of a country for computer education and training, and that, unless the problem is taken in its totality, computer introduction in these countries will greatly suffer.

## References


Hebenstreit, J. (1975) *Informatics Education in Developing Countries* in Pipe, G.R., Veenhuis, A.A.M. (Eds) *National Planning for Informatics in Developing Countries.* North Holland Publishing Co., Amsterdam.

Miligi, I.S. (1978) *Attitudinal Problems and Computers in the Sudanese Public Service* in Proceedings of International Conference on Cybernetics and Society, Tokyo.

Miligi, I.S. (1979) *Towards a National Policy for the Application of Computer Technology in Developing Countries.* Dirasat in Economics and Business, **15**, 1, 2, Libya.

Pipe, G.R. and Veenhuis, A.A.M. (Eds) (1975) *National Planning for Informatics in Developing Countries.* North Holland Publishing Co., Amsterdam.

United Nations (1973) *The Application of Computer Technology for Development.* Second Report of the Secretary General, United Nations. E.73.II.A.12. New York.

# Education and Training in Informatics in Africa

***R.J.P. Scott***
*University of Nairobi, Kenya*

## INTRODUCTION

**Reports on Computer Technology for Development:** For the past decade international bodies have discussed ways in which computers will affect development. Particular emphasis has been given to the application of computer technology to development. The first report published by the United Nations in 1971 is an important document which is still relevant today in discussions of the subject.

Since its publication a second report has been prepared by the United Nations (1973) in which 30 recommendations were made. In 1971 and 1974 two conferences in Jerusalem on Information Technology took place with two large volumes of papers published. In 1978 the SPIN conference was held in Spain, (Inter-Governmental Conference on Strategies and Policies for Informatics). This conference made 40 recommendations (UNESCO-IBI, 1978). In all these conferences the recommendations endorse the enhancement of social and economic development with the effective utilization of computer applications. Nearly a decade has passed since the first UN report was prepared; in the countries of Africa, computer development has either been stagnant or haphazard and certainly not planned according to the numerous recommendations put forward at the various conferences on Information Technology.

**Current Applications:** To date most applications of computers in this region are for the routine clerical tasks and employ standard data processing procedures using large data files. In some instances these procedures are totally magnetic tape based and still involve the use of punch cards as the primary input medium. At the first Computer Managers Conference in Arusha in 1979 (EAMI, 1979) the ranking of the top ten applications processed at the installations represented were found to be applications of standard accounting functions. It should be emphasised that most government computer centres in the region process a considerable number of activities using a computer which in relation to the development of the nations concerned must be regarded as advanced uses of data processing when compared with their counterparts in the developed countries at the same relative stage of development. As stated in Scott (1971) "There is one question which occurs to most people and this is: why use computers when there is such unemployment in emerging countries? Most government departments and commercial organisations are relatively new in structure and can use a computer at an early stage in their growth. The organisations will have the advantages of up-to-date reports and files on which decisions for future development can be made, yet at the same time avoid the 'clerical mess' that will require a computer solution at some later stage."

James Clayson (1980) argues that one of the main reasons why methods of operations research are not used extensively is because there is no facility for local training in such techniques. He cites examples where such techniques have been successfully implemented by direct importation from outside but warns of possible antagonism if not supported by all the workers directly or indirectly involved. The same can apply to data processing techniques.

Two regional meetings have been held: the first Computer Managers Conference in Arusha in September 1979 and the UNESCO sponsored meeting of Computer Centre Directors in Africa in Arusha in April 1980 (UNESCO, 1980). To date one book has been published on the use of computers in Africa (Taylor and Obudho, 1977).

**Levels of Computer Activity:** Julian Bogod (1979) proposed that the different levels of computer activities in a country can be classified as initial, basic, operational, and advanced.

From the above it appears that Kenya is at the Basic level or moving towards the Operational level. There are now approximately 100 computer mainframes in Kenya in a population of 16 million persons. One major problem is still staff recruitment; for example the Institute of Computer Science of the University of Nairobi has had 3 vacant posts of Senior Computer Programmer and Senior Systems Analyst for 30 months.

**Formulation of National Education and Training Policies:** In order to be able to quantify data on the existing application of computers in a country an important research project must be established. Such a project is outlined in the last section but must include manpower projections. It is surprising to note that even today many senior government officials are not aware of the extent of computerisation in Kenya.

The following section will outline some of the background issues related to computer education and training in the region. The second section will categorise the main courses now required for new entrants to the profession. The third section will deal with advanced workshops and the education of management and users. It is my contention that all of these issues must be dealt with on a national basis without further delay.

## BACKGROUND ISSUES RELATED TO EDUCATION AND TRAINING

*Subject Nomenclature*: The use of the terms computer science, informatics, information processing are discussed in some detail (IBI, 1976). At the University of Lagos there is a Department of Computer Sciences. At the University of Nairobi an Institute of Computer Science teaches and researches computer science and data processing subjects, while a diploma in computer science has started this year.

*Estimates of training needs*: Much has been written on the needs for national training plans for informatics (United Nations, 1973; IBI, 1976, 1978) but little can be done without quantitative data on existing computer installations. It should be possible to maintain and update a directory of computer installations and related staff requirements for the whole country. From this data educational and training projections can be made. Such data was scheduled for publication in Kenya (April 1981) from a research project funded by IDRC in Canada.

*Practical Oriented Training Requirement*: At present the type of education and training required as a priority should be practical rather than academically oriented. The installation of many smaller mainframes in the 1980's leads to a different requirement for training which will involve an installation staff running a computer custom-programmed by the vendors.

*Centres for Education and Training*: If education and training are to be successfully provided proper equipment (hardware and software) must be available as well as a proper training environment. With so many packaged courses available each institution must be equipped with overhead projectors, slide projectors, slide making equipment, video player, etc. Such equipment has recently been donated to the Institute of Computer Science by DAAD.

*Retention of Training Staff*: Considerable difficulty is experienced in the retention of data processing staff by the public sector. In particular a university or similar teaching institution faces a severe problem in retaining local staff after a considerable investment of time and money in training. The only realistic solution is to provide for public employees in this field to be paid a professional supplement to equate their remuneration with that obtained in the private sector, or to actively encourage consultancy work to be undertaken. Whilst local staff development is being undertaken international aid agencies will be still required to assist in the finding of skilled expatriate staff.

*Aptitude Testing*: Most computer vendors and private computer training firms still provide for a fee the ritual of the aptitude test as the key to entry to computer careers. Most of such tests are imported from the USA and Europe and little work on their current relevance in the region has been undertaken.

*Private Computer Training Schools*: One of the observations made in Arusha (British Computer Society, 1980) was concern at the exploitation of the general public by some private computer training companies. Control of their activities by national governments was implored to prevent uninformed members of the public from being given unsuitable training which may not be acceptable by potential employers. Until alternative sources of training are available on a sufficient scale this may well be counter-productive.

*Examination and Assessment*: Examinations and assessment of existing data processing specialists becomes a major problem when new formal courses are introduced for new entrants from formal schooling. For the orderly growth of the profession some ways of assessment for experience should be devised. It may be appropriate for a computer society or association to evolve to become a professional society with a full-time secretariat to undertake this role. Experience to date throughout the world has not been without considerable difficulties.

*Pan African Association of University Departments/Institutes or Institutions*: A most fruitful exchange of ideas and developments can be maintained with the establishment of a pan African association of University Departments/Institutes or Institutions involved with information processing training. It is the kind of association that UNESCO should be able to support and enable an annual meeting to take place. Regional conferences and seminars could be arranged from this forum and exchanges of experiences and material initiated. A recommendation to this effect was made at the UNESCO-sponsored meeting of Computer Centre Directors in Africa held in Arusha, April 1980 (UNESCO, 1980).

*Coordination of Training and Education on a Regional Basis*: The international aid agencies should be requested to provide scholarships to local personnel from the region to attend existing or proposed courses at Institutions in the region. The duplication of resources should be discouraged at all costs.

## EDUCATION AND TRAINING FOR NEW ENTRANTS TO THE PROFESSION

*Postgraduate*: The main developments that have taken place are at the postgraduate level. The postgraduate diploma course in computer sciences at the University of Lagos is sponsored by the University of Lagos, UNESCO, and OAU/STRC. The University of Nairobi commenced a postgraduate diploma in Computer Science in 1981 at the Institute of Computer Science, initially for 11 students, where the British Government has provided an ICL 2950 computer system with a terminal network under its capital aid programme to Kenya. The course comprises the following subjects: Foundations of Computer Systems, Information Processing, Programming Methodology, Software Systems, Principles of Accounting and Management, Business Systems Analysis and Design, Assembler Language Programming, Computer Technology, Operating Systems, Scientific and Engineering Computing, Computer Management, Database Management, Advanced Programming, Simulation, Computational Statistics, Mathematical Programming, Computer Assisted Engineering Design; and a computer systems project. Both of the above courses have compulsory and optional subjects. The Lagos course seems more scientifically oriented whilst the Nairobi course is more business/data processing oriented.

*Undergraduate*: At a considerable number of Universities undergraduates from the Faculties of Engineering, Science, and Commerce receive service courses in the foundations of computing and at least one programming language. At the University of Nairobi a modular structure to these courses is being evolved. A considerable number of course syllabuses for Computer Science curricula have been published (CACM, 1968; The Computer Journal, **18**, 4). Work should be undertaken to evolve more courses in information processing subjects at the undergraduate level eventually leading to joint degrees. All undergraduates should receive at least one course in information processing. The most suitable position of the unit in a university may be independent from existing faculties to demonstrate the multi-disciplinary nature of the subject (Scott, 1979).

*Post A-Level (Higher) School Course*: A considerable number of students from the region have attended courses equivalent to the Scottish Higher National Diploma in Computer Data Processing at Aberdeen College of Commerce. The need for this level of course was mentioned in the observations in Arusha in 1979 (EAMI, *op. cit.*). The most suitable venue for this course may be at a national technical college or university. The Kenya Polytechnic in Nairobi has just taken delivery of 14 PET microcomputers and expects to start courses using this equipment shortly.

*Post O-Level (Form 4) School Course*: At least one such course was mounted in Zambia at the Zambia Institute of Technology in Kitwe as a 2-year Programming Certificate course. A good syllabus already exists in the Scottish National Certificate in Computer Operations. The need for this level of course was mentioned in the observations in Arusha in 1979. In Kenya a schools computer project has been started at Starehe Boys Centre with support from local computer installations and vendors on an experimental basis. It is currently planned to introduce an experimental A-Level course at the same institution.

*School Courses*: Attention should be drawn to the work already undertaken (British Computer Society, 1980) to keep so called 'Computer Studies' courses up to date. A text published in Nairobi in 1971 was aimed at the schools population (Scott, 1971).

*Systems Analysis & Design Training*: The East and Southern African Management Institute in Arusha has been running an Information Systems Course for several years. A well documented and updated course with an examination scheme has been available for many years from the National Computing Centre in the U.K. There is a demand for more of these courses to be given in the region at least once a year.

*Data Entry and Data Control Personnel Training*: Two areas where key personnel are involved is in data entry and data control. In most instances these personnel have only attended a skills course on data preparation. It is time that all new entrants to these positions are given formal training on the principles of information processing.

*Vendors Courses*: The current practice is for the majority of new entrants to the profession only to be able to attend the courses run by the computer vendors usually at a regional centre rather than on a national basis. These courses are usually well designed but have the obvious disadvantages of being product orientated.

## COURSES FOR EXISTING PERSONNEL AND EDUCATION OF MANAGEMENT AND USERS

*Advanced Workshops*: A framework for advanced courses on data processing subjects on a national basis should be established. A consultative committee would devise an annual programme of short fulltime courses and "workshops" of an advanced nature for computer professionals and information processing specialists in the country.

Such a consultative committee is expected to be drawn from the University, government, private industry, vendors, international aid agencies, and the local computer association. In Kenya it is proposed that each academic member of staff of the Institute of Computer Science would give at least one such course in his/her specialization in each annual programme of courses. At least three professionals from outside Kenya would be invited each year. Courses that are expected to be mounted in 1980/81 include: Structured Programming, Microprocessors, Operating Systems, Financial Modelling, Distributed Processing, Statistical Computations and Databases.

*Video Recorded Teaching Packages*: There are now available some courses that are documented and video recorded to enable students to follow the material at their own pace. The National Computing Centre in the UK has such courses available which should be reviewed.

*Part-time Extra Mural Courses*: Where a public establishment exists a series of part-time, (evening) courses on an extra mural basis are usually well attended and provide an important input into the computer education of users.

*Government Institutes of Administration*: It is most likely that national governments have a training centre for public administrators. It is most important that courses for senior administrative personnel be provided at all levels in order to obtain maximum use of government computer centres.

# ESTABLISHMENT OF COMPUTER RELATED RESEARCH PROJECTS AND PRIORITY COMPUTER APPLICATIONS FOR DEVELOPMENT

Below are enumerated a number of important research projects and priority computer applications for development, most of which will be funded on a national basis and run, either at a national teaching or research institution, or at the appropriate government department.

## Computer Related Research Projects

*Computer Utilisation Study*: An important project already mentioned which should be undertaken on an on-going basis is on computers and development policy in each country. Such a project should maintain an inventory of computer installations, manpower, utilisation statistics and applications processed. A directory of computer utilisation can be published to increase general awareness of computer utilisation and to facilitate an exchange of experience within the country. The information should be updated on an annual basis.

*Telecommunications for Data Transmission*: Most telecommunications facilities are in the monopoly hands of a semi-state post and telecommunications corporation. Most of them in the region only provide data transmission using dedicated lines with their own supplied modems. It should be possible to develop the use of dial-up facilities for data transmission on one institution's telephone network in conjunction with the telecommunications corporation.

*Microprocessor-Based Products*: The exploitation of the microprocessor in product development is now taking off worldwide. National universities research projects should be established by departments of Electrical Engineering, Physics, and Computer Science. In this way prototype products designed for local conditions can be investigated.

*Microprocessors*: With the availability of word processors to streamline clerical work, their introduction needs to be studied with regard to the socio-economic effects on traditional urban employment opportunities.

*Library automation*: Considerable difficulties exist in the maintenance of manual library records in nearly all national university libraries. Considerable progress in the automation of libraries has been undertaken in Europe and America. It should be possible to undertake a general project to find a feasible solution to this problem. Once the national university library is automated all accessions within a country are the next priority.

*Application of Portable Microcomputer Systems*: The uses of the new, relatively cheap, portable microcomputer in the fields of computer assisted instruction, medical diagnosis and farm management for the rural population needs investigation as these are applications that can benefit different aspects of development.

### Priority Computer Applications for Development

Some of the scientific applications of computers are often linked to research projects from other disciplines. These projects are often undertaken at one of the university computer installations because only at such sites are any suitably qualified technical applications staff available to give proper advice and guidance.

It is apparent that considerable research that could use computers in this way is not being undertaken due to the shortage of technical manpower. An urgent need is to establish or strengthen full-time technical and scientific systems analysts and programmers at university and central government computer installations.

An example of the projects currently identified at the University of Nairobi are:

(a) access to and analysis of Landsat satellite data recorded on magnetic tape.
(b) conversion of rainfall data on magnetic tape to disc.
(c) establishment of a research profile database of all on-going research projects in the country.
(d) assembly of gynaecological records at the teaching hospital over the past 10 years to form a database for research.
(e) processing of statistical records of infant mortality in Nairobi.
(f) provision of experimental software to enable surgeons to diagnose the need for surgery.
(g) access to census data for planning.
(h) selection criteria for small business computers.

In the above research projects and technical applications the most important requirement is the co-ordination of effort, avoidance of duplication, and communication of work already undertaken and in progress.

## WORKSHOP ON NATIONAL EDUCATION AND TRAINING IN INFORMATICS

Once the up-to-date data on usage of informatics within the country is available it should be appropriate to hold a national workshop to devise education and training plans for the country. Such a workshop is proposed for Kenya to draw together all the independently active institutions and interested groups later this year in co-operation with the Directorate of Personnel of the Government of the Republic of Kenya.

## CONCLUSIONS

DESPITE THE CONSIDERABLE amount of talk of strategies for the development of informatics in developing countries, action is still being hindered by a lack of resources. Assistance in the form of personnel, graduate training programmes, proven training courses and teaching aids, capital investment for hardware and buildings are still in demand and will continue to be in demand in a country like Kenya if its move from a basic level of data processing is to be achieved in the next decade.

Research projects must be undertaken to publicise the relevant applications within a country and most important of all an updated inventory of computer applications and manpower projections should be maintained.

Once that is done sufficient national interest can be generated by all parties concerned to attend a workshop on how these manpower needs and research activities can be provided and managed.

## References


Bogod, J. (1979) *The Role of Computing in Developing Countries.* British Computer Society Lectures. Series No. 2.

British Computer Society (1980) *Syllabuses for the Future.* Schools Sub-committee Report.

Clayson, J.E. (1980) *How Relevant is Operational Research to Development: The Case of Kenyan Industry.* J. Operational Res. Soc., 31, No. 4, London.

CACM (1968) *Recommendations for Academic Programs in Computer Science.* Communications of the ACM.

EAMI (1979) Report on First Computer Managers Conference.

IBI (1976) *Guidelines for a National Policy for Education in Informatics.* Spin 101, IBI Green Series.

IBI (1978) *Considerations on Education Policies in Informatics.* Spin 201, IBI Green Series.

Report on Degrees in Data Processing. The Computer J, **18**, 4.

Scott, R.J.P. (1971) *Computers for Developing Countries.* EAPH, Nairobi.

Scott, R.J.P. (1979) *The Role of a Computer Installation in a University in a Developing Country* in *Overseas Universities,* ODA, London.

Taylor & Obudho (1977) *The Computer and Africa: Applications, Problems and Potential.* Praeger, New York.

UNESCO (1980) *Informatics: A Vital Factor in Development.* UNESCO, Paris.

UNESCO-IBI (1978) *Strategies and Policies for Informatics.* Final Report.

United Nations (1971) *The Application of Computer Technology for Development.* First Report of the Secretary General. United Nations, E.71.II.A.1, New York.

United Nations (1973) *The Application of Computer Technology for Development.* Second Report of the Secretary General. United Nations, E.73.II.A.12, New York.

# Advanced Technical Training in Developing Countries—Some Wider Considerations

*K.T. Ryan*
*Trinity College, Dublin*

## INTRODUCTION

It is a fallacy to equate the training needs of an established company in the industrialised world with those of enterprises, however sophisticated, in the developing countries. Training is itself part of the development process. Very often its main objective is to replace expatriate with local employees. Successfully achieving this requires putting training in a much wider context than might usually be the case.

This paper is a review of some of the circumstances that impinge on advanced training in developing countries. It is based mainly on experience with computer staff training in Africa, principally in Zambia, but is intended to be of wider interest. No specific training methods or techniques are described. Instead the objective is to show the training problem in all its aspects, whether technical, economic or organisational. If it appears at times to emphasise what is obvious, it is because this emphasis has been needed in practice. Similarly, if its scope seems broader than its title, it reflects a need to illustrate the wider context so that difficulties and obstacles are more easily anticipated.

## TECHNICAL ASPECTS

Economic development can be defined as the process of raising the sustainable GNP above a certain per capita level. At each stage the GNP is made up of a subsistence element and a surplus available for either investment or consumption. This is the case in both advanced and developing countries. In the latter however the surplus resources being allocated are usually scarcer. Accordingly, great care must be taken if limited finance and personnel are to be used for the greatest good of society.

Leaving aside foreign aid, the rate of economic development is principally determined by the proportion of current production which is *not* consumed. This investment is used to provide for future production. Thus the greater the proportion that is consumed immediately the slower the rate of economic growth. In a country which is primarily agricultural this relationship is clearly seen. The individual or co-operative farm is a microcosm of society as a whole. A farmer can either consume his cattle or breed them—he can't do both. In an economy based on extractive industry (e.g. oil or minerals) the relationship may be obscured. Mineral resources are non-renewable. (Agriculture, in contrast, is a growing asset). Therefore the surplus that is available for investment in a mining economy is essentially temporary in nature. The need for maximum growth and hence minimum consumption is all the more urgent in those cases (e.g. copper mining) where the world market price is subject to great fluctuation. However, for as long as there is a high rate of profit the surplus available can be used to promote long term growth which will build an economy that can survive during harder times.

We shall consider only the case of a mining (or oil rich) country which has had a history of colonial exploitation. Political independence must always precede economic independence for such a country. It is unlikely, however that the attainment of political independence will of itself lead inevitably to economic independence. Generally, the colonial administration will try to maximize the profit extracted from the colony and will restrict investment to the minimum necessary to ensure reliable and efficient production. Civil and military authority, infrastructure as well as educational and health facilities are provided for this reason. There is little to be gained by the imperial power from the advancement of local employees. Their loyalty to the colonial government is always questionable and their usefulness is restricted to their own country of origin. Until such time as the world labour market dictates otherwise, it is easier and more profitable to fill all but the most unskilled jobs, with imported personnel.

Political independence is an opportunity to change this state of affairs. It is no coincidence that almost every newly independent country, whether in Africa or elsewhere, has made education its top priority in the post-independence period. The appropriateness and the cost/benefit of such education has lately come under considerable scrutiny. Investment in general education does not guarantee economic growth. It may produce a literate and partly trained workforce but it may simultaneously produce people whose abilities or inclinations are unsuited to the country's development needs.

One aspect of the manpower training (or retraining) problem is the need for expertise in the more advanced sectors of industry. There are many types of foreign "experts" however and decisions as to their relative usefulness are early pre-requisites in any evaluation of training needs. These decisions themselves require expert knowledge and this is usually not available within the developing country. The need for expert advice can be met by suitable outside consultants, chosen either through the open market or through an international aid-giving organisation. It is important to realize that the expert's professionalism is more important than any commitment he (or she) might have to the host country. Naturally an ideological antipathy would be a hindrance to his work but it is his professional abilities that are being hired, not his politics. Once suitable technical consultants have been engaged they must be given their terms of reference. Because of the host country's lack of technical knowledge it may take some time to define realistic and acceptable goals. The division of responsibility between the government and the outside consultants should be quite clear and explicit. The government sets the objectives; the consultants, once they accept that the objectives are attainable, specify the steps necessary to achieve the objectives. A consultant's typical report in the training area would specify at least:

(a) Manpower needs, current and projected.
(b) Available local manpower, current and potential.
(c) Proposed training to optimise use of local personnel.
(d) Resources required for (c).
(e) Recruitment of expatriate staff.
(f) Evaluation of training progress.

The second last point (e) has a major bearing on the type and quality of training that will be offered subsequently. If there are not enough experienced supervisory staff, on-the-job training is impossible. As some form of apprentice-ship is desirable in almost all technical occupations, expatriate line staff have a major training function. Here we must distinguish between the "accelerated" training plan of a newly independent country and the normal "progressive" training of any well managed enterprise. In the former instance the primary objective is to combine the minimum of experience with the maximum of training. Training can be looked upon as packaged or compressed experience. There is a shortage of technical research in the area of "accelerated" training, but suffice it to say that rapid growth industries and war-time military experience provide empirical evidence that rapid staff promotion is feasible.

It should be obvious that the more rapid and radical are the changes that are desired, the more this is going to cost. A consultant's report may specify a range of options showing the relative costs/benefits of different training schemes. The choice between these schemes is however a *political*, and not a technical, one. Once a particular scheme has been chosen it must be adequately staffed and equipped. Competent technical training staff are in short supply. It is preferable to offer employment conditions that will attract a number of well qualified staff from among whom a selection can be made.

The broad outlines of planned training facilities should also be produced by the consultants. Decisions as to the number and size of training centres will reflect the relative priorities of general and specific training. The nature of the technical disciplines to be taught may be such that particular instructional media are more appropriate than others. If the equipment and facilities required are of a highly specialised nature it may be better to provide them *before* recruitment of staff. On the other hand, it is better to leave as much as possible to the discretion of the training manager so that his individual preferences and methods are reflected in the physical facilities he will use.

Periodically the government should request and obtain a clear statement of objectives, and achievements. It must ensure that the objectives set are precise, quantifiable, and observable *even by the technically naive.* Unless there are extenuating circumstances, failure to reach stated targets should result in some or all of the following actions:

(a) Dismissal of the training manager and/or staff.
(b) Reappraisal of the training plan by government.
(c) Recall of the consultants if this is felt necessary.

Since the training staff will be judged solely by results, it is essential that they have all the necessary authority and freedom to carry out their programmes. The managerial and social problems that this may entail are considered below.

## ECONOMIC ASPECTS

THE OBJECTIVE of foreign investment is to maximise the profits of the investing company. In the colonial situation the imperial government can count as its profits not just the taxes or duties imposed on industry and employees but also the retained profits of those companies and *most especially*, the repatriated earnings of non-local employees. The manipulation or control of profits and taxes are a macro political problem and outside our present scope. It is however relevant to consider the cost to the economy of using expatriate labour in technical areas. Multinational companies are aware of the benefits of employing expatriates overseas. However many of these benefits do not accrue to a nationally owned enterprise in an underdeveloped country. Variety and flexibility of work that may attract and keep personnel do not exist unless the state enterprise has a close relationship with an external multinational company. On the other hand many of the costs are only too apparent.

(a) Technically qualified expatriates must usually be offered a standard of living higher than they would get at home and, unless the host country is very rich, much higher than that of local employees of the same status.

To maintain this standard of living will definitely involve heavy foreign exchange payments for imports and may also cause a structural bias in the emerging local economy. A small example: if superior (but foreign) tomatoes are imported to suit expatriate tastes the local producer of edible but lower quality tomatoes is not encouraged.

(b) Expatriate savings consist entirely of foreign exchange, either currently or when they eventually leave the country. A shortage of "hard" or negotiable currencies is a frequent problem for under-developed countries. In view of what we have seen earlier about the scarcity of investment capital, this makes expatriate labour a particularly expensive commodity.

(c) There are many overheads in having expatriate staff, some of which are often ignored or underestimated by "host" governments. Besides the obvious expenses such as providing superior housing, paying for travel and baggage, mounting recruitment campaigns etc., there are such "hidden" costs as those of disruption due to frequent staff changes or the additional furniture and house maintenance required due to quick turnover of occupants.

(d) One definite but difficult to quantify cost remains. It is the social cost to the host country of having an alien elite whose life-styles and tastes are inappropriate, and most probably unattainable, for the vast majority of the local population.

Notwithstanding all of these costs, expatriate staff are often a net benefit and are occasionally essential. They are so expensive, however, that a positive cost/benefit should be shown for each position, stating clearly the need for a non-local employee. It is in this context that we must consider the costs and benefits of competent professional training.

The cost of a line employee, whether local or expatriate, is a current cost to be met out of current income. The occasional education and training necessary to enable the employee to do his *present* job satisfactorily, is likewise a *current* cost. But the cost of training a local employee to fill a position currently occupied by an expatriate is an *investment* cost. At a later date it produces a saving both in internal and foreign exchange terms. The money so saved is available for reallocation. In long-term planning therefore it is better to spend as much as possible on training in the near future, if the overall expense to the developing country is to be minimised. An example may help to make this clear.

Let us consider a typical technical position, using costs that I have taken from an actual company.

| | | | |
|---|---|---|---|
| Local employee: | total annual cost | 5,000 | |
| Expatriate employee: | total annual cost | 16,000 | (5,000) |

(Foreign exchange element shown in parentheses represents gratuities, remittances, air fares, etc.)

Let us suppose that the following training resources are required annually for a full five-year period to train 10 local employees for this position:

| | | | |
|---|---|---|---|
| 2 × Training Staff | annual cost | 66,000 | (12,000) |
| Training equipment etc. | annual budget | 10,000 | (6,000) |
| Overseas Courses | annual budget | 10,000 | (10,000) |
| | | 86,000 | (28,000) |

Therefore the five years of training will cost 430,000, of which 140,000 is foreign exchange.

Having replaced 10 expatriates by 10 newly trained local employees the annual saving will be:

| | | |
|---|---|---|
| 10 × expatriate employees | 160,000 | (50,000) |
| 10 × local employees | 50,000 | |
| | 110,000 | (50,000) |

After four years this saving will total 440,000 of which 200,000 is foreign exchange.

The entire training cost can thus be written off in four years and in addition there is a continuing reduction in foreign exchange costs. If foreign currency is very scarce in the developing country this saving is even more important. It is the reduced foreign exchange cost that helps provide the additional investment capital necessary for longer term development. The reduction in the social costs mentioned above will also be significant. Of course the danger remains that the "host" country becomes accustomed to the expatriate life-style and a local elite is formed whose inclinations are inimical to the long term growth of the economy.

## ORGANISATIONAL ASPECTS

WITHIN A PRIVATE COMPANY the purpose of training is to contribute to profitability by increasing productivity, job satisfaction and general awareness. Training's position with respect to line management is almost self-defining. If a training policy conflicts with management requirements, the training policy changes. Even within nationalised industries in developing countries, training is evaluated on its contribution to reduced costs or higher staff morale. Such an industry may not have a profit motive but, generally speaking, its social or political goals are defined outside the training area. A developing country which wishes to achieve accelerated promotion of local employees in technical areas, has set a social or political objective whose realisation is part of the training function. As we have seen, one reason for wishing to localise as quickly as possible is that the optimum profit for the company (even if nationalised) is not necessarily the optimum level for the country as a whole. The disparity may be due to the different costs, in local and foreign currencies, of the expatriate and local employees. There are also the social and political aspects of unemployment, national morale, ideology etc. The wish to localise is not enough, however. Having set a "higher" goal the government must consider how it is to be attained notwithstanding the established line authority. It is not enough to obtain the consent, however genuine, of management. Yet even this consent may be absent, or at least lukewarm, especially in those cases where the management feels itself threatened. A typical situation therefore is one where an external training section is given a set of objectives, the achievement of some of which may be opposed by management. Such conflicts are unfortunate but in achieving "accelerated promotion" they are almost inevitable. Tact, diplomacy, and a willingness to compromise, will minimise the head-on collisions and a measure of maturity and experience of the training staff will be their ability to maintain harmony *without* sacrificing their overall goals. Given that the training staff are employed specifically to achieve these goals however, and given that they are of the calibre and competence recommended by the outside consultants, their opinion as to the need for a course of action should overrule that of management unless a clear and unambiguous case can be made to the contrary.

Within the under-developed countries of Africa there is a social or racial overtone to any localisation effort. Expatriates tend to be European or Asians. Training aims generally to replace them with black Africans. In addition, if there is a history of overt racial discrimination, its memory may aggravate the organisational stresses of rapid local promotions. "Reverse discrimination" may be alleged, and in a way that is exactly what should be happening. The crucial difference lies between the long-term career of the local and expatriate employee. The former is entitled to anticipate spending his full working life within his own country, achieving as high a level as his country's development and his own abilities will allow. An expatriate is, or ought to be, a *short-term* employee. No expatriate should look for, or be offered, a lifelong career within a single company. If he is not contributing directly to his own obsolescence, then someone else ought to be. This "someone else" will usually be a training person whose usefulness, in turn, is also of limited duration. "Discrimination" in favour of local employees is thus the meaning of rapid training. Rationality is no antidote to prejudice however, and the merit of the training function does not guarantee its success. It is essential that the element of racial conflict be faced openly and dealt with decisively. Management, both local and expatriate, should take every opportunity to demonstrate and encourage racial harmony. If their own prejudices make this impossible for them, then be they local or expatriate, they are unfit for management in the developing country. The presence of some black expatriate employees might also help to soften the racial boundary.

## CONCLUSION

TRAINING LOCAL EMPLOYEES to replace expatriates is an essential part of economic development. Quite apart from political, social or ideological goals it can be justified in cost/benefit terms. However this activity is qualitatively different from the normal DP training function and it is accelerated. It may cause some disruption and resentment. Management's clear perception of these differences and difficulties can ensure the setting of realistic training goals. Outside help, especially consultants, may be sought in choosing a strategy towards these goals. Firm commitment by the indigenous management can then guarantee their achievement.

# The Need for a National Information Centre on Industrial Organisation in a Developing Country

*Ural Sözen*
*Ankara Academy of Economic and Commercial Sciences, Turkey*

## INTRODUCTION

One of the main problems of production is idle capacity. The reasons for this idle capacity in developing countries are drastically different from the ones prevailing in developed countries. But the techniques for fighting this illness in developing countries are not much different from the ones used in developed countries. Therefore in most of the cases, developing countries fail to cope with this problem.

The cause of idleness in production activities in developing countries is structural in nature. Since the sector activities are incomplete, a firm must bear the production activities where otherwise there is no possible way for them to obtain the necessary services.

During the development processes an industrial sector might develop by itself. But in most cases the sector does not find this opportunity. As a result there are duplications in investments and unnecessary production capacities, which in turn increase the amount of idleness as a whole.

We do think it is inevitably necessary for developing countries to adopt, what we would like to call, a central organising unit to assume the responsibilities of coordination. In this paper we take Turkey as an example and propose a model for such a centre. This topic was dealt with at a conference held in Istanbul in 1971 jointly organised by National Productivity Center and UNIDO (MPM, 1973).

## SITUATION IN TURKEY

In a 1976 survey of Turkish manufacturing industry 12.9 per cent utilised less than 39 per cent of their capacity, 31.2 per cent utilised between 40 and 59 per cent, 37.9 per cent utilised between 80 and 100 per cent. By a rough estimate, 45 per cent of manufacturers utilise 60 per cent or less of their capacity (State Institute of Statistics, 1977).

According to a 1977 survey 15.1 per cent of manufacturers utilised less than 39 per cent of their capacity, 29.11 per cent utilised between 40 and 59 per cent, 37.1 per cent utilised between 60 and 79 per cent and only 25.7 per cent of them utilised between 80 and 100 per cent. These ratios do not show much difference between 1976 and 1977. Again, 45 per cent of manufacturing industry utilised less than 60 per cent of capacity (*ibid*, 1978).

In the 1978 study, 25.2 per cent of manufacturing industry utilised less than 39 per cent of its capacity, 34.8 per cent utilised between 40 and 59 per cent, 31.5 per cent utilised between 60 and 79 per cent and only 10 per cent utilised over 80 per cent of capacity (*ibid*, 1979). In 1978, the situation seems to be somewhat different, but this difference is due to the economic bottleneck Turkey entered in 1977. Difficulties in transfers and especially the inability to provide industrial raw material inputs in addition to the existing structural difficulties caused the situation to change slightly.

## IDLE CAPACITY CREATING FACTORS

Usually idle capacity is caused by maintenance and production necessities, material procuring difficulties, market fluctuations, labour and employment problems, financial bottlenecks, doubts in reaching optimisation, faults in the control mechanism, and selection of technology (Kobu, 1979; Buffa, 1971; Levin *et al.*, 1972; Louis, 1963; Riggs, 1976). The same factors have been quoted in the questionnaires used in Turkey since 1976. In Turkish industry, the main cause of idle capacity is raw material deficiencies. Labour problems, financial bottlenecks followed by insufficient demand are the other idle capacity creating factors. Table 1 shows the idle capacity creating factors which were stated between the years 1976–79 from the evaluation of the questionnaire. Of "other" factors in the table, energy deficiency predominates.

**Table 1. Idle Capacity Factors in Turkey**

| Years | Material Deficiencies | Labor Problems | Financial Bottlenecks | Insufficient Demand | Other |
|---|---|---|---|---|---|
| 1976 | 35.9 | 12.8 | 13.5 | 15.6 | 22.2 |
| 1977 | 37.5 | 11.7 | 12.7 | 17.8 | 20.3 |
| 1978 | 46.36 | 9.6 | 14.8 | 17.9 | 11.4 |

Source: State Institute of Statistics, Manufacturing Industry 1976–79.

## THE PROBLEM CONCERNING UTILIZATION OF IDLE CAPACITY

IN THE 1979 programme of the fourth five-year development plan, concrete proposals are made especially for the utilisation of idle capacity in the state economic enterprises. Some measures are designed to prevent idle capacity arising (DPT, 1979).

State economic enterprises had entered other areas of activity that were connected to their operations. For that reason investment duplications took place. The measures in the 1979 programme envisage the utilisation of idle capacities of the state economic enterprises by manufacturing goods for each other.

At the base of these precautions are the Co-ordination Department's August 1979 report which suggested that, in order to increase productivity among the state economic enterprises, they should utilise each other's idle capacities. The difficulty of such an application lies in gathering information about the machine powers and capacities and information concerning the monthly loads of each enterprise. In the report it is stated "It should be ensured by legislation that the state economic enterprises evaluate the publications of the (information) center and provide the necessary coordination for full utilisation of the total capacity" (DPT, 1979).

## THE REASON PROPOSED FOR IDLE CAPACITY IN DEVELOPING COUNTRIES

BEFORE PROCEEDING with the development of a model for developing countries, one must consider whether there is a special reason for the occurrence of idle capacity in these countries.

Thompson (1967) stresses the importance of obtaining capacity balance in organisations. Furthermore, he states that the lack of balancing activities inevitably results in the emergence of bottlenecks or idle capacity problems. When a firm comes face to face with the decision of integration, it definitely should consider the problem of balance prior to the final decision. Thompson suggests that the criteria for decision should ensure that the capacity of the marginal organisational unit results in maximum utilisation. In other words, each new organisational unit should be added if, and only if, its full capacity is utilised.

This proposition should not be considered as an abstract judgement. Raw materials, resources and market demand, above all, affect the main capacity. These, as a result, influence the added capacity. For this reason, capacity determination should not be considered as an independent variable.

Buffa (1971) on the same subject, states that "Balance is the central problem in designing a production or assembly line. This is not to minimise the other problems of physical positioning of equipment, material handling devices, design of special tools and workplace layout, for in many instances solutions to these problems will contribute to the balance of the line. Balance refers to the equality of output of each successive operation in the sequence of a line. If they are all equal we say that we have perfect balance and we expect smooth flow. If they are unequal, we know that the maximum possible output for the line as a whole will be dictated by the slowest operation in the sequence, often called the bottleneck operations".

Although they do not explicitly state the emergence of sector structure, Thompson and Buffa both accept it as the underlying assumption. Both of them assume that the firm evaluates its means and market expectations, and according to the availability of an opportunity, acquires new capacity that satisfies the balance restriction. The system rationale is based on the availability of choice alternatives. The firm adopts new capacity only if this application is profitable.

In the developing countries, however, such an opportunity of choice is not present. In many instances the firm is forced to produce goods that can be easily imported in the case of developed countries. For sector development is still at introductory stages.

However in some sectors, the main product, because of its "pace maker" characteristic, provides the opportunity, for sector evolution. Most of the time, in the sectors that have oligopolistic and other imperfect competition conditions production of the main product organises the sector and as a result, the firms that comprise the support system gradually emerge.

The automobile industry has these characteristics. As a matter of fact, with the establishment of Fiat and Renault factories in the 1970's in Turkey, the development of supporting sector units were realised in a space as short as ten years. Nowadays, the quality of spare parts produced by this supporting sector can compete in world markets.

In summary, in developing countries idle capacity continues to exist for firms, because sectoral organisation is incomplete at the beginning and because, even after completion of this organisation, the investments of the firms concerned are already consummated.

There are other reasons of a social and economic nature which motivate the firm to acquire a capacity for the services it needs. But the main problem here stems from the fact that in sectors involving a large number of firms, firms tend to import technologies incompatible with the existing technology or that, although they can purchase certain services from other firms in the same sector, this they fail to evaluate. This is because they are inadequately informed. Thus, on the one hand, these technologies incompatible with the sectoral organisation, and tending to create idle capacity, are introduced into the sector; on the other hand idle capacities occur, even when there is no nominal incompatibility, because of duplications.

## THE NECESSITY FOR AN INFORMATION CENTRE

IT IS ABSOLUTELY NECESSARY for developing countries to establish some mechanism that will support the formation of sectors. This should not be interpreted to mean that a custodian establishment must be imposed upon the sector. The proper solution is the formation of a body within the sector that will accumulate, store, classify and dispatch information concerning financial opportunities, market situations, technology, idle capacity and conditions of utilisation, demand projections etc. The contemporary development of means particularly in the field of information processing render such an enterprise possible.

## A PROPOSAL FOR THE CASE OF TURKEY

IN THE CASE OF TURKEY the information needed for such a centre can easily be collected. This information already exists in rough form in various institutions relating to Turkish industry.

These institutions are cited below according to the information they have. The idle capacities in computer time which could be used are also cited.

(1) Türdok: Tubitak (Turkish Scientific and Technical Research Association) is concerned with the subject of technology transfer and provides a service to interested firms.

However, the Türdok computer centre files only data that is related to newly created or adapted technologies. Data related to the users of the technologies or related technologies are not collected by this unit. For this reason, the information in its existing form does not provide solutions to the problems presented in this paper.

(2) Odalar Birliği (Union of Chambers of Commerce and Industries): As a result of the supplementary industry strategy this institution has collected capacity reports on 22,000 firms. These reports contain information about the technologies utilised, capacities, raw material needs, resources of these firms, market potential and the potential demand for the products. However, the information is kept in separate files for different firms.

The Union of Chamber of Commerce and Industries research in order to increase the use of capacity. However, they base their studies on traditional factors. Moreover, the capacity reports are not utilised in these studies.

(3) The Ministry of Industry and Technology: The Ministry of Industry and Technology has not conducted research on this subject.

(4) Computer utilisation: The Public Sector in Turkey owns 139 computers of which 72 are large scale, 14 are medium and 53 are small. Of these computers a total of 359 hours of computer capacity is created. Of this capacity, only 100 hours is utilised and 259 hours is idle.

The private sector owns a total of 206 computers of which 96 are large, 27 are medium and 83 are small. On these computers a total of 359 month/hours of computer capacity is created. 133 hours of this capacity is utilised and 226 hours is idle.

A total of 42 per cent of public sector computer capacity is used for services, 12 per cent is for production and 18 per cent is for education. In the private sector 40 per cent is used for services, 54 per cent is production, and only 0.05 per cent is for education.

Fifty per cent of the total computer capacity is in Istanbul, 45 per cent is in Ankara and 5 per cent is in other provinces.

## CONCLUSIONS

(a) In developing countries, the idle capacities of enterprises is a problem related to sectoral organization.

(b) Because of incomplete sector formation the enterprises are involved in manufacturing areas unconnected to their basic operations.

(c) In some sectors (especially in oligopoly situations) firms contribute to sector formation.

(d) Generally, and especially in competitive sectors, it is not possible for sector formation without intervention. If no aid is provided there will be idle capacity generated.

(e) By establishing a supportive central information system at the beginning, the centre could generate information and publish it in the fields of marketing, finance, technology and machinery.

(f) There is sufficient disorganised information and data capacity to warrant such an information centre. All that is required is to organise it.

(g) In the case of Turkey, the information required is mainly concentrated in the Union of Chamber of Commerce and Industries where such a centre could be organised.

## References

Buffa, E.S. (1971) *Basic Production Management.* 2nd edn, Wiley, New York.

DPT Koordinasyon Dairesi Tetkik ve Tahlil Şubesi (1979) Kamu Makina Fabrikalarında Verimliligin ve Üretimin Araştirilmasına Ilişkin Çalişma Grubu Raporu. Agustos.

Kobu, B. (1979) *Uretim Yonetimi.* 2nd edn, Arpaz Matbaacilik, Ankara.

Levin, R., McLaughlin, I. *et al.* (1972) *Introduction to Operations Management.* McGraw Hill, New York.

Louis, J.R. (1963) *Production Analysis and Control.* 2nd edn, Haddan Craftsmen Inc., Scranton, PA, U.S.A.

MPM (1973) *Atil Sinai Kapasiteden Ihracat için Yararlanma Olanaklari Semineri.* 123. Ankara.

Riggs, J.L. (1976) *Production Systems: Planning, Analysis and Control.* 2nd edn, Wiley, New York.

State Institute of Statistics (1977) *Manufacturing Industry: Employment Expectations.* Ankara.

Ibid (1978).

Ibid (1979).

Thompson, J.D. (1967) *Organisations in Action.* McGraw Hill, New York.

# Informatics and the Irish Economy: An Overview

*Eamon Lalor*
*National Board for Science and Technology, Ireland*

## INTRODUCTION

This paper aims to present a brief overview of the importance of the electronic information technologies (including electronics, computers, telecommunications, software etc.) in the Irish economy.

## THE IRISH ECONOMY

For an international audience it is worthwhile sketching briefly recent developments in, and the present state of, the Irish economy. Although towards the bottom of the scale of its OECD partners, Ireland is, in world terms, a relatively wealthy country. There are, however, a number of features—the openness of the economy, the relative size of enterprises in all sectors, the sectoral breakdown in employment, recent trends in industrialization and the demographic pattern, for example, which make it unique.

The fact that the economy is both small and open means that Ireland is never immune from worldwide developments. Because the domestic market for products is small, industry is heavily dependent on selling abroad, but because the Irish output is small in world terms, the selling price for the products is largely determined by world conditions. Recession in the economies of our trading partners inevitably has a significant effect on the Irish economy. On the other hand the openness of the economy has helped in its rationalisation, since competitiveness is in many cases a necessity not only for growth but for survival.

Thirty years ago the country's economy was largely based on agriculture. More than two fifths (43 per cent) of the workforce were employed on the land. About half that number ((22 per cent) were in industry and the remaining one third (35 per cent) working in services. Employment in agriculture has continually fallen since then to the point where it is much less than that in industry today (20 per cent as against 32 per cent). About half the working population is now employed in services (48 per cent).

In addition to the increase in employment in manufacturing in this period, output from the sector has increased dramatically—to the point where manufacturing output is now more than double agricultural output—and this growth in output has fuelled the sustained growth of the whole economy.

The work of the Industrial Development Authority has resulted in the rapid build up of a modern manufacturing sector, which more than compensates in employment terms for the decline of traditional industries while significantly raising output generally. Much of this new industry is in high technology areas such as chemicals and electronics. Much of the output of these firms is exported and this has enabled the economy as a whole to grow steadily.

The Irish demographic pattern is unusual, with a high dependency ratio per worker. The population of Ireland is about 3.5 million. Of these some 2.3 million are not in the workforce being either too young or too old, in full time education, incapacitated or not seeking work for some other reason. Of the 1.2 million remaining, about 10 per cent are unemployed leaving 1.1 million in employment.

A very large fraction of the population is young and has not yet entered the labour market. There will, in fact, be large numbers of young people becoming available for work each year for the foreseeable future. This poses both an opportunity and a challenge. If productive employment can be found for these new entrants the economy can grow significantly with sizeable consequential increases in per capita incomes. The problem will, however, be to find enough new jobs.

In summary, therefore, the Irish economy is small and open. It has a modern, growing, exporting manufacturing sector which is fuelling the growth of the entire economy and it has a growing, mostly young population.

## THE IRISH ELECTRONICS INDUSTRY

Within a generally growing manufacturing sector the electronics industry has shown the most remarkable development. The activities of the IDA were to a great extent responsible for this development. As these activities are described in another paper in these Proceedings (Hanna, p. 134), the discussion here will be limited to a brief overview.

In 1974 the IDA identified the electronics sector as one which at the time was relatively isolated from the impact of the recession and which would expand rapidly in the following years. A specific development strategy was prepared and has been actively pursued since then. The strategy included the identification of key technologies and the selection of target sectors.

The key technologies include microchip fabrication, and the IDA identified world leaders in this field and promoted their involvement in Ireland. Companies in this area presently manufacturing here include Analog Devices, Mostek and Fujitsu.

The target sectors are those with high stable growth rates and strong technological bases. High risk sectors, for example, certain consumer electronics and components which are labour cost sensitive and have short life cycles are avoided. These target sectors include; data processing equipment (mini-computers, terminals, printers), business electronics (e.g. word processing equipment), industrial electronics (e.g. process control equipment, scientific instruments) some consumer products, automotive products and electronic components.

The IDA has achieved a remarkable success in this area to date. There are now some 85 companies in this industry in Ireland including a number of the world technology leaders. These companies include DEC, Amdahl, Prime, Apple, NEC, Wang, Technicon, Westinghouse, etc.

Employment in this sector has more than doubled since 1973 and now stands at 15,000. It is expected to reach 30,000 by 1985.

Exports of electronics products have increased by a factor of 16 since 1973 and now stand at £450 million per annum. This constitutes a remarkable 20 per cent of all manufactured exports. A figure of £1,250 million is projected for electronics exports in 1985 and this would constitute a significantly larger fraction of all manufactured exports at that time.

This is clearly a remarkable success story. Prudence, however, dictates that we take a closer and critical look at certain aspects of these developments. The growth of this industry results from substantial foreign investment in Ireland—notably from the U.S. and Japan. This investment is welcomed. The question must, however, be asked whether in the highly competitive environment which has developed worldwide for mobile investment, does Ireland have the capacity to sustain existing investment of this kind and continue to attract further such investment?

Questions can also be asked about the quality of technology transfer involved given that most of the Irish subsidaries of multinationals in this sector do little or no research and development in Ireland—although there are significant exceptions to this. It is also true that relatively little of the marketing of the products of these companies is undertaken from Ireland. Furthermore, there is a relatively small volume of sourcing of raw material, components and sub-assemblies by these firms from Irish sources.

The overall question then is: is this industry developing in such a way as to ensure self sustaining growth generating spin-off activity and being constantly refuelled by local innovation leading to new products which will be continually required by the rapidly changing marketplace? This surely must be the objective.

Present industrial policies have served Ireland well in the last decade or so. The changing world economic environment has, however, dictated that these policies be reviewed. The Government has, therefore, requested the National Economic and Social Council to undertake a wide-ranging and fundamental review of Irish industrial policy. The NESC report is expected to be published by the summer of this year. It is to be hoped that the study will identify opportunities for enhanced development of self-sustaining indigenous enterprise working in parallel with and perhaps linked with foreign investment. It would certainly seem that such opportunities must abound in the electronics sector.

## THE APPLICATION OF MICROELECTRONICS

IT WOULD BE A MISTAKE to think that the contribution of electronics to our economy is limited to the output of the electronics manufacturing sector itself. The fact is that the use of electronics is becoming so widespread and its effects so significant that it is having an important impact on most if not all industrial sectors both manufacturing and services.

The National Board for Science and Technology (NBST) has undertaken a study of the implications of microelectronic technology on the Irish economy in the decade of the '80s (NBST, 1981). This work is being reported elsewhere in these Proceedings (Cochran p. 245) and will, therefore, be only briefly touched upon here.

Microelectronic technology has advanced with extraordinary rapidity recently. The greatly increased power and reliability of integrated circuits together with the much reduced size and cost have resulted in the scope for appropriate and cost justified applications growing explosively.

It is clear that there is considerable scope for the application of microelectronics in many sectors of the Irish economy. In some sectors the effects will be more far reaching than others, and results will be seen in some sectors earlier than others. But it is abundantly clear that the use of microelectronics will be a growing factor in the competitiveness of Irish industry and this, given the openness of the economy and the fierceness of international competition, gives urgency to the task of ensuring that the technical possibilities are understood and that the technology is available to potential users.

There will be a need in Ireland to raise the level of awareness of the technology amongst industrialists across the whole spectrum of enterprise from the small firm to the large. The State will have a role to play in this area and a programme to this end will be established.

There will be a need to ensure that sufficient competence in applying the technology in a wide variety of industrial and commercial environments exists to meet the growing demand for such expertise in the years ahead. There would appear to be an important role and opportunity for the private sector here with, perhaps, appropriate State support.

It is worth emphasising here that this technology is not just another technology. It is all pervasive in its use and effects. It is a general enhancer of the productive process. Its timely application is critical in maintaining and enhancing the competitiveness and efficiency of the economy generally.

## INFRASTRUCTURE

In this section we shall deal with two different kinds of infrastructure vital to support and sustain informatics development in Ireland. These are firstly Telecommunications and secondly the national capacity for Education, Training, Research and Development.

### Telecommunications

It is possible to consider telecommunications as a branch of the electronics industry on the one hand and a component of the service sector on the other. Such an approach would, however, miss its central role in the economy.

Telecommunications is being strongly influenced by electronics nowadays with new digital switching and transmission techniques being widely adopted. But, given the increasing convergence between information processing and telecommunications, the existence of an excellent public telecommunications network is itself a prerequisite for the healthy development of informatics in the economy and indeed of the economy itself.

The existence of an excellent telecommunications network facilitates social and economic development generally and its absence retards such development.

Let us briefly review recent developments and future plans for Irish telecommunications (see Byrnes, p. 153 for further details).

The report, recently published, of the Posts and Telegraphs Review Group established by the Government to advise on appropriate forms of organisation to run the Posts and Telecommunications services, documents the poor present state of public telecommunications in Ireland. There has been an extensive period of unsatisfied demands for telephone and other services, and the quality of the service provided has been inadequate.

On the other hand this time will be seen with hindsight to have been a watershed period for Irish telecommunications. A number of important decisions with far reaching consequences have been taken. The timely decision of the Post Office engineers to specify digital switching equipment to meet the future needs of the network is particularly important. This digital (wholly electronic, software controlled) exchange equipment will be the most modern of its kind available and will facilitate in due course the provision of advanced facilities and new services on the network. Paradoxically the present underdeveloped state of the network will—in the period of rapid growth ahead—allow the early development of one of the most modern networks in Europe as the capacity of the digital equipment installed overtakes and exceeds that of the existing electromechanical equipment in the mid to late eighties.

This future Irish digital network should be well able to deal with the general flow of information whether that information be in the form of voice, data, text, or still or moving pictures. This will strongly facilitate Ireland's transition to the information society.

The Government decision to accept the advice of the Review Group and to create a State sponsored body, An Bord Telecom, to run the telecommunications services along commercial lines, and the decision to inject substantial capital into telecommunications development over the next five years, are also much to be welcomed and will facilitate these developments.

The statutory Bord Telecom will be established shortly and will have considerable scope to adopt imaginative new approaches to the development of this vital national service and of the manufacturing sector which supports it. Significant opportunities will exist for cooperation between An Bord Telecom and private industry for their mutual benefit.

## Education, Training, Research and Development

One of the attractions of Ireland for foreign industrial investment is the presence here of a well educated workforce. Our system of primary, secondary and university institutions has for many years provided a liberal education of a high standard in world terms.

In the last 20 years with the rapid growth of technology intensive industry we have seen the creation of Regional Technical Colleges (nine, distributed throughout the country) and National Institutes for Higher Education (one in Limerick and one in Dublin) which provide education and training primarily of a technological nature and for the most part explicitly oriented towards the requirements of industry.

The traditional educational establishments—schools at second level and universities—have been slow to adapt to the rapidly changing environment of technology based industry. It is, therefore, to be welcomed that the Government White Paper on Educational Development recently published, recognises the needs for a balanced educational curriculum which provides for improved facilities for scientific and technological education.

In the evolution of our educational institutions which will continue throughout the eighties, it will be important to keep a proper balance between a liberal and technical education on the one hand and within technical education between broadly based education and narrow specialization on the other. Both temptations—to concentrate exclusively on narrow specialization and to ignore it altogether—should be resisted equally.

The Universities and the National Institutes for Higher Education are the seed beds from which our electronics graduates will spring. They are the source of our future innovative capacity in this area. They need to be continuously and increasingly supported and they need to adopt flexible institutional arrangements in addressing national priorities.

A number of interesting developments of relevance to the information technologies is presently taking place. A Microelectronics Applications Centre, supported by a number of State institutions has been established on the campus of the National Institute for Higher Education, Limerick (NIHE). A Microelectronics Research Centre has been established at University College Cork. This is and will increasingly become a centre of excellence in semiconductor research. A National Microelectronics Laboratory is soon to be established on a third level campus in the Dublin area to address the research, development and training requirements of the growing Irish semiconductor fabrication industry. In addition, new institutions dealing with the research, development and training requirements of the telecommunications sector—both service and manufacture—are presently being specified and a software innovation centre is under discussion.

It will be important that each of these specialised institutions has an active and mutually enriching interaction with the existing third level educational establishments.

## CONCLUSION

Ireland is a small open economy dependent for its continued growth on competitive exports. The economy is largely sustained by a modern, growing, increasingly high technology manufacturing sector with a particularly strong electronics industry which has grown mostly as a result of foreign investment.

The electronic information technologies can contribute to the continued health of the economy through: further growth of foreign investment in this sector; the acceleration of the growth of a self sustaining indigenous industry in this area; the development of an advanced competence in microelectronics applications throughout Irish industry and the development of an excellent telecommunications network.

In order to ensure that this growth is organic and self sustaining, the scientific and technological infrastructure of Education, Training, Research and Development institutions will have to be strengthened to meet the growing and changing requirements. This will be particularly true for the indigenous component of the sector. Developments in this regard have started and will continue throughout the 1980s.

## References

NBST (1981) *MicroElectronics: The Implications for Ireland.* National Board for Science and Technology, Dublin, Ireland.

SECTION 5

# The Implications of Microelectronics for Productivity and Employment

The potential economic effects of microelectronics and the implications for employment are examined. Discussions range from a global overview of the electronics sector itself to an examination of the likely impact in other fields such as the garment industry and consideration is given to the likely effect of different economic environments with particular emphasis on the case of developing countries.

# Comparative Advantages in Microelectronics

***Juan F. Rada***
*Centre for Education in International Management*
*Switzerland*

## FOUR BASIC ASSUMPTIONS

THE INITIAL POINT of departure for analyzing comparative advantages in microelectronics is to make certain explicit assumptions. First, technology changes conditions but does not determine comparative advantages. Many other factors such as industrial structure, skill endowment, managerial practices and the overall structural characteristics of any given economy play a crucial role. The role of technology in conditioning comparative advantages is nevertheless beyond doubt. The majority of studies, from different angles, and using different methodologies, tend to agree on this point. Therefore, comparative advantage, particularly in manufactured goods, is to a great extent a function of R & D intensity and the scientific/technological environment of any given country (US Dept. of Labor, 1980; Schmoranz, 1980).

Secondly, the entire industrial infrastructure is becoming increasingly abstract, based upon technologies and guided by science. Because of the importance of technology, comparative advantages have become increasingly man-made rather than determined by historical circumstances or geographical position. (This is not to diminish the importance of these factors, but to put them in the right perspective.)

As technology becomes more abstract, it also becomes more pervasive. The digitalisation of information relates in many important ways to human intelligence functions and the still infant processes of genetic engineering are related to basic mechanisms of evolution and life.

Furthermore, because of this abstraction technology is being increasingly embodied into tangible goods. (It has been estimated that the amount of technology embodied in Japanese exports between 1962 and 1977 has more than doubled). The 'information society' seems to be one where increasing amounts of information are incorporated into material goods and the process of their production. Unfortunately, the present statistical base is incapable of providing a clear assessment of this shift, except via occupational categories of the labour force or rather old definitions of research-intensive products and science-based industries.

Thirdly, the all-pervasive characteristics of information technology will affect to different degrees all industries and productive activities. This comprehensive technology simultaneously affects:

Production { Products
Processes }
Structured and unstructured office work
Services
Information flows

This takes place because of the digitalisation of different types of signals which is now possible due to the development of the "basic" components: the microprocessor and the microcomputer. Current technological developments will affect all areas where information exchange takes place, whether these activities are of an electric, electromechanical, mechanical, electronic, hydraulic, pneumatic or intellectual nature.

Finally, production (in the general sense of transformation of nature and not only manufacturing) is becoming more and more a product of capital rather than labour. In a more theoretical vein, this seems to respond to the overall tendency of capital to become as independent as possible of the factor that conditions its reproduction, whether it be labour, materials or other constraints.

The above is nothing but the continuation of a historical trend which was first clearly evident in agriculture and then in large scale continuous production such as petrochemicals. This in itself raises a number of issues in relation to comparative advantages, particularly in terms of capital requirements, even if a technology reduces unit cost as is the case in many microelectronics-related applications.

This growth in capital intensity produces greater demands on management, skills, marketing, etc., as a result of the need for the full utilisation of equipment, shorter product and design cycles and the introduction of data processing equipment in heavily manned areas such as the office. The types of knowledge and skills required for these changes are far from clear, except at the software and system level. What can be said is that these 'intangibles' or 'invisibles' will grow in importance. (Introducing a speculative note here, one can easily imagine that in the long term, one could be talking about 'management intensive' products as hardware incorporates more skills and becomes more economical. At the moment, we have no way of measuring this to anything approaching a reliable indicator. This is further complicated by the varying 'management styles' which characterize different countries and cultures.)

These four elements provide the overall framework in which the question of comparative advantage in microelectronics needs to be considered. It is also valid in understanding the impact of other advanced technologies on the international division of labour, since these technologies share some of the characteristics described. (The reference is to information technology, biotechnology and new materials technology.)

## RESEARCH AND DEVELOPMENT

ELECTRONICS is significant because of the *convergence* characteristic of the industry. The importance of this industry now and in the future is a foregone conclusion. In quantitative terms in 1975 in the OECD area, those industries producing electrical and electronics machinery, equipment and supplies, including computers, data processing supplies and accessories, spent about 13 billion U.S. dollars on R & D and employed 225,000 research scientists and engineers. This represents about 30 per cent of all industrial R & D resources in the OECD area. The industry became the biggest R & D performer of the industrial sector, overtaking the chemical and aerospace industries.

Table 1 gives a breakdown of expenditure by "main performers".

**Table 1. R & D in the Electrical Group (Including Computers) 1975**

| | R & D Expenditure | | RSE (a) | |
|---|---|---|---|---|
| | $ millions (c) | % | FTE (b) | % |
| United States | 7,325 | 57 | 119,500 | 54 |
| EEC Countries | 3,945 | 30 | 56,500 | 25 |
| Other | 1,691 | 13 | 47,900 | 21 |
| (of which Japan) | (1,345) | (10) | (43,900) | (20) |
| TOTAL | 12,960 | 100 | 223,900 | 100 |

(a) Research Scientist and Engineers (RSE)
(b) Full Time Equivalent (FTE)
(c) At Current Exchange Rates (1979)
Source: OECD, "Trends in Industrial R & D in selected OECD Member Countries, 1967–75", OECD, Paris 1979, p. 32.

**Note:** Figures on R & D, particularly international comparisons, are always qualified because of methodological difficulties. For full details consult the source.

Many other areas not accounted for in the electrical groups, such as in aerospace and machinery, are heavily based on electronics and considerable R & D efforts are undertaken. In addition, production support, in-company training, management development, and other items which are part of the "technological environment" are not accounted for, although they may have a significant impact on results.

For instance, it has been consistently argued that Japanese innovative capacity is as much related to R & D efforts as to management style, particularly in relation to quality control and "zero defects" programmes with shop floor participation. Japanese R & D efforts are modest by comparison to other OECD countries, but nevertheless produce very substantial results. Additional weight is given to the "technological environment" argument and the socio-cultural conditions for innovation when one considers that the USSR is the second biggest R & D spender and employs the greatest number of scientists and engineers, but has considerable difficulties in commercialising innovations on the home market let alone compete in the international one.

Military budgets are also crucial and not necessarily accounted for in R & D budgets given due, for instance, to disclosure rules in countries such as the USA (the 10-K form). Military expenditure was important to the initial stage of the industry (more as a market rather than in R & D projects). Although this expenditure had declined in the 70s, this is no longer true. Today, only 7 per cent of the integrated circuits are destined to the military market in the USA, against 70 per cent in 1965. Nevertheless, an increasing percentage of the research budget of the US Department of Defense will go to electronics, from about 37 per cent in 1979 to 46 per cent in 1989. The budget of the Department of Defense is the most important of the U.S. Federal budget, followed by NASA and the Department of Energy.

Already a substantial $250 million program for VHSIC (Very High Speed Integrated Circuits) has been launched by the Department of Defense with expected results by 1986. The aim is to produce circuits 100 times more powerful than those available today. The first contract has been signed with IBM (Mouy, 1980). Adding to this support about three-quarters of NASA's budget of about $5,000 millions is spent on R & D. Between 1961 and 1975 about half of the patents commercialized as a result of NASA contracts were in the electrical machinery category (including computers) and the item electronic components figures as the single most important item in terms of patents commercialized (Hertzfeld, 1980).

Since 1975 (the date shown in Table 1) governments have increased their support to the electronic industry. In addition to the case of the USA which has been described above, a quick run down shows the following (figures at current exchange rates). (Mackintosh Consultants, 1979; General Technologies Systems Ltd, 1979)

**Japan**

1971–1979 US$1658.3 million of government financial support to the information industry as a whole, including computer industry (hard and software), technical systems for public welfare, and preferential loans.

**Federal Republic of Germany**

1979–1981 US$552 million from the Federal Ministry of Research and Development.

1975–1979 US$37.5 million in the area of Very Large Scale Integration—VLSI.

**United Kingdom**

| | | |
|---|---|---|
| 1978– | US$150 million | IC industry support |
| | US$100 million | INMOS |
| | US$155 million | Applications |
| | US$ 50 million | Education |
| | US$ 75 million | Retraining |
| **Total** | US$490 million* | |

(* Not totally committed)

| | | |
|---|---|---|
| 1981– | US$430 million | Support of ICL |

**Italy**

1977–1981 US$600 million direct grants
US$280 million low-interest loans
US$250 million in increased government expenditure

**Total** US$1,130 million*
(* Not totally commited)

**France**

1979–1984 US$144 million for the support of the IC industry

The figures do not refer of course to total expenditure. They only indicate direct support to the industry, specifically in the area of components. Furthermore, they are part of a comprehensive *strategy* in the field of electronics and informatics.

In addition to these national efforts, the EEC is investing US$60 million to increase the Community's share in the world production of integrated circuits from the current 6 per cent to 12 per cent in 1984–85. (Commission des Communautés Européennes, 1980).

Private sector investment in R & D is equally if not more important, depending on the countries. Table 2 shows the situation up to 1975.

Around 1975, R & D efforts accelerated in electronics due to fast technological change and keen competition. Thus, the ten most important U.S. independent semiconductor companies (i.e. those companies selling on the open market) have more than doubled their R & D budgets, from over US$200 million in 1975 to over US$500 million in 1980 (estimated). These figures include expenditures by the companies and exclude programmes from outside contractors such as the U.S. Government (Business Week, 1977; *ibid*, 1978; *ibid*, 1979).

**Table 2. Percentage Sources of Funds for R & D in the Electrical Group (Including Computers) 1975**

| | Private | Government | Abroad | Total |
|---|---|---|---|---|
| Japan | 98 | 2 | – | 100 |
| FRG | 84 | 14 | 2 | 100 |
| United States | 62 | 38 | – | 100 |
| France | 58 | 30 | 12 | 100 |
| United Kingdom | 47 | 44 | 9 | 100 |

Source: As for Table 1.

For the Japanese industry as a whole, R & D expenditures in integrated circuitry grew from about US$100 million in 1974 to US$200 million in 1978. The amount of people working in the field increased from 3,800 to 5,300 in the same period. (B.A. Asia Ltd., 1980).

An element which I have not considered here, essentially because of lack of data, is investment in basic scientific research which is becoming increasingly important to maintain the lead in electronics.

## CAPITAL INVESTMENT

THE NEXT CRUCIAL ELEMENT in generating advantages in electronics is capital investment. Electronics can no longer be seen as a labour-intensive, low-capital investment, except in some assembly operations which are becoming less important as the industry reaches for global markets and economies of scale. The graph below provides an illustration of current trends in semiconductors (see Figure 1).

It is acknowledged today that a minimum fabrication facility, in order to be truly competitive, will cost about US$ 30 million and will need approximately US$ 40 million of output per year. These figures do not consider R & D, marketing and so on. (Heikes, 1980).

The combined effect of growing R & D and capital expenses has resulted in business and thus technological concentration. Most of the pioneers of the semiconductor industries have been obliged to sell all or part of their companies to finance necessary growth. The usual corporate means to raise funds through public stock offering or internal profits were not adequate. For all its innovation and growth, the industry's cash generative levels are not sufficient to pay for its next generation developments, particularly considering the risks involved. In fact,

**Figure 1. Capital expenditure by semiconductor manufacturers (USA and Japan)**

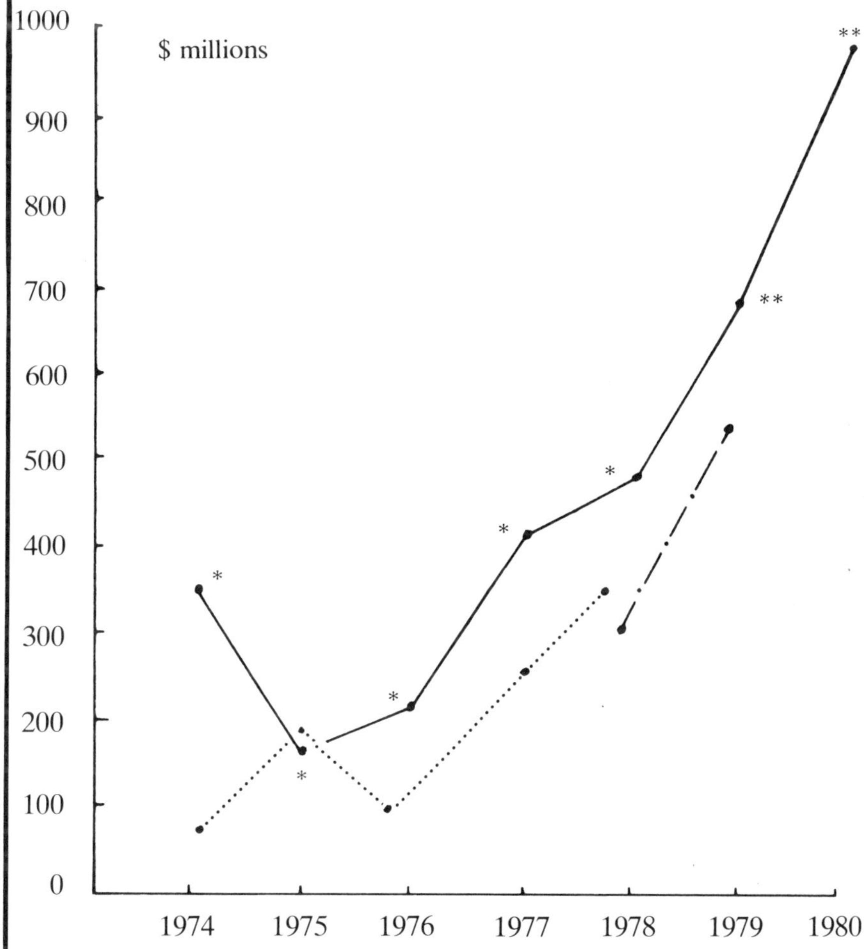


— USA * 5 main companies ** 9 main companies.
· · · JAPAN Integrated Circuitry equipment investment (All producers).
–·– JAPAN Plant and equipment; 9 main producers.

Source: For the USA: Li, M. (1979), *The Electronic Industry—How the investment Banker sees it.* Financial Times Conference: Tomorrow in World Electronics, March, p. 76–77 and Business Week (1980), July 21, p. 189.
For Japan: BA Asia Ltd. (1980).

the rate of profit in the industry has diminished through a continuation of growing cost, fierce competition and uncertainties about the production process, particularly with regard to yields per wafer.

As a rule of thumb, it would seem that the lower the prices, the less likely that non-vertically integrated companies can compete. This, of course, is not

**US Semiconductor firms—Acquisitions and Independent Companies**

*Estimated* 1979

| US Films | Integrated Circuitry Sales | Affiliation |
|---|---|---|
| 1. Texas Instruments | 680 | Division of T.I. Inc. |
| 2. Motorola | 425 | Division of Motorola Inc. |
| 3. Intel | 400 | Independent |
| 4. National Semiconductor | 320 | Independent |
| 5. Fairchild | 305 | Acquired by Schlumberger |
| 6. Signetics | 250 | Acquired by Philips |
| 7. Adv. Micro Devices | 160 | Acquired by Siemens (20 percent) |
| 8. Mostek | 155 | Acquired by Unit. Technologies |
| 9. RCA | 145 | Division of RCA |
| 10. Intersil | 140 | Acqu'd by Gen. Electric/Northern Telecom |
| 11. Harris | 100 | Division of Harris Corp. |
| 12. American Micro Systems | 95 | Acqu'd by R. Bosch GmbH (25 per cent) |
| 13. Rockwell Semiconductors | 85 | Div. of Rockwell Intern. |
| 14. General Instruments | 80 | Division of G.I. Inc. |
| 15. Synertek | 50 | Acqu'd by Honeywell |
| 16. Analog Devices | 40 | Acqu'd by Standard Oil |
| 17. Monolithic Memories | 35 | Acqu'd by Northern Telecom |
| 18. Siliconix | 30 | Acqu'd by Lucas Industries/ Electronic Engineers |
| 19. Solid State Scientific | 22 | Acqu'd by UDO Adolf Shindling |
| 20. Databit | 19 | Acqu'd by Siemens |
| 21. Zilog | 15 | Acqu'd by Exxon |
| 22. Microwave Semiconductors | — | Acqu'd by Siemens |
| 23. Litronix | — | Acqu'd by Siemens |
| 24. Unitrode | — | Acqu'd by Schlumberger |
| 25. Electronic Arrays | — | Acqu'd Nippon Electric Co. |
| 26. Spectronics | — | Acqu'd by Honeywell |
| 27. Interdesign | — | Acqu'd by Ferranti |
| 28. Micropower Systems | — | Acqu'd by Seiko |
| 29. Mos Technology | — | Acqu'd by Commodore Intern. |
| 30. Precision Monolithics | — | Acqu'd by Bourns |
| 31. Semtech | — | Acqu'd by Signal Co. |
| 32. Western Digital | — | Acqu'd by Emerson Electric |

necessarily applicable when processes rather than products change. Business concentration could on the other hand, lead to the creation of market niches due to the loss of flexibility of larger units. This proposition is by no means a clear-cut one inasmuch as one of the features of current changes is the potential for flexible manufacturing systems able to "individualise" products within the same production line.

The above-mentioned trend can be clearly seen in the list of acquisitions of US firms by US, European and Japanese corporations (see p. 206).

This concentration has important implications for developing countries in terms of transfer of technology.

Today, advanced electronic technology—mainly concentrated in U.S. and Japanese firms—is obtainable for the most part only through co-production, joint ventures, and cross-licensing, although more than one channel is generally used. Joint ventures have been extensively used by countries which were lagging behind in technology (e.g. France and the UK) while others, such as Japan, have mainly used cross-licensing, as they were already at a more advanced state.

Experience has shown that transfer of advanced electronic technology to developing countries rarely takes place (Kim, 1980). The crucial technologies are in the design of circuits and wafer fabrication and this section of the industry will remain in the advanced countries for the foreseeable future with a few notable exceptions in the lower end of the market. There are several reasons for this:

(a) The high capital investment involved in design and wafer fabrication makes manufacturers hesitant to transfer this process to a developing country. Wafer fabrication, which is a batch production, is generally centralized to justify economies of scale.

(b) The need to maintain a close relationship between R & D, production and marketing.

(c) Maintenance of quality which is felt could not be obtained in offshore installations.

(d) Decreasing importance of labour cost as the industry becomes more dependent on capital equipment, R & D, marketing and managerial skills.

## TECHNOLOGY

UNDERLYING THE TRENDS in terms of R & D, capital investment, and technology transfer are the trends in the technology itself which condition the entire structure of the industry. Three elements are of great importance here vis à vis developing countries. All three are closely interconnected.

(1) Increasing levels of integration transfer value-added in a number of electronic and other products to the manufacture of components.

(2) The growing embodiment of technology due to the level of integration and trends in software.
(3) The erosion of the relative importance of low-labour cost due to automation and integration.

The replacement of mechanical, electromechanical, older electronics and other types of components by integrated circuits, microprocessors and microcomputers has sharply altered the production processes and products.

The alteration of products has an important consequence: the transfer of a greater part of the value added to the component manufacturer. This happens because the electronic component becomes a much more important part of the product. This is most applicable to mass-produced 'chips' which contain a standard program (e.g. TV games), but increasingly the value added will depend on the application, i.e. how the circuits are put to work. The cost of integrated circuitry for a watch or a TV is decreasing rapidly; design, quality and flexibility will increasingly become the areas where most of the value added will be created, coupled with the capacity to design systems (e.g. home entertainment systems integrating audio, video and games with memory capacity).

Traditionally, the manufacturing process could be simplified schematically as the production of materials into components followed by their assembly into sub-systems which in turn are assembled into products. The final stage is testing, and sales.

In general, the later the phase the more labour intensive the process becomes.

The use of microelectronics somehow reverses the process as the use of the integrated circuits allows sub-systems and component assembly to be replaced by component manufacturing and the final stage of product assembly becomes the assembly of components. The manufacturing of components is becoming less and less labour intensive as automation takes over.

The following simplified schematic view compares the old structure and the emerging one (Figure 2).

As the semiconductor industry pushes towards higher levels of integration ($10^6$ elements per chips is on the horizon) it will find itself in the computer area, forcing the end-equipment manufacturer to be concerned mainly with the software of particular applications and human-machine interface. This is because hi-fi equipment, car electronics, radio, TV, telecommunications, controls, etc., will ultimately be designed around some type of computer as most functions will be digitalised.

The software and human-machine interface areas are where some developing countries could play an important role, provided they possess a degree of specialisation and forceful government policy. It must be understood, however, that an important level of dependency will continue to exist at the component level. This should encourage developing countries to pool resources in order to reach economies of scale on a regional basis, even at the software level. Software is a complex skill, but at the same time remains an extremely labour-intensive activity which demands close contact with the end-user for most applications.

**Fig. 2. Participation of the components sector in the manufacturing of equipment**

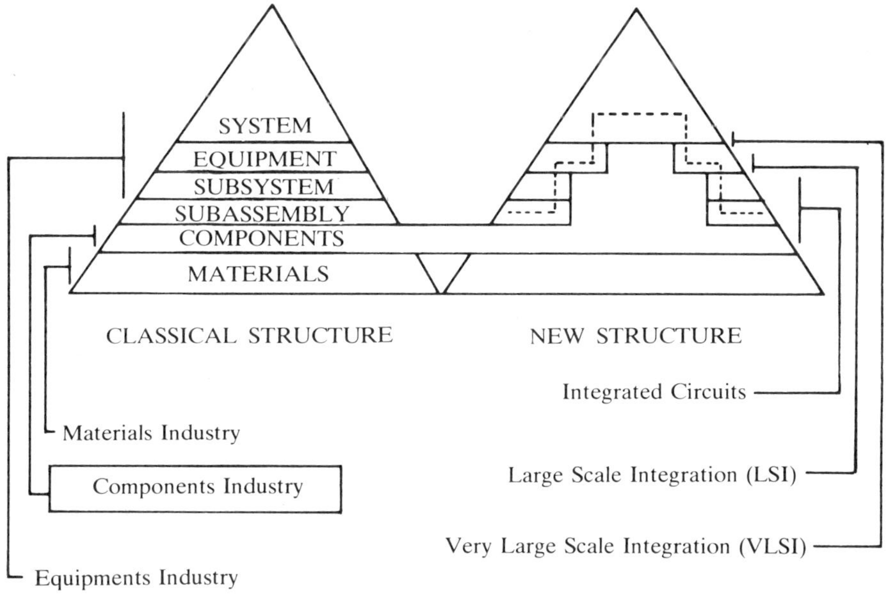


The embodiment of technology takes place by further integration of components and by current trends in software. Today, the electronic watch is essentially composed of five parts (battery, quartz crystal, LED, integrated circuit and case), which carry out the functions performed before by dozens of parts. The examples are many, TV sets, hi-fi equipment, machine tools, sewing machines, cameras, etc. In the case of machine tools, the future will see the commercialization of machinery packages, of which the hardware will be only part of a total system.

In the case of software, the trends are already discernable. At the hardware level, the main technological emphasis of the 1970s was on increasing the number of components per chip. In the future it will be to extend this integration to the system level and this of necessity includes software (see Figure 2). This means moving the software back to the hardware, where it first started. Software that cannot be incorporated into hardware will become more modular and transportable. Products will become more user-oriented rather than designer-oriented.

Some time will pass before this trend fully materialises, but the movement is in that direction. Thus component manufacturers are striving to sell systems (rather than components) and systems vendors, whether they are sophisticated software houses or non-vertically integrated computer manufacturers, are attempting to manufacture components or design and assemble their own boards.

This trend is only natural since hardware (components) represents a decreasing percentage of total system cost and thus a diminishing source of revenues. Table 3 illustrates the trend.

**Table 3. Percentage of Large Microcomputer System Cost**

| | Hardware | Software Development | Software Maintenance |
|---|---|---|---|
| 1970 | 55 | 24 | 21 |
| 1975 | 45 | 30 | 25 |
| 1980 | 35 | 33 | 32 |
| 1985(E) | 30 | 35 | 35 |

Source: Creative Strategies Intern. (1981).

For a complete stand-alone microcomputer system (CPU + CRT + mini-floppy disk + Printer) selling from US$5000 to US$10,000, the cost of components before assembly into a microcomputer board would be between US$62–US$112, while the assembled board would be worth US$250–US$1,000 (Creative Strategies Intern, 1981). These elements are pushing the industry into further vertical integration, inter-links and embodiment of technology.

## OFFSHORE INSTALLATIONS AND LABOUR COST

ANOTHER IMPORTANT CONSIDERATION IS the erosion of the importance of low labour cost in the assembling process of components, products and systems.

Although this question goes far beyond the electronic industry and covers much of the manufacturing, I shall limit myself to the electronic sector and the use of offshore installations (i.e. foreign subsidiaries).

Offshore installations were initially created to make use of low-labour costs particularly for the assembling and testing of electronics components and later on for products. The use of these installations and their role in company strategies varies greatly, depending upon the position in the industry as well as nationality. Thus, U.S. firms are encouraged to use offshore installations because custom regulations tax only value added abroad. Japanese companies have used them mainly to supply local firms (national or transnationals) rather than to export to Japan.

From the point of view of some developing countries, this was seen as an opportunity to create employment, increase exports and obtain some technology.

An illustrative case is the Republic of Korea, which has enjoyed spectacular growth of the electronics industry and has been regarded by some as a model to be followed by other developing countries. South Korea has relied heavily on imported technology tied to foreign investment and has been criticised on a number of occasions precisely because of the vulnerability of its economy to outside decisions, particularly in relation to the type of technology, its domestic diffusion, and product cycle of foreign investment.

In 1972 Korea exported US$177 and imported US$142 million worth of electronic goods. By 1978 exports were up to US$1,386 million and imports to US$1,000 million. Of total exports, components represented US$614 million or 44 per cent (integrated circuits 27 per cent, semi-conductor devices 6 per cent and other parts 11 per cent) (Electronic Industries Association of Korea, 1979).

The above would seem to imply that the industry is particularly vulnerable to changes in the component sector. The Chairman of the Electronics Industry Association of Korea has spoken about the situation of the industry in 1980 with a "tone of despair rather than hope" due to the speed of technological development, lack of technology transfer, and severe competition in the international market which makes it difficult to obtain a market share of high-technology products (Kim, 1980).

Although the above makes reference to a specific situation, it is also applicable to the electronics industries of the developing countries as a whole. The fact that most products are becoming highly technically oriented in electronics is due to the characteristics of the component. Those products which remain labour intensive in their assembly have been divested of an important part of their value added due to the change in components. This is the case, for instance, of television sets where the number of parts has been halved since 1970 because of the use of more sophisticated components which obviously eliminate many assembly operations.

This has permitted automation to a level never achieved before and in addition TV sets in the developed countries are produced in large quantities. As a result, "the manufacturing cost of a TV set in Korea and that of the U.S. are now practically comparable with each other. Rapid advancement of industrial technology is eliminating labour-intensive portions of the electronic industry; this tends to make it harder for Korea to earn enough foreign currency to import expensive new technology" (Kim, *op. cit.*).

The continued use of offshore installations will depend very much on what type of product is manufactured. Thus, for those products where the selling price and margin are low, the assembly time long and the relative weight of labour in the product high, off-shore installations will be maintained for some time. However, in those cases where the product incorporates expensive raw materials, is produced in long series, and has a short assembly time, it may be profitable to repatriate the installations in the Far East. With automation and robotics, the cost of investment and exploitation remains basically the same, regardless of location (d'Argoeures, 1980).

In the case of semiconductor components, there is a growing agreement in the industry that further substantial investment in offshore installations cannot be justified. The essential reason again is the automation of the testing and assembling process. However, some offshore assembly investment is taking place, (e.g. Motorola is investing US$40 million in Sri Lanka). Because the lead time for building highly-automated capital-intensive plants is very long, continued use of offshore installations ensures a supply of some components, which are scarce in the international market. Increased salaries in Southeast Asia are also forcing some companies to search for different locations, while those staying in countries such as Singapore are increasing automation of assembly (Eng Fong, 1977).

The strategies of each manufacturer are different and are constantly being modified due to the characteristics of the industry. For Japanese producers the use of offshore plants represents only about 10 per cent of production and in some areas is declining (discrete semiconductors). Japanese automated bonding equipment represents a labour efficiency factor (in terms of people) of perhaps ten to one. This has helped to compensate for growing salaries in Japan. Another factor of equal importance to Japanese producers is the concern for quality which they feel cannot be maintained through the use of offshore installations (BA Asia Ltd, 1980).

This element is growing in importance and will force other producers to follow similar considerations. Part of the explanation for the Japanese semiconductor producer's rapid penetration of the US market can be found in the quality of their product. Thus a test run by Hewlett-Packard on 300,000 random access memories (RAM) from three Japanese and three American companies showed the Japanese integrated circuits to have consistently less defects. Not one device from Japan failed on inspection of incoming new parts, whereas the failure rate for US companies ranged from 0.11 to 0.19 per cent. In field operations, Japanese chips showed failures at rates from 0.01 to 0.019 per cent, while the rates for American firms were 0.059 to 0.267 per cent (Science, 1980).

Japanese firms attain higher quality by a very high degree of automation and a managerial system which involves shop floor workers in quality control methods, particularly the widespread use of "zero defect" groups.

In the case of other producers such as ITT Semiconductors, the assessment is the following:

> "The relative advantages of increased mechanization versus offshore assembly are reviewed next and the conclusion comes down firmly in favour of the former, permitting closer customer contact, elimination of logistics problems, and upgrading of local or national technology, without which complete business transfer to low labour cost areas will occur. The conclusion is that the successful European semiconductor manufacturer must have complete production cycle in one manufacturing location, worldwide marketing, fast introduction and close co-operation with end user industries, e.g. computer, entertainment, telecommunications and automotive." (Roessle, 1979).

Other assessment coincides particularly because of increasing capital cost (Crémieux, 1980).

There are three main reasons, however, why a massive repatriation of offshore assembly in this sector is unlikely in the medium term, although the amount of investment is and will diminish further while new plants will be set up in the OECD area. First, under financial stress companies prefer to maintain offshore installations and invest at a slower pace in new plants. At the same time, there are supply problems which extend the life-expectancy of the offshore installations. What happens is that reduced savings and profit is not necessarily a loss and, although profit levels may not justify a considerable new investment, neither do the reduction of profits justify repatriation. Second, labour legislation in offshore locations often permit terms of employment that are not tolerated in the developed countries. This is especially important for firms which are very sensitive to drops in demand. An illustrative example is the retrenchment in Singapore during 1974–75 recession, where for the electronic industry as a whole there was a decline of 25 per cent to 30 per cent of the total labour force. Four major US semiconductor companies laid off almost 5,000 workers in about 6 months (Eng Fong, 1977).

Third, there is great pressure by the governments of Southeast Asia to maintain facilities and up-grade technology. Direct government intervention and guarantees are an important element and explain, for instance, the agreement between American Micro Systems and Gold Star of South Korea, which could eventually allow the joint company (Gold Star Microsystems Inc.) to compete in portions of the semiconductor market in the Korean and Southeast Asian market.

For other sectors of the industry, such as suppliers of equipment, crystals, chemicals and so on, there is no question that they will remain in the high technology areas.

The overall conclusion from the trends described is that there is a rapid erosion of the comparative advantage of developing countries in the electronic field due to a large extent to rapid technological change.

The main problem here is that we are not talking about an isolated technology or industry but about an all-pervasive, *convergence* industry—a basic industry of the future.

## PARTICIPATION OF DEVELOPING COUNTRIES IN THE PROCESS OF 'INFORMATISATION'

One must establish the relative level of participation of developing countries in the process of "informatisation" in order to assess how this process will affect them and what their absorption capacity will be. Tables 4, 5 and 6 give a purely quantitative picture and do not account for the incorporation of electronics into items such as transport equipment or machine tools which are internationally traded under specific categories.

**Table 4. World Microprocessor/Microcomputer Forecast Sales (Chips, Boards and Systems) (US$1000 m)**

| | 1981(E)* | Percentage of Total | 1983(F)† | Percentage of Total | 1985(F) | Percentage of Total |
|---|---|---|---|---|---|---|
| Developed Countries (USA, Japan & Western Europe) | 1.69 | 98 | 2.79 | 97 | 4.84 | 96 |
| Other countries (a) | 0.04 | 2 | 0.11 | 3 | 0.20 | 4 |
| Total | 1.73 | 100 | 2.90 | 100 | 5.04 | 100 |

Market Share—Producers

| | 1981(E) | Percentage of Total | 1983(F) | Percentage of Total | 1985(F) | Percentage of Total |
|---|---|---|---|---|---|---|
| USA | 1.40 | 81 | 2.20 | 76 | 3.68 | 73 |
| Japan | 0.21 | 12 | 0.43 | 15 | 0.86 | 17 |
| W. Europe | 0.12 | 7 | 0.23 | 8 | 0.40 | 8 |
| Other Countries (a) | – | – | 0.04 | 1 | 0.10 | 2 |
| Total | 1.73 | 100 | 2.90 | 100 | 5.04 | 100 |

(a) A big percentage here refers to the Centrally Planned Economies, including the USSR.
* (estimated).
† (forecast).
Source: Derived from data of Creative Strategies International, 1981.

**Table 5. Value of Data Processing Equipment (US$1000 m)**

| | 1978 | Percentage of Total | 1983(F) | Percentage of Total | 1985(F) | Percentage of Total |
|---|---|---|---|---|---|---|
| Developed Countries (USA, Japan & Western Europe) | 110 | 83 | 180 | 82 | 250 | 80 |
| Other countries (a) | 22.5 | 17 | 40 | 18 | 61 | 20 |
| Total | 132.5 | 100 | 220 | 100 | 311 | 100 |

(a) A big percentage here refers to the centrally planned economies, including the USSR. They account perhaps for more than half of the "other countries" category. Thus "developing countries" are considerably less than what appears in this category.

Source: Derived from data of Diebold Europe, 1979.

**Table 6. Worldwide Telecommunications Equipment Markets* (US$1000 m)**

| | 1980(E) | Percentage of Total | 1985(F) | Percentage of Total | 1990(F) | Percentage of Total |
|---|---|---|---|---|---|---|
| Developed Countries | 36 | 90 | 53.5 | 89 | 75.4 | 86 |
| DevelopingCountries | 4 | 10 | 6.7 | 11 | 12.1 | 14 |
| Total | 40 | 100 | 60.2 | 100 | 87.5 | 100 |

* *Includes*: telephone, telegraph, telex data communications, satellite communications, mobile radio and radio telephone, radio paging, cable television.

Source: Derived from data of Arthur D. Little Inc.

There are two important aspects which can be derived from this purely quantitative check. First, the participation of developing countries in "informatisation" (components, computers and telecommunications) is indeed very small. Second, as time passes the gap in amount of investment in these fields widens. The same is true for market share in production of microprocessors/microcomputers, chips, boards and systems.

An additional element which I shall mention in passing is the question of the relative skill endowment of the developing countries.

On the human resource side, it is sufficient to say for instance that Korea had 0.4 researchers per 1,000 of the population in 1978, compared to 2.6 and 2.4 in 1977 for the U.S. and Japan, respectively. Expenditure in R & D per researcher is about $21,500 in Korea, compared to $47,560 in Japan and $80,680 in the U.S. (Kim, 1980).

I take Korea as an example not because it is a "typical" developing country (if such a thing exists), but because it has been acknowledged as an "electronic miracle", a "Japan in the making" and belongs to the NIC's category.

The diffusion of information technology in developing countries resembles the general pattern followed by the West, but has its own characteristics. The most visible aspect of this diffusion is the growing use of computers, but other types of 'intelligent' equipment are being introduced through trade.

There are several kinds of practical factors that condition diffusion and act as brakes. One important factor is labour costs, particularly in direct application of computers for conventional procedures. The low cost of labour makes the equipment less competitive and amortisation takes longer. The lack of standardisation in a number of activities and of modern managerial practices has made software requirements more demanding and, therefore, the total cost of installations more expensive than they would have been otherwise. "Software packages", even in the banking sector, are difficult to implement. This situation has led, paradoxically, to well-developed software applications in some countries. Additionally, companies need to spread the costs of infrastructure, service, and operations over relatively few pieces of equipment which also contributes to increased costs (Rada, 1980).

The single most important factor concerning diffusion of the technology is undoubtedly government action in a wide range of fields, from imports, telecommunications, and transfer of technology policy to industrial strategy. An increasing number of countries have adopted centralised systems to process import requests, monitor applications and purchase equipment for government needs. These types of measures cover data processing equipment especially, and do not affect other types of electronic or "intelligent" devices. The situation in each country varies considerably.

## EMPLOYMENT

THE EMPLOYMENT EFFECTS of information technology have been at the centre of the current debate about its impact in developed countries.

The short and long term effects in developing countries are different due to the structure of industry and services. The development of information technology creates jobs at the manufacturing and end-user level by allowing the production of new equipment and development of services. The employment created in manufacturing, assembly, peripherals and related industries is only valid in very few developing countries where most of the equipment and ancilliary goods are imported. Thus, no compensation effect takes place within the national economy.

In the use of equipment many studies exist which refer to direct computer utilisation which affects a very small proportion of the total labour force and can only serve as a general indicator of possible developments. (International Labour Office, 1972; Government of India, 1972; Prasad and Verma, 1976).

Despite the qualifications, all the studies mentioned agree on the fact that there is a loss of potential employment which will have different effects in developing countries with high unemployment rates and growing educated unemployment.

More important in terms of employment are the general development prospects in the light of what one might call the "external effects", i.e. the conditioning of developing countries' possibilities due to the technological push of the advanced countries, such as the erosion of comparative advantages.

Some industries will choose developing countries as sites for establishing industries based on the new technology. The decision will be based less on low labour cost than on tax advantages, lack of labour resistance, lower start-up costs and cheaper capital.

It is doubtful, however, that this type of modern installation will create significant employment or contribute to the general development of the economy inasmuch as the diffusion of microelectronics in developing countries will remain a sophisticated technology for most products and processes.

At the same time, the use of the technology by developing countries will widen the gap between technology and its application, because entire processes will have to be imported on a turnkey basis. The net result could be islands of high-technology within economies characterised by low-productivity and artisanal practices. This dichotomy and the link between the two areas represents one of the existing dilemmas of industrialisation.

For some developing countries the combination of high technology and low labour cost, particularly of educated labour, could be one of the opportunities to develop an embryonic informatic industry and create employment. What is of interest for the purpose of microelectronics is the possibility of using it to enhance these activities and thus the question of software is crucial. This depends on the country's current endowment of software skills, its capacity to keep software specialists at home. The question of "brain drain" here should not be overlooked since the recruitment policy of large hardware and software companies is as global as all their other operations. Developed countries are already sub-contracting software development on developing countries, especially for routines most applicable to their conditions.

The development of the capabilities in this area will largely depend on public policy, since in many cases the purchase of skills outside the country is more economical than developing national skills.

## CONCLUSIONS

CURRENT TECHNOLOGICAL CHANGES and the advent of a comprehensive information technology call for a re-conceptualisation of development strategies beyond mere implementation of informatic policies. This re-conceptualization should consider that informatics will profoundly affect the productive infrastructure and the international division of labour.

The process of informatisation is narrowing the policy options of developing countries with regard to industrialisation; whether this is looked at from the quantitative or qualitative side of the Lima Target. The latest estimates show that 9 per cent of the industrial capacity of the world will be in the developing countries by the year 2000, but this estimate does not consider the possible impact of technological change. An interesting example of how options are narrowed is the case of the electronics industry itself, particularly the component sector which is becoming the 'heavy industry' of the future. The R & D capital requirements, pace of technological change, and the nature of the changes imply that only total vertically integrated companies will be able to compete and they will require substantial public sector support to do so, directly or indirectly. At the same time as the level of integration increases, so does the amount of technology embodied in hardware and software. The assembly of different functions is transferred to the manufacturer of components, decreasing the number of manual operations, while increasing the possibilities and flexibility of automation.

These processes partly explain the fact that offshore installations are becoming less attractive. Although they have never been a good mechanism for the transfer of technology, they created incentives for the development of local skills. There has never been any question of transferring the highly technical stages of production to developing countries and now even the assembly operations are being automated. Strategies vary considerably from company to company and this explains, to a degree, the differences in the use of offshore facilities.

Developing countries should consider the following three elements in elaborating public policy towards the electronic industry (Rada, 1981):

(1) Renew efforts in the area of science and technology, research and develop-

ment. They should not merely reproduce the efforts of the developed countries but base their actions on an assessment of long-term comparative advantages. The use of interactive links in this field could be of great benefit.

(2) Search for specialized areas in the field of informatics in hardware and software (market niches) rather than attempt the development of the entire "information front". Here regional agreements and horizontal links are especially important. It is unlikely that with a purely national base a developing country would be able to master the required capital, R & D, and human resources for a stake among the giants. This is even true at the software level, where the trends are towards systems which demand tailoring of hardware.

(3) At a national level adopt policies of selective application of the technology bearing in mind the need to spread dependency and reinforce long-term possibilities. These elements should be guided by a clear assessment of long-term comparative advantages with the understanding that they are man-made and based on science and technology.

This search of necessity requires a consideration of alternative development strategies and styles and consequently a more comprehensive body of knowledge to understand the links between advanced and less advanced technologies. This covers not only the area of informatics but also of other advanced technologies, such as bio-technology and materials.

A final general consideration is that the developing countries should evolve a common science and technology strategy, geared to their needs and aspirations. This will allow them to command and direct technological change, rather than just react to it with controls and half solutions.

**References**

BA Asia Ltd. (1980) *The Semiconductor Industry in Japan*, 1980. Hong Kong.

Business Week (1978) *R & D Scoreboard* 1977. Business Week, July 3.

Ibid. (1979) *R & D Scoreboard* 1978. Business Week, July 2.

Ibid. (1980) *R & D Scoreboard* 1979. Business Week, July 7.

Commission des Communautés Européennes (1980) *L'Europe Face aux Technologies de l'Information.* Dossier de l'Europe, 1980/3 Mars.

Crémieux, A. (1980) *Impact Économique de Circuit Intégré*, in Analyse Financière, N. 42, 3 Trimestre, p. 47.

d'Argoeures, T. *et al* (1980) *L'Avenir de l'Electronique Européenne Grand Public.* in Analyse Financière, N. 42, 3 Trimestre, p. 36.

De Benedetti, F. (1979) *The Impact of Electronic Technology in the Office.* Proc. Financial Times Conference: Tomorrow in World Electronics, London, p. 130.

Electronic Industry Association of Korea (EIAK) (1979), Seoul.

Eng Fong, P. (1977) *The Electronics Industry in Singapore: Structure, Technology and Linkages.* ERC Research Monograph Series N. 7, Singapore.

General Technologies Systems Limited (1979) *Netherlands Microelectronics Study.* Middlesex.

Government of India (1972) *Report of the Committee on Automation.* Ministry of Labour, New Delhi.

Heikes, R.R. (1980) *Industry Strategies: The United States View of the European Markets.* Proc. Financial Times Conference—World Electronics: Strategies for Success, Monte Carlo, p. 55.

Hertzfeld, H.R. (1980) *The Impact of NASA Research and Development Expenditure on Technological Innovation and the Economy.* In *Economic Effects of Space and Other Advanced Technologies.* European Space Agency, Paris p. 90, table 1.

International Labour Office (1972) *Automation in Developing Countries.* Geneva.

Kim, W.H. (1980) *Challenge to U.S. Domination: The Promises of Technology for Newly Industrialised Countries.* Proc. Financial Times Conference—World Electronics: Strategies for Success, Monte Carlo, p. 99.

Kurt Salmon Associates (1979) *The 1980's—The Decade for Technology?—A Study of the State of the Art of Assembly of Apparel Products.* A report to the Commission of the EEC, London, p. 3.11.

Mackintosh Consultants (1979) *Microelectronics in the 80's: A Digest of International Scenarios and Government Strategies for the Microelectronics Industry in the 80's.* Mackintosh Publications Ltd., London.

Muoy, P. (1980) *Le Rôle des Marchés Militaires dans le Développement des Industries Electroniques Américaines.* Analyse Financière N. 42, 3 Trimestre, p. 47.

Prasad, K. and Verma, P. (1976) *Impact of Computers on Employment.* New Delhi.

Rada, J. (1980) *The Impact of Microelectronics and Information Technology: Reference to Argentina, Brazil and Bolivia.* UNESCO.

Rada, J. (1981) *Some Issues and Possibilities Posed by the Unfolding Information Revolution.* Development Dialogue, February.

Roessle, H.F.L. (1979) *Prospects for the European Semiconductor Industry.* Proc. Financial Times Conference: Tomorrow in World Electronics, London, p. 97.

Schmoranz, J. (1980) *The Information Sector in International Trade.* OECD, DSTI/ICCP/80.17 Paris.

Science (1980) May 19, p. 585.

US Department of Labor (1980) *Trends in Technology—Intensive Trade.* Bureau of International Labor Affairs.

# The Consequences of Microelectronics for Employment and the International Division of Labour

***Øyvind Kvåle***
*Resource Policy Group, Norway*

## INTRODUCTION

MICROELECTRONICS is one of the factors which lie behind the improvements in efficiency and changes in the production conditions of the various geographical areas, but as this article shows, the employment consequences of the new technology are far from unambiguous.

It is important to realize that both the fate of the labour force and current material standards of living are to a great extent determined by the economic system, economic policy, and the power of the consumers—not by microelectronics.

It is illusory to believe that we can choose whether or not to make use of microelectronics. To a limited extent, we can influence where and when, but the introduction of microelectronics is a consequence of our economic system.

Microelectronics helps form the basis for new products, it makes production processes more efficient, and it provides the basis for an altered international division of labour. Nevertheless, it is important to realize that microelectronics only plays a subordinate role in the major processes of transformation which the world is experiencing today. The most important factors behind the current transformations are the increased amounts of know-how which a number of countries have acquired, and the effects of changes in the price of oil.

An increased amount of know-how is an important prerequisite for establishing traditional industrial production in a number of new industrialised countries.

The old industrialised countries are having trouble maintaining this kind of production, e.g. automobile production, steel production, or shipbuilding.

It is obvious that the enormous increases in oil prices will give rise to a major redistribution of revenues between various countries. This must necessarily alter the conditions for the global structure of consumption. It is difficult for people to adapt their consumption to changing conditions. The large gap which exists in many countries between their own production and consumption requires either large investments from abroad or external loans. These major "recycling" problems create unwanted dependence and uncertainty. (These transformation problems can be registered as problems with stagflation and the balance of trade.)

In general, efficiency is quite desirable in a poverty-stricken world. Of course, an increase in efficiency is not without its problems. We can lose in some areas what we gain in others, e.g. by rationalising manpower out of productive activities. For the world as a whole, however, material affluence can only increase if the population increases its efficiency. In this context, microelectronics can be a tool for making the world materially richer. In order to ensure that the increase in efficiency will result in increased production, while maintaining the desired demand for manpower, economic policy will be completely decisive.

On a more detailed level, we find that shifts in the international distribution of revenue, economic growth, and an altered product spectrum will change the demand for various goods and services. The freedom which we have when it comes to choosing which goods and services we will consume can be in contrast to the range of products which industry is capable of producing. Therefore, the power of the consumer determines the fate of the labour force. Without a restructuring of industry, we will quickly reach a situation where too many people are producing things which too few really want.

It takes time to bring both the domestic structural changes and the international division of labour which are continually promoted by our economic system. Many people will neither endure nor accept transformations which are as rapid as those created by the technological innovations and the world's changing production conditions. The result will be that an increased material standard of living will be attained by forcing the weakest and oldest workers off the payroll.

If people's transformation problems become greater when the degree of occupational change and geographical displacement increases, then microelectronics' employment consequences will be far from negligible. Without a doubt, microelectronics will serve to increase changes in the production processes. Seen in isolation, this will worsen the difficulties which people have in adapting.

On the other hand, the potential of microelectronics to make the service sector more efficient will contribute positively to reducing the rate of transfer, which we observed in earlier times, between important sectors in the society. It is the latter structural changes which give rise to the greatest occupational and geographical transfers. In this way, microelectronics can help to reduce the worst difficulties of transformation for people in the labour force.

# MICROELECTRONICS—A CONSEQUENCE OF AN ECONOMIC SYSTEM

IN TODAY'S WORLD, mankind is not asked for his opinion about microelectronics—whether it is desirable or undesirable. To some extent, we are faced with the question of when it will come and where. There are two important reasons why microelectronics will be introduced:
(1) the introduction is a consequence of a decentralised decision-making structure;
(2) for the older industrialised countires, microelectronics is important in order to maintain the unequal distribution of incomes in the world.

## Microelectronics—A Consequence of a Decentralised Decision-Making Structure

The economic system which prevails over large parts of the world today is characterised by a relatively decentralised decision-making structure. Economic profits, income, and existing jobs are determined at a disaggregated level. The power of the market is considerable. The fact that both income and survival are closely tied to the actions taken by a firm, provides incentive for constantly trying to improve productivity. Microelectronics can be used to improve existing production processes in order to lower costs, or it can be used in radical innovations of completely new processes and products, which, at first, are not very price sensitive.

In this case we have stated that the possibility of increasing profits was the most important incentive, but even without this motive, it is obvious that it would be difficult to try to ignore the process of change. Even if one does not undertake anything at all, the changes in efficiency will tend to result in lower prices, so that those who do not improve their efficiency will have declining profitability. These hardpressed firms could avoid closing down by gradually accepting a lower, or maybe even a negative wage development. Everyone knows that in practice this sort of strategy is extremely difficult to carry out.

However, the most important thing is that product competition is often not essentially a question of prices. The primary factor involved in the choice of a method of production is often the visual and qualitative demands which are placed on the finished product.

Microelectronics, and especially the miniaturisation which it brings about, naturally gives rise to a number of new products. Moreover, microelectronics will allow many traditional goods to be simplified and improved so that previous products are displaced. If we refuse to accept microelectronics in either consumer goods or production equipment, the spectrum of products which we can produce will become narrower and narrower. We should also keep in mind that the successful introduction of microelectronics in the vulnerable sectors will often indicate things which will be repeated in the protected sectors. There will often be a great deal of competition in individual areas within the protected industries. Moreover, we often see that the authorities and large interest organisations will promote initiatives to make these industries more efficient. Low efficiency here could easily result in high costs for the vulnerable branches.

## Microelectronics—A Consequence of the desire to maintain the gap between the Rich and Poor Countries

Changing conditions in the international division of labour will be a threat to the unequal income distribution which has prevailed up to now. Older industrialised countries can no longer be certain that in the future they will still be able to exchange a few of their man-hours for many from other countries.

Two conditions are especially important in explaining the structural changes in production and income distribution between the various countries:

A number of raw materials which previously were easily accessible and inexpensive have become scarce and expensive.

Technological knowledge is spreading over the entire planet. The recently industrialised countries are the result of a higher level of education and more favourable national economic conditions.

Western Europe and the USA have previously attained high levels of income and consumption by purchasing inexpensive raw materials from underdeveloped countries, refining these materials in their own countries, and re-exporting some of the finished products back to the underdeveloped countries. Their high standard of living was a result of high efficiency and favourable trading terms.

In recent years the development in this traditional price pattern has followed a different course—a number of raw materials have become scarce and expensive. In particular, it is the oil rich countries that can now exchange only a few of their own manhours for many from other countries.

The advent of the new industrialised countries means that the value added for traditional industrial production will tend to decrease. A lower wage level and high efficiency in some parts of industry have put pressure on the price of a great number of traditional finished products.

The traditional view of the structure of world trade is that the underdeveloped countries are only qualified to produce traditionally labour intensive products. In recent years the growth in global steel production stands out as an example that the new industrialised countries can also participate in new and relatively capital intensive production (Figure 1). In the course of a few years, the new industrial countries have more than doubled their share of global steel production. Correspondingly, the old industrialised countries are losing their traditional dominant grip on this economic and strategically important sector.

In Europe, the attitude for the most part seems to be a belief that we can uphold both income and employment levels. In some countries we find that

**Figure 1. The distribution of steel production**

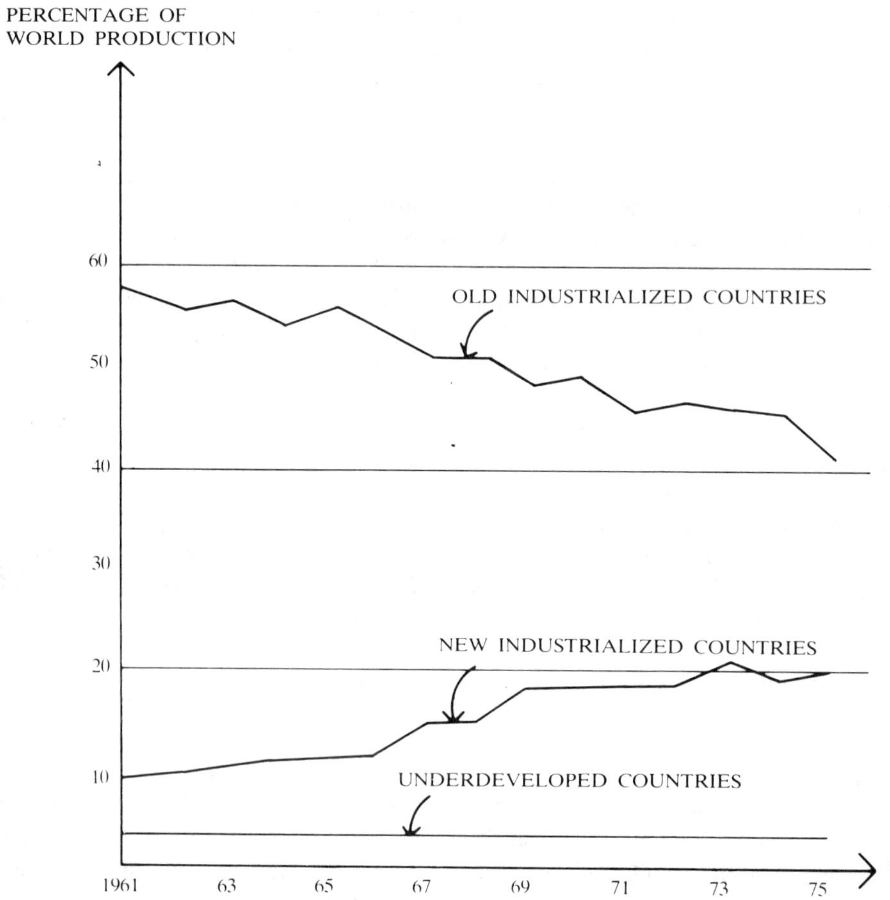


upholding the level of income clearly has the highest priority. Two strategies are often tried:

General efficiency improvements in order to reduce the effects on private incomes of the redistribution of incomes which is funneling greater wealth to the producers of raw materials.

An attempt to utilise the high level of education/knowledge in various countries to instigate radical product innovation and the production of special products to a greater extent, and at the same time concentrate less on working with standardised products and production processes. Products which are unique and production processes which few have mastered are much less price sensitive, so that there is little competition from the new industrialised countries, at the same time as the absence of price sensitivity results in a neutralisation of the increases in prices of raw materials.

It is obvious that both points make it very difficult for these countries to reject either the introduction of microelectronics or many of the structural changes that have to come. In the struggle to maintain a high income level in the old industrialised countries, both employer and employees have common interests. These interests become clearly apparent in the pressure for growth in real disposable income.

In the structural transformation which is now occurring, large transfers of technology are being made to industries in underdeveloped countries. The OECD countries are well paid for their information intensive industries, which are performing development work when production information is transferred to underdeveloped countries.

For underdeveloped countries the critical point will often be the transfer of information. Those who are in possession of this knowledge have a very strong strategic position and a corresponding position of power. The pattern can be clearly seen in the increasing number of joint ventures and daughter companies between the OECD countries and the rest of the world. Such situations are and will be major power problems for the authorities in underdeveloped countries.

The spread of technology is closely connected with the level of knowledge. In almost all of the underdeveloped countries, a noteworthy rise has occurred in the level of education. A higher level of education permits more complicated production processes.

The higher level of knowledge in the world makes it less important to distinguish between labour intensive and capital intensive countries. The low wage levels and, perhaps equally important, the relative absence of restrictions on the exploitation of manpower and the environment are important competitive factors in the underdeveloped countries.

We will see major changes in those products which are labour intensive, because new technology will radically change many production processes. In many cases the most advanced technology, with microelectronics, will be so efficient that there will be little or no possibility of using less capital intensive technology unless one is willing to accept next to nothing in the way of wages.

The most important features of industry in underdeveloped countries that produce for export will be production in large batches and the fact that the products are standardised, in other words, production processes which are easy to manage, but which will often be capital intensive with microelectronics incorporated into the product and the product processes.

## MICROELECTRONICS AND THE DEMAND FOR MANPOWER

THUS WE HAVE TO EXPECT that the applications of microelectronics will continue to expand, because they are promoted by the economic system. What then will be the consequences of the fact that microelectronics brings about improvements in efficiency? In a global context, two possibilities arise:

The possibility of rationalising away manpower.

The possibility of increasing production.

The question as to whether production will increase in step with the efficiency increases is a question as to whether the potential producers can find a market for their increased production.

When weak demand causes the production potential to be slack, this is not necessarily a sign of a generally slackening demand for goods and services. Assertions of this sort may seem downright comical to underdeveloped countries, and even in industrialised countries they seem irrelevant. The demand for reduced taxes and better health services indicates that there is also a strong desire for greater consumption in these countries.

On the other hand, there may well be saturated consumption for some individual goods, because production may become too great in relation to the consumers' wants. This happens quite often when new products enter the market. We can imagine, for example, that other countries might offer new goods so that those which we have traditionally produced can no longer be sold. As a result, we try to produce something which we do not want for ourselves. Without some sort of readjustment, we run the risk of ending up uncompetitive as regards both price and quality.

With a broader and broader spectrum of products and a steady renovation of products, it becomes more and more difficult to judge what can be sold. Product developers will need a lot of technical knowledge and will be situated to a greater extent in the vicinity of their customers.

Since the need for goods and services apparently is present, the economic policy becomes important in order to assure full employment. The nation-states may well regulate the magnitude of their domestic demand, but they are dependent on their trading partners when it comes to determining the level of foreign demand. In this way a strong relationship of dependency develops between different countries. If we are to utilise all the available production capacity in the world, then the total demand of all of the countries put together must be equal to their total production capacity.

One of the major problems today is that some countries want to operate with a surplus in their balance of trade. They want to produce more than they consume. This can give them a feeling of greater freedom of action in the event that something unexpected should happen. Some of these surplus countries also have problems because extra increases in demand will create bottlenecks and thereby inflation.

The trading nations that have a large deficit in their balance of trade are forced on the defensive. They have to balance their foreign accounts with large loans or foreign investments that create dependence and uncertainty. The result is that the growth in demand for all countries becomes weaker than it should be.

In this perspective the new industrialised countries are not a threat to the exports and employment of the older industrialised countries. All of the new industrialised countries utilise all of their export revenues for imports. On the whole, these countries never provide the older industrialised countries with balance of trade problems. As long as we have a low standard of living in the underdeveloped countries, it is rather obvious that they will import an amount which corresponds to the quantity which they are able to export to the older industrialised countries. In this way they stimulate economic growth in the older industrial countries.

The current remedy for problems of inflation and the balance of trade seems to be a restrictive financial and monetary policy. Demand is reduced in the hope that the participants in the economy will feel greater concern for their own jobs and thus moderate their expectations of higher incomes, and at the same time the country's import bill is lowered. But if most people refuse to give up their standard of living, the result will be stagflation.

Development is directed towards easing the situation for manufacturers in the hope of increasing the rate of innovation and easing structural transformations. To some extent these incentives may work, but social unrest, as a result of the inequalities which arise, can rapidly eliminate the gains.

This strategy might possibly promote the international division of labour, but it can only improve an individual country's balance of trade to the extent that that country is more successful in this respect than its trading partners. As long as every country's supply-side economic policy is equally successful, the problems with the balance of trade will remain. But improvements in efficiency can cause domestic conflicts over distribution to abate because there is more to be divided.

From this point of view, microelectronics will be a less important factor contributing to unemployment. It is economic policy and society's ability to re-organise that will determine the demand for manpower.

## THE CONSUMER'S POWER IS THE FATE OF MANPOWER

DEVELOPMENT in the labour market, however, is also dependent on what happens with the individual products. As consumers we have, to a great extent, real freedom to choose what we will buy with our income. Therefore, in periods of real growth in incomes, the private consumer will have decisive influence over which production processes will expand and who will be unemployed.

If we compare the efficiency improvements in production which are brought about by microelectronics with the income effects from the same efficiency improvements, we can deduce the following structural changes:

> If demand increases in correspondence with efficiency of production, the need for employees will remain constant. If we have increases in demand in areas where the improvements in efficiency are also large, the employment consequences will be small.

The most serious situation arises if the efficiency improvements come in sectors where the development in demand is weakest (even negative).
If microelectronics are introduced on a "broad front" so that the efficiency improvements are evenly divided, the consumers' behaviour alone will dictate the employment consequences.

From the current discussion we can see that it may be incorrect to assert that it will be disastrous for the microprocessor to automate or improve the efficiency of work for completely new groups (in the tertiary sector for example). If the consumption in these sectors (both public and private) is constantly in a state of rapid growth, it is conceivable that efficiency improvements might be very advantageous. Efficiency improvements can prevent the transformations between sectors from proceeding as rapidly as they have done in the past.

In Figure 2, we can see how employment has been distributed between primary, secondary, and tertiary sectors, as the income per capita has increased (the average for the OECD-countries). The characteristic feature here is the primary sector's reduced significance and the corresponding growth in the tertiary sector. The structural changes have been a consequence of low demand and large increases in efficiency in the primary sector together with a reverse situation in the tertiary sector, i.e. rapid growth in demand and sluggish increases in efficiency. If

**Figure 2. Structural changes between main sectors within a country when income increases**

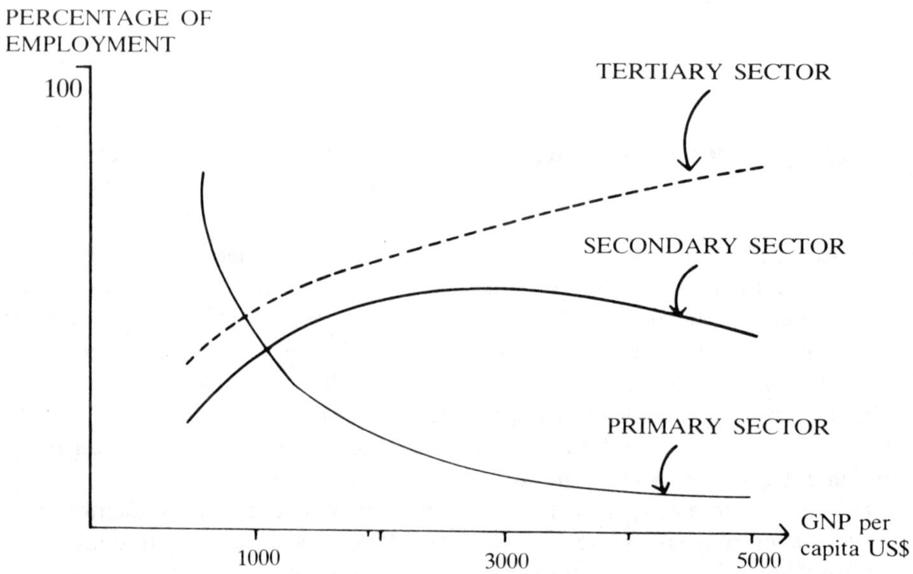

the efficiency increases in the tertiary sector had been greater, the structural changes between the sectors would have been less.

Even though microelectronics can help to uphold the demand for manpower within the main sectors of society, some individual industries will nevertheless be ruined. Microelectronics undoubtedly plays an important role as a basis for new products. If the consumer begins to prefer something new, other producers will be left with a product which we, as consumers, no longer want. In addition, continuous changes will be taking place within the actual production processes. Microelectronics will promote these transformations.

It is not so easy to shift manpower over to the types of production which are in demand. The employees who will have problems will vary somewhat according to the type of transformation, but the most established (socially and geographically), those who stand to lose social status, and those who have the least professional flexibility are among the vulnerable groups.

Studies in Norway indicate that of those who become unemployed, 10–20 per cent will be permanently expelled from the labour market.

## CONCLUSIONS

In the future microelectronics will be an important factor behind a continued increase in efficiency and production potential. If production increases more rapidly than the increases in efficiency, then the demand for manpower will increase.

There is undoubtedly a tremendous need for a higher material standard of living throughout most of the world. The major obstacles to increased growth in world production are the difficulties inherent in making structural transformations. The need for transformations and the problems involved in carrying them out can be detected in a low demand for manpower, inflation, and problems with the balance of trade.

The forces which are currently promoting the demand for increased structural transformations are not primarily microelectronics. Important driving forces behind this pressure are altered conditions for the international division of labour and the changes in the international terms of trade, where the changes in oil prices are the dominant factor.

However, microelectronics will increase structural transformations within established production processes and increase the access to products. These changes can expose the people involved to a great deal of stress; many will not be able to maintain their jobs. Seen in isolation, this will result in increased unemployment.

On the other hand, microelectronics' capacity to make the service sectors more efficient will decrease the traditional structural changes between the main sectors in the society. Moving manpower *within* the main sectors, e.g. the secondary sector, probably involves less transformation stress than movements from one sector to another, e.g. from the primary to the tertiary sector. The fact that microelectronics' potential for efficiency improvements will affect most workplaces may decrease the transformation problems for employees.

## References

LeFrance, J.D. (1980) *The Future for European Business—Strategies for Growth.* Long Range Planning, **13**, August.

Lundgren, G., Inganäs, O. and Mölleryd, B. (1980) *Automatisering och internationell konkurrenskraft—en framtidsstudie om strategival i svensk verkstadsindustri.* Styrelsen för Teknisk Utveckling, nr. 170–171, Stockholm.

Moene, K.O. (1979) *Klassisk eller keynesiansk arbeidsløshet.* Memorandum fra Sosialøkonomisk Institutt, Oslo.

Nilsson, J.E. (1981) *The Impact of Microelectronics on Economic Development: A Comment.* GRS-342, Oslo, March.

Sharp, M. (1980) *Technology and Growth. The Challenge of Long-Term Structural Change.* Futures, October.

Sleigh, J., Boatwright, B., Irwin, P. and Stanyon, R. (1979) *The Manpower Implications of Microelectronics Technology,* Her Majesty's Stationary Office, London.

# Employment and Skills

*Ernest Braun*
*University of Aston in Birmingham, United Kingdom*

## MICROELECTRONICS REVISITED

MICROELECTRONICS has been talked and written about so much that a special explanation is required for a further talk about this ubiquitous subject. The word ubiquitous provides the clue—for microelectronics is not only an ubiquitous subject for discussion, it is indeed an all pervading technology. It is impossible to ignore a technology which not only dominates telecommunications, computing, automation, control and all forms of data processing, but also plays a major role in domestic entertainment and is continually finding new applications.

The essence of microelectronics is that it can store and manipulate items of information—be it words, numbers or measurements. The result of the manipulation may be displayed, transmitted or used to control machinery or processes. All these things were possible with the old type of electronics too, but microelectronics equipment occupies a fraction of the space previously required, uses a fraction of the power, operates at enormous speed, is capable of unbelievable complexity and, above all else, is incredibly cheap. The combination of these qualities has made the technology irresistible. To obtain so much capability at such a low price is enough to ensure ready customers for microelectronics and the question of whether or not they need the new capabilities does not arise. If so much is available for so little—everybody wants it, very few can resist.

Clearly developing countries must ask themselves how they should deal with this technology. In particular, they need to decide on the type and extent of any support to be given to the manufacture and/or use of microelectronics and of equipment using microelectronics.

## THE MANUFACTURE OF SEMICONDUCTOR ELECTRONIC COMPONENTS

SEMICONDUCTOR ELECTRONIC COMPONENTS range from the relatively simple diode rectifier right up to the enormously complex very large scale integrated (VLSI) circuits, such as the most recent memory chips and microprocessors. The design and manufacture of VLSI circuits require very large-scale resources in scarce engineering skills and in capital. Even some of the well established leaders of the industry in the USA have recently run into cash flow problems and sold out to firms with larger financial resources.

The microelectronic industry is extremely competitive, yet linked by numerous networks of personal acquaintance and of licensing agreements. Even giant firms cannot afford to develop everything from scratch and often are content to serve as second sources for some components. No firms can manage without the use of somebody else's patents and there is a brisk trade in licences, partly for fees and partly as barter.

Like all manufacture, only much more so, the manufacture of microelectronics is dependent upon a multitude of specialist suppliers of goods and services. In fact the industrial systems shows a web of interdependencies and synergetic relationships. The absence of an extensive network of competent suppliers of goods and services within the industrial systems forms one of the great obstacles to industrial development. In the case of microelectronics, supplies of extremely pure chemicals, encapsulations, wires, microscopes, furnaces, clean rooms, computer aided design and test facilities, are just some of the dozens of items the manufacturer needs.

Because of the extreme sophistication, competitiveness, dependence upon scarce skills, capital intensity, and dependence upon synergy, it would be sheer folly for a small developing country to try to establish an indigenous microelectronic manufacturing industry. To invest in such an enterprise would be tantamount to pumping money down the river. The cost would be enormous, the chance of success pretty close to nil.

A different and much more feasible strategy is to invite a foreign manufacturer to establish a facility in the developing country concerned. The manufacturer knows his technology and his markets and any investment is much more likely to bring a return. On the other hand, the manufacturer is interested in so-called offshore facilities only for very special reasons and under very special conditions. Availability of a skilled, docile and cheap labour force, (albeit only quite a small labour force is required), tends to be one attraction. Penetration of new markets otherwise closed for political reasons, availability of grants and tax incentives, and various other trade-offs are further common incentives.

Whether the country concerned has much to gain depends upon a range of policies. If a substantial proportion of supplies can gradually be shifted to home manufacturers and if training of native personnel is achieved, then the presence of a foreign manufacturer may provide some stimulus for the growth and development of an industrial system. Very careful policies are required to achieve the desired strengthening of the industrial infrastructure.

## THE ELECTRONICS INDUSTRY

QUITE APART FROM the manufacture of active semiconductor circuit elements, e.g. rectifiers, transistors, microprocessors and memory chips, there are large industrial sectors making other parts of electronic equipment and making the equipment itself. There is more, much much more, to making electronic equipment than the manufacture of semiconductor components. To name but a few products of the electronics industry which incorporate semiconductor circuit elements, we have telecommunications equipment, radio, television, home entertainment, computers, word processors, industrial measurement and control equipment, numerically controlled machine tools.

Some of the above industries are as complex as the microelectronics industry itself; others are relatively simple and can form an important item in the industrial strategy of a developing country.

There can be two main planks in a policy of building up an electronics industry. One is import substitution on items such as radios, electronic control equipment, electronic watches, etc. The other is to use ingenuity to put together well designed original items, perhaps new electronic toys, new measuring devices, new electronic controls, or whatever, and attempt to capture an export market. The beauty of electronics lies in the fact that ingenuity and cheap labour can add very high value to imported components. The condition of success is, unfortunately, a harsh one: the availability of first rate entrepreneurs with first rate technical ideas.

Neither of the above options is particularly easy and especially the latter is precisely what every entrepreneur in technically developed countries, with all the benefits of a complete industrial infrastructure, is trying to do.

This leaves the possibility of simply trying to attract electronic assembly work from firms in advanced countries. There are numerous examples of this happening, but how great a benefit it brings to the country concerned will largely depend on details of the agreement; for example, on whether the firm employs a significant number of people, whether it gradually shifts to a domestic purchasing policy and whether it provides training.

## USE OF MICROELECTRONICS IN MANUFACTURING INDUSTRY

THE PROBLEM of introducing microelectronics into manufacturing industry is complex and multi-faceted. Four questions must be tackled:

(i) What can microelectronics do to aid manufacture directly?

(ii) How can microelectronics be introduced?

(iii) What effect does the new technology have on skills, employment, and competitive positions?

(iv) How relevant is the experience of developed countries to developing ones?

Clearly no more than a sketchy attempt at answering these questions can be made here.

(i) Apart from purely administrative uses, such as payrolls, invoicing and bookkeeping, (which will be discussed later), microelectronics has several functions directly linked with the manufacturing process. They fall broadly into four groups: a) the management of production, e.g. scheduling of the flow of work, keeping and dispensing of stores, maintenance scheduling; b) process control, e.g. the measurement and control of temperatures, flows, curing times, composition, mixing; c) design, e.g. computer aided design; d) control of machinery, e.g. numerically controlled machines, programmable handling machines (robots), automatic transfer machines.

The range of potential applications is clearly very great, ranging from the relatively trivial introduction of a digital micrometer to replace a traditional one, right up to the vastly complex introduction of a computer controlled automated production line.

(ii) The introduction of any new manufacturing equipment into an established plant producing an established product, requires a constellation of circumstances.

Such manufacturing innovation is normally carried out in response to the identification of a weakness in the chain of linked operations constituting the manufacturing systems. A weakness may consist of many different sub-optimal conditions. Examples are: labour productivity which is too low—at prevailing wage rates—to allow competitive pricing of the product; unavailability of required skills in the labour force; frequent breakdowns in machinery; excessive energy consumption; low utilisation of raw materials; unsafe working conditions; inadequate quality of product. It is obviously a vital task of management to monitor the manufacturing system continually in order to identify weak links in it.

Once a weakness has been identified, solutions to the problem must be sought. In some cases the appropriate solution may consist of the purchase—or even development—of new equipment. In many cases such new equipment will contain some microelectronic components. To quote just a simple example: the loading of a press for the manufacture of extruded metal parts can be a hazardous and unpleasant occupation. The solution to the problem may consist of introducing better safety features on the press and perhaps providing some noise reducing facilities. On the other hand, there may be a case for purchasing a robotic device which will carry out the loading operation automatically.

Which of these solutions is appropriate in a given situation depends on a multitude of factors, such as the relative cost of the two solutions, the abundance or scarcity of labour for loading presses on the one hand or skilled labour to maintain robotic machinery on the other, the spatial layout of the presses and adjacent machinery, access to spare parts.

It is obvious from even this single simple example that the appropriate solution to an identified manufacturing weakness depends upon a host of circumstances—a constellation of interdependent factors.

Similarly, the successful implementation of the identified solution depends upon a constellation of circumstances. Decisions must be made about technical matters such as where and how to obtain the machinery, where and how to install it, what suitable maintenance arrangements to make. Of at least equal importance, decisions concerning labour must be made. The introduction of new equipment is generally most successful if full discussions with the workforce have preceded the appearance of the equipment on the factory floor. New wage rates may have to be negotiated and new manning levels may require a decision. Finally, it may be necessary to acquire new skills either from outside or by appropriate training.

(iii) These questions, at long last, bring us to the problem area that this section is dealing with. However, they put the problem of skills, training and employment in the correct context, as one aspect, albeit a vital one, of manufacturing innovation and of industrialisation.

Skills have become an emotive issue. There can be no doubt that skills are closely linked with technology, in the sense that people and their skills in interaction with machinery produce the goods and services that society requires. It is impossible to divorce the typist from the typewriter, the pilot from the aeroplane, the turner from the lathe, the car mechanic from the motor car. In that sense—as long as the above remains true—it is clear that skills must change as technology changes. Just as machinery is constantly evolving and changing, so the skills of the people interacting with it have to change. The skilled bowman went out with the bow, the wheelwright with the wooden wheel, the thatcher with the thatched roof.

Changes of skill represent no problem other than that of training and of temporary mismatches if technologies change faster than people's skills. Real problems do arise out of two considerations: when new skills become simpler and less satisfying than the old ones they replace, and when new machinery runs largely without the benefit of skilled human intervention. The two problems can really be boiled down to one: the problem of reduced skill requirements. Reducing the question of skills to this single denominator means ignoring the difficult issue of whether craft skills are healthier and more necessary to human well-being than mental skills. In other words, can we really compare the skill of a turner with that of a programmer of a numerically controlled machine?

Although the above question may be of great cultural importance we must set it aside except in the context of training. Our question therefore is not whether new technology, and particularly microelectronics, causes changes in skill requirements—because there can be no doubt that it does that—but whether it reduces the total of skills required. Put this way, the question is not very helpful. An answer is difficult to obtain, highly uncertain, and dependent on too many other variables, such as total employment. Even if an answer were available, its policy implications would not be at all clear.

An associated question of equal importance and greater utility concerns the distribution of skills. Technology alone does not uniquely determine what skills its users shall apply—there are various ways in which the total of required skills can be distributed. A numerically controlled machine tool may serve as an example. Such a machine can be operated in two distinct ways: We can have an unskilled machine minder who only feeds material to the machine and switches it off in case of a malfunction. The machine minders have to be supported by maintenance staff, machine setters and programmers. Alternatively, we can have a machine operator who combines the roles of minder, setter, programmer and maintenance worker. Such an operator too will need support at a higher level of skills, e.g. a designer-programmer and a maintenance engineer. The two cases are, however, very different. In one case we have many totally unskilled and bored machine minders backed by a few highly skilled programmers and maintenance workers. In the other case we have mostly fairly skilled machine operators, backed by a few very highly qualified personnel.

Empirical research has shown that the latter case leads to a much more satisfactory operation; but it may not always be feasible. The determining factor is training. To operate a system of distributed skills, an adequate system of training must always be available. While it is possible to rely on external training, such as that provided by the education system, for the highest skills, distributed skills can only be effectively acquired by learning on the job. The principle of giving people opportunities and incentives to learn while doing a job ought to be firmly embodied in any good personnel policy.

The fundamental belief behind this principle is that people are able and willing to expand their range of skills gradually, building new layers upon those of existing knowledge. It must remain questionable, of course, whether complex mental manipulations can be learned by people with little basic schooling; but these detailed questions can only be answered in a specific situation.

The next complex problem area we must tackle is that of total employment. Two arguments prevail. One is that excessive increases in labour productivity cause redundant labour; the other that inadequate labour productivity causes lack of competitiveness and thereby loss of jobs. Thus we come to the conclusion that labour cost per unit of product must be optimal, which in this context means assuring competitiveness and reasonable growth of the firm.

It may come as a disappointment to some that I am not making any statement of the nature: 'electronics destroys jobs, therefore destroy electronics'; or the opposite 'electronics enhances productivity, use it wherever technically possible'. My view is that the use of microelectronics or any other modern machinery must be dominated by three major considerations: (1) The survival and reasonable growth of the firm through the production of competitively priced goods at competitive quality. (2) The employment of as many people as possible at socially acceptable wages. (3) The improvement of working conditions and the widest possible distribution of skills.

(iv) All that has been said so far is based on experiences and knowledge gained in developed countries. I have tried to consider the relevance of this knowledge to developing countries because I think that the fundamental considerations are the same everywhere and detailed considerations cannot be generalised.

The other major area of application of microelectronics not discussed so far is in administration. By the very nature of the technology—the processing of information—administrative tasks are the ones to which it can be applied with the greatest ease. Space permits only one or two general observations.

The first principle must be to use computer and word processor facilities in all those cases where a genuine improvement in services can be obtained. It is difficult to measure efficiency in administration as there are no good measures of output; nevertheless most people recognise the difference between a good and a bad administrative service.

The question of employment in administration is not clear-cut either. Some argue that as computers and word processors are introduced, so fewer and fewer people will be required in administration. Others argue that the range of possible administrative and, even more important, information services can be expanded almost indefinitely. If this happened, then employment could be maintained despite increased productivity.

The question is the more difficult as there is no direct competitive criterion as in the case of goods. Clearly, wherever a competitive product or service is administered, the administration does fall into different categories.

The nearest one can come to an answer would be to say that economic policies should be such as to allow an expansion of those services for which there is real demand—whether in the private or public sector—and the computers can make services cheaper and thereby increase demand for them. On the other hand, any proliferation of unwanted and unnecessary administration is clearly a waste of total resources.

## CONCLUSIONS AND POLICY RECOMMENDATIONS

(1) The manufacture of semiconductor circuit components is too complex to be contemplated in isolation. Foreign competent firms should only be encouraged to undertake such manufacture if they are prepared to make real contributions to the industrial infrastructure and to training.

(2) Similar considerations as in (1) apply to the electronic industry in general, although a greater element of import substitution and domestic entrepreneurship may be possible.

(3) The introduction of microelectronics and other modern machinery into manufacturing processes should be governed by three main considerations: competitive prices and quality of production; maximum employment; widest possible distribution of skills and training.

(4) The introduction of new machinery is a complex process requiring suitable constellations of circumstances and considerable managerial skills.

(5) The introduction of computers and word processors into administrative and information services should be undertaken in such a way as to maximise the provision of useful services, thereby ensuring improved economic activity and guarding against redundancy of personnel.

# The Informatics Sector in the Context of Industrial Innovation Policies

*Dennys Watson*
*Commission of the European Communities*

## INTRODUCTION

THIS PAPER does not describe the approach of any single country towards integrating the informatics industry into an industrial strategy; still less does it attempt to prescribe some ideal policy for achieving this goal. It reviews rather the general question of strategies to stimulate innovatory industries, some of the factors characteristic of the informatics sector, and determinants of long-term success which policy-makers should take into account.

## THE INDUSTRIAL DEVELOPMENT PROCESS

INDUSTRIAL DEVELOPMENT is not, for any country, just a one-time step, involving decisions by the private or public sector to purchase and install plant and equipment, to arrange transfer of the necessary technology, to recruit and train management, technologists and other workers, and then start producing goods. Even in the apparently simple case of import substitution, there are changes in technology and the market, more or less rapid according to the product sector but continually taking place, and customers have a habit of demanding the latest and best. So a firm, or an economy, that is not positively committed to industrial innovation will usually decline: the exceptions, such as firms making hand-crafted traditional goods, or operating in a non-competitive siege economy, have no relevance to a conference focused on informatics!

This commitment to industrial innovation involves an on-going process of analysis of current and potential changes in the market and in relevant technologies, of appraisal and choice of investment opportunities (including R & D projects, patents and licences, as well as plant), and adaptation of management and workforce to successive changes in products and manufacturing processes, so as to remain competitive in world markets. Public authorities have the power to influence this innovative process in various ways, even though it is primarily within the firm that the process takes place. Some of those ways are more or less obvious and transparent: policies and attitudes towards foreign investments (inward and outward), trade and tariff policies, fiscal measures and other support for R & D, public purchasing practices, provision of education, training and re-training facilities. Other ways may be less immediately obvious: the regulations covering banking and stock exchange transactions may affect ease of access to necessary capital of smaller firms: norms, standards and test procedures may be established in ways which either encourage or inhibit innovation and technology transfer: mobility of key personnel: and, in general, government policies can have a critical effect upon the encouragement and acceptance of risk-taking and of necessary change by managers, workers and the general public.

## THE INFORMATICS INDUSTRY

INFORMATICS is a sector in which changes have been extremely rapid, both in the technologies employed and in the market: and there is no reason to suppose that future changes will be any less rapid. For many parts of the sector, the initial investment costs of entry are at an increasingly high threshold: and, because of the competitive pace of development, the ongoing annual investment costs in R & D alone are correspondingly great. So the stakes are high and the risks of failure are all too apparent. Against this, one can balance the undoubted advantages of direct involvement in a sector with great potential growth still to come, and whose technologies have application in a wide range of other industries and services. The concept of the informatics industry as 'an integrated industrial sector', referred to in the conference literature, would seem to embrace the manufacture and marketing of a very wide range of products, from VSLI through mainframe computers to peripherals and software, with at least some product design and development capacity, if not applied research. But this very ambitious goal is one to which developing countries seem unlikely to aspire, at least in the short term. The problem then is how to have some share of the benefits, without excessive costs and risks.

## THE INFORMATICS SECTOR IN THE EEC

WITHIN THE EUROPEAN COMMUNITIES, informatics is considered as part of the integral concept of 'New technologies of information' (in French, 'télématique'), which was the subject of a document submitted by the Commission to the European Council in November 1979. The point was made that certain activities in this field could no longer readily be tackled by any one Member State alone, owing to the threshold of effort required to remain competitive in world markets: and it spelt out a range of measures needed: social and educational, the creation of new markets, the promotion of industry and technology. These include, amongst others, the creation of a pool of studies concerned with the impact on employment of the new technologies, the Community's four-year Informatics Programme (promoting the application and industrial development of computing via calls for tender of suitable projects for support, standardisation and public procurement policies), continued promotion of data-bases and a European information industry via proposals for a third action plan for Scientific and Technical Information and Documentation, as well as proposals in the field of education and training. In the broader context of innovation policy, the Commission submitted a document "Industrial Development and Innovation" to the European Council held in Luxembourg in December 1980, and is now responding to the European Council's request to examine ways of eliminating the fragmentation of markets, and improving incentives to innovation and the dissemination of knowledge.

Mention is made here of initiatives within the European Communities, to show that even countries that have long been considered as industrialised are greatly concerned with the international challenges of the informatics industry, and the need for innovation as the basis for economic revival. It may be argued that such wide-scale efforts are not needed for a firm or country which is content with a more modest sub-sector of the market, such as some peripheral equipment, or specialised software: and of course there is an element of truth in this. However, in such a technologically dynamic field, if the peripheral equipment designer and manufacturer or the software house does not have a continuous and effective interface with the microchip producer and the mainframe manufacturer, he is likely to be left behind in development; ways have to be found of overcoming the problems of physical remoteness, and at least some local capacity for developing the 'de-bugging' equipment and programmes must be built up.

## SUMMARY

To SUMMARISE, the informatics sector has undoubted growth potential and other attractions; but it is a highly competitive, fast-changing sector in which the stakes are high and the risks should not be undertaken blindly. In order to succeed in the long-term, a country needs good educational and training facilities, an adaptable workforce, an environment that encourages flexible innovation and openness to technological transfer: and, for the firms concerned, if they are independent, the financial resources to make repeated and substantial investments.

# Study of the Impact of Microelectronic Technology on the Irish Economy

***Robert Cochran***
*National Board for Science and Technology,*
*Ireland*

## INTRODUCTION

TOWARDS THE END OF 1979, the National Board for Science and Technology (NBST) perceived that the area of microelectronics and its applications was worthy of special attention. In particular, the electronics industry was one of the fastest growing sectors in the Irish economy, and one which could offer considerable employment over the coming decade. Furthermore, it was seen that the application of microelectronic technology in all sectors of business—both industrial and commercial—would be a major factor in international competitiveness in the years ahead.

There was also a popular belief that the application of the technology, particularly in certain industries, would cause large scale unemployment, and the NBST wanted to examine and report on this issue. It also recognised the fact that the pervasion of microelectronic technology into all aspects of commercial and domestic life would have far-reaching effects on the whole of society.

Thus it was clear that the technology would be an important factor in Ireland's economic and social development over the coming decade and on to the end of the century. It was also clear that the experience in Ireland might not be typical of experiences in more highly industrialised countries, and that studies, policies and programmes developed in and for such countries would not necessarily be appropriate for Ireland. The NBST therefore decided that a major study was desirable which would provide a balanced perspective from an Irish viewpoint. The terms of reference for this study were:

(1) To examine the trends in microelectronic technology and its applications;
(2) To assess, in the context of the Irish economy over the period to 1990 the new industrial and employment opportunities arising and the potential impact on the nature and level of employment in existing manufacturing and service industry; and
(3) To report conclusions and recommendations.

It is important to note that the study embraces all sections of the economy within its terms of reference and also that it aims to establish the positive benefits and industrial opportunities for Ireland from this new technology while not ignoring the structural and personal adjustments that will be needed in certain areas.

It is also worth noting that the project team was assisted by a consultative committee representing both sides of industry as well as the relevant Government Departments and State Agencies. In addition, some financial support was provided by the Directorate-General for Regional Affairs of the EEC.

## BACKGROUND

It is clear that a study such as this cannot be wholly concerned with technology. The technology must be seen in the context of the overall state of the economy and of industrial development. Thus in order to gauge the effects of microelectronic technology on the Irish economy, certain things about this economy as it stands must be determined. In particular, to predict the impacts on employment, the occupations engaging the workforce and their proportions need to be ascertained, along with the trends that already exist in each sector of the economy, and any forecasts that have already been made for them and the basis for these forecasts.

It is also necessary to establish the state of the technology in relation to the various occupations/sectors identified. There is also a need to know not only what is possible with the technology but also the availability of currently commerically viable systems, and reasonable estimates of trends and forecasts for what will be available commercially up to 1990. It is also necessary to establish how fast the available technology will be accepted in the marketplace and to identify the factors which either accelerate or decelerate this acceptance.

Other aspects which could have a bearing on the conclusions are the openness of each sector of the economy to external competition, and the opportunities for growth and employment made possible by the development of new goods and services.

In devising an appropriate methodology, consideration was given firstly to the general approach to be adopted. There are in practice three such approaches:

(a) Firstly one can approach it from a 'macro' point of view. It seemed clear however, that the development of econometric models has not progressed sufficiently to take account of rapid technological change of the kind being considered. In particular, such models tend to use highly aggregated data and rely heavily on historical time trends, both of which are serious disadvantages in attempting to study future changes. A similar line of reasoning also indicated that input-output analysis was not feasible, despite its superficial attractiveness.

(b) Secondly, one can take the opposite approach and deal with the problem on a 'micro' level. This essentially involves undertaking detailed case studies in isolated firms and attempting to draw some generalisations from these investigations. The project team felt it important that such case-studies be undertaken in order to gain a better understanding of the interplay of all the relevant factors which influence the diffusion and application of technology at plant level. It was however, considered that this approach would not be a satisfactory way of examining the impact of this technology on the economy as a whole.

(c) The third approach, which was the one adopted, attempts to steer a middle course between the other two. This 'meso' approach involves dissecting the economy into its component sectors and examining each of these in some detail. At this level one can get sufficiently detailed data to permit an understanding of the nature of the products and processes involved in each sector, as well as the nature of relevant variables such as manpower, degree of capital intensity, competitiveness, openness to external influences, etc. It may be objected, with some validity, that this approach ignores linkages between sectors, or the influence of imports and exports etc. The projects team was very conscious of this, and attempted to collect data to address these questions as far as practicable.

## METHODOLOGY

In line with the sectoral approach adopted, the economy was first broken down into its three main sectors of agriculture, industry and services. These were then further subdivided as outlined below.

### (*a*) *Agriculture*

Agriculture currently accounts for over 20 per cent of the labour force, which is the highest such proportion in the EEC. Employment in this sector has been dropping steadily and is expected to continue doing so for some time in the future. Some of the questions which were considered to be relevant to the examinations of the impact of microelectronics were:

Does automation on farms depend on farm size?

How will external competition from possibly highly automated farms affect the market for Irish produce?

Is there a difference between 'family farms' and farms run as business enterprises?

The analysis of this sector for the study was based on published material, along with detailed discussions with the Agricultural Institute (the state research body for agriculture) and contact with other research bodies and equipment suppliers around the world.

### (*b*) *Industry*

This sector employs over 30 per cent of the labour force, 20 per cent being in the manufacture of products while the other 10 per cent are in building and construction, utilities and the extraction of raw materials (mining etc). For the purpose of this study, special investigations were commissioned on the following sectors within manufacturing industry.

Food
Drink and Tobacco
Chemicals and related products
Engineering and Electronics
Printing and Paper
Textiles
Clothing and Footwear
Construction.

The other parts of manufacturing and related areas were examined by means of discussions with key people and by analysis of available data from a variety of sources.

### (*c*) *Services*

The services sector of the economy covers a diverse range of activities but together they account for almost 50 per cent of the labour force. These activities can be categorised as

Distribution/Commerce and Transport
Public Administration and Defence
Personal and Recreational Services
Professional Services
Insurance, Banking and Business/Financial Services
Communications.

While all parts of the services area were being considered, special studies were undertaken as part of the project of those areas where it seemed *a priori*, that the technology was most likely to have an impact. These were the areas of distribution and transport, office and administrative work, and banking. It was recognised also that the first two have important horizontal components which spread into the manufacturing areas as well.

## FINAL REPORT

At the time of writing, almost all the preliminary investigations, collection of data, and the collection and examination of revelant literature has been completed. At present, this material is being synthesised into the final report of the project.* This is not an easy task, involving as it does the reconciliation of material from different sources (with different approaches and assumptions). In addition, there is the task of putting the sectors together in a way that makes sense in the context of the economy as a whole, taking account of the interactions between the sectors as far as is possible and trying to reflect the overall dynamics of the economy. It is not simply a matter of looking at existing activities, postulating the introduction of microelectronics and assessing the direct effects on these activities alone.

The report now taking shape will have the following general structure:

Section 1 will give the background to the study, describe its methods, and give a brief description of the findings, conclusions and recommendations. In addition, this section will include a general review of the development of the technology.

Section 2 will describe employment in Ireland today, in all sectors, and discuss the economic factors and trends relevant to it.

Section 3 will describe the state of technology in the various sectors and will discuss the scope for application of microelectronics, sector by sector in the economy, and then describe the factors which will accelerate or retard the uptake of the technology in each sector.

Section 4 will discuss the changes that are foreseen in each sector; e.g. the replacement of labour by capital, the expansion of business, structural changes. It will attempt to quantify the jobs at risk in each sector in the short, medium and long term. It will also describe the possible new industries that may arise.

Section 5 will examine ways of dealing with problems and of maximising opportunities offered by the technology. It will also deal with the educational and training requirements that will be necessary. These, and the other recommendations likely to be discussed, are outlined in more detail later.

Section 6 will briefly look beyond 1990 and at the trends to the end of the century. It will try to assess the implications for planning in the 1980s of the developing information society, and will identify areas needing further study. It will also indicate follow-up actions that need to be taken.

* Now published: see reference NBST (1981).

## POLICY ISSUES RAISED

An important aim of the report is to adequately address the policy issues which arise from the analysis of the information collected. Examples of the issues to be considered are:

(*a*) *Education and Training*

It will be necessary to ensure that the educational system is geared to produce people with the right qualifications, in adequate numbers. Examples of possible problem areas are electronics engineers and technicians, computer professionals, maintenance craftsmen etc. It is also essential that adequate exposure (in both a general and a specific way, as appropriate) to electronics and computing is available in all relevant second and third level courses.

(*b*) *Retraining*

Provision needs to be made to cope with structural changes in employment. This will involve all skill levels: semi-skilled, craft, technicians, engineers, managers etc. Also there is a need to consider in-service training for engineers and others to keep them up-to-date.

(*c*) *Promoting Awareness*

Raising the awareness within industry of the possibilities of the technology for improving products and processes is vitally important but detailed considerations of the most appropriate mechanism to achieve this is required.

(*d*) *Infrastructural Requirements*

This will include ensuring that adequate R & D facilities exist, and the role of bodies such as specialised Application Centres. Also the role of the telecommunications service needs to be addressed, both in relation to its basic services and also in relation to the new range of potential services.

(*e*) *Consultation*

The most beneficial and effective utilisation of the technology will occur when workers affected by the technology have sufficient participation in the process of change.

### (*f*) *Planning*

Ensuring that an adequate focus for the continued monitoring of and planning for the changes due to the technology is established at government level will be considered. This is necessary if the technology is to be utilised towards the achievement of national economic and social objectives.

## CONCLUSION

THIS PAPER has concentrated on the background, rationale and methodology of the study while leaving aside the technology itself, on the assumption that developments in the technology itself are reasonably well known or are adequately dealt with elsewhere. Also ignored in this paper are the actual results of the study, since the analysis of the information is not complete at the time of writing. The final report of the study will however be published in June 1981 and will be available on request from the NBST then (NBST, 1981). The approach outlined here will, it is hoped, prove useful to others who may be considering undertaking similar studies elsewhere.

### References

NBST (1981) *MicroElectronics—the Implications for Ireland*. National Board for Science and Technology, Dublin.

# Microelectronics and the Garment Industry: Not Yet a Perfect Fit

***H.J. Rush and H.K. Hoffmann***
*University of Sussex, United Kingdom*

## INTRODUCTION

CURRENT NORTH–SOUTH TRADE relationships and the underlying structural characteristics which determine the resultant international distribution of benefits set the context into which microelectronics based innovations (MBIs) will be introduced. Any speculation on possible scenarios regarding the impact of microelectronics on the Third World must take these contextual factors into consideration.

One of the most significant of these factors is the phenomenon of Third World industrialisation efforts which have led to fairly significant increases in the export of manufactured products. Rates of growth of manufactured exports averaged around 26 per cent per year during the 1970s.

These aggregate growth rates, although impressive, do not necessarily reflect the situation of individual nations since developing countries are by no means a homogeneous group. Different countries have experienced varying degrees of success in increasing the value and volume of their manufactured exports. This is largely due to differences between countries in terms of the structure of their economies and the orientation of their industrialisation strategies. Underlying these differences are factors likely to be crucial in determining the ability of these countries to respond to changes in trading conditions brought about by the use of MBIs. These factors include the diversity of industrial activities and the size of the internal market; the degree of foreign control, (particularly in sectors where exports are important); the percentage of value added in exports; the degree of integration with local suppliers of intermediates, capital goods, and technical services; the level of indigenous technological capability; and the location of markets for the exported products.

It is difficult at this stage to assess how the combined weight of these variables will influence the impact of microelectronics on individual countries or on the Third World as a whole. Detailed case studies with a strong data base drawn from field work within these countries are required. Nevertheless, a reasonable first approximation of some aspects of the problem can be developed from the careful monitoring and analysis of the process of innovation and diffusion of MBIs within the advanced industrial economies. The first manifestations of how microelectronics will affect international comparative advantage will be highlighted by the rate and extent of the diffusion process in those developed countries which are either competitors or markets for the Third World.

A major study along these lines is underway at the Science Policy Research Unit, University of Sussex and is focussing on the MBI innovation/diffusion process in the garment sector in a select number of developed countries. The garment sector was selected for study because of its importance to both developed and developing countries in terms of trade and employment (see Tables 1 and 2). Moreover, the production process in this sector is highly labour intensive.

A common argument in the literature is that such processes are particularly susceptible to cost reducing innovations based on microprocessors which would allow full automation and dramatically reduce labour inputs. Private sector firms in the developed countries are expected to be the first to introduce these innovations. It is alleged that this would improve their competitive position *vis à*

**Table 1. International Trade in Clothing Among Selected Market Economy Countries (US$)**

| | 1969 | 1977 |
|---|---|---|
| Hong Kong | 620,713 | 2,935,834 |
| Korea, Republic of | 160,691 | 2,023,675 |
| India | 26,916 | 333,579* |
| Singapore | 27,825 | 210,879 |
| Mexico | 5,639 | 32,848* |
| Brazil | 737 | 85,172 |
| Thailand | 788 | 81,827* |
| Tunisia | 620 | 70,406* |
| Colombia | 916 | 39,666* |
| Philippines | – | 79,999* |

* 1976 data.

Source: United Nations (1978) *Yearbook of International Trade Statistics.* Vol. II, New York.

**Table 2. Employees in Clothing Industries and their Percentage in Total Employment in Selected Manufacturing Industries**

| | 1975 | 1976 | 1977 | 1978 |
|---|---|---|---|---|
| *Africa* | | | | |
| Cameroon | 6963<br>21.8% | 7011<br>21.4% | 6844<br>19.8% | 7084<br>17.2% |
| Mauritius | 7574<br>36.5% | 11484<br>42.9% | 13675<br>44.0% | 14282<br>46.0% |
| Kenya | 4284<br>4.0% | 4785<br>4.1% | 4911<br>3.9% | 5011<br>3.8% |
| *Americas* | | | | |
| Colombia | 38715<br>8.5% | 45503<br>9.7% | N/A | N/A |
| Costa Rica | N/A | 26086<br>29.4% | 32883<br>32.1% | 32143<br>31.0% |
| Chile* | 19671<br>9.6% | 18998<br>10.2% | 19494<br>10.6% | N/A |
| Honduras | N/A | 16430<br>25.2% | 16919<br>24.8% | 17562<br>24.4% |
| Puerto Rico† | 33834<br>30.0% | 34743<br>29.3% | 34980<br>28.9% | 33963<br>26.5% |
| *Asia* | | | | |
| Hong Kong | 192532<br>30.5% | 239941<br>32.0% | 231739<br>30.4% | 237557<br>30.1% |
| Korea, Republic of | 152100<br>11.4% | N/A | N/A | N/A |
| Singapore | 17739<br>9.4% | N/A | N/A | N/A |

* Including footwear.
† wage earners only.

Source: ILO Bureau of Statistics (1980) *ILO Second Tripartite Technical Meeting for the Clothing Industry*: *General Report.* Geneva.

*vis* Third World producers and deprive developing countries of crucial export markets.

Given that the apparel industry shares some similarity with footwear, textiles and other labour intensive sectors, it was felt that a detailed examination of developments in the industry would provide valuable insights at a more general as well as at the specific level. In the next section we briefly review the nature of the production process (prior to the introduction of microelectronics) and some of the structural and institutional characteristics of the industry. In the following section, innovations in sewing technology containing microelectronics are described; while in the final section some concluding observations are presented on the implications of these changes for the Third World.

## STRUCTURAL AND INSTITUTIONAL CHARACTERISTICS OF THE INDUSTRY

THE MANUFACTURE OF GARMENTS typically involves a sequence of activities where an operative is required at the interface between material and machine at each stage. (The discrete activities include: design, grade and cut pattern, plan optimised lay, lay and inspect for faults, mark, cut, label and bundle, transport to sewing station, assembly, inspect, press and finish, inspect, pack.) although the basic steps in the process are the same for all garments, the tremendous variety of wearing apparel that is produced to meet the demands of the fashion conscious consumer in the developed countries in fact imposes widely differing operating parameters from garment to garment. In some cases, such as jean manufacture, production runs are long. Only a relatively few pieces of material need to be sewn together to make the jeans, the sewing tasks are straightforward and style changes are comparatively few. In ladies' clothing, however, the situation is precisely the opposite; there are many style changes, short runs, complicated sewing tasks are required to accommodate design changes, and there are frequently many pieces to be assembled.

In the face of these conditions, the industry historically has relied upon highly skilled operatives and reliable, inexpensive, all purpose sewing machines which can be cheaply and quickly adapted to the different sewing requirements for each type of garment. The rate at which innovations have been introduced into the industry has been much slower than in other sectors. The basic sewing machine design, which still predominates, is almost identical to the industrial machines of 50 years ago!

There are many reasons for this slow rate of technical change, the major one being the continuing ability of highly skilled manual operatives to respond quickly and efficiently to the technical demands of rapid style changes. However, there are other factors—the industry is highly fragmented, very undercapitalised and, apart from the large firms, is burdened with archaic management practices. Equally as problematic as these institutional obstacles are some significant technical barriers to innovation which centre around the problems of handling the limp fabrics which make up garments. As we shall see since the average rate of investment in R & D in the apparel industry is very low (about 0.05 per cent of sales) and the capital cost of overcoming these problems very large, prior to the advent of (micro)electronics there has not been much progress towards their solution.

As a result of these structural and technical characteristics of the industry in the developed countries, the level of technology in use by Third World manufacturers is roughly on par with that employed in the advanced industrial economies—certainly the gap between best practice techniques and the average production methods in use in developing countries is much smaller in the garment sector than in other sectors. Consequently, relative labour costs have been the main determinant of competitive position although quality factors have so far prevented developing countries from competing internationally in some lines of apparel. Due to low wage rates, Third World manufacturers have enjoyed an increasing degree of comparative advantage in a number of high volume sub-sectors as wages have risen in the West.

Although technical change has, as stated above, been relatively slow in the industry, during the 1950s, 1960s and early 1970s the introduction of electronic controls allowed for continuous, if modest, increases in productivity for specific sub-processes now carried out by specialised machines. Due to the relatively high capital costs of these specialised machines, now costing thousands of dollars as opposed to the average cost of $600 for a standard sewing machine, the comparative advantage of less developed countries has remained intact. The differential in labour costs (see Table 3) has remained so that a large number of international firms have located an increasing proportion of assembly in low-wage countries, through sub-contraction, joint ventures or wholly owned subsidiaries.

The introduction of microelectronics in the late 1970s raised the expectations of some observers, both inside and outside of the industry, that the formidable

**Table 3. Comparison of Wage Rates USA/UK/Far East**

| | Labour Rate Per Hour | Effective Labour Cost/Hour after allowing for Prod/Organisational and Efficiency differences |
|---|---|---|
| USA | £1.00 | 100p |
| UK | £0.40 | 44p |
| Japan | £0.20 | 26p |
| Hong Kong | £0.10 | 16p |
| Taiwan and South Korea | £0.05 | 10p |

Source: M. Saibel, Singer Corporation, in Brooke, P.G., *Mechanisation of Garment Assemblies*, Clothing Institute Journal, **XXII** 135.

technical obstacles to automation would be removed. It was felt that the microprocessor, with its vast information processing capacity and inherent flexibility had the capacity to facilitate radical technical changes at the sub-process and systems level.

While the changes that were widely predicted have yet to advance beyond the first generation of innovations, the awareness and interest of the industry in advanced industrial nations has certainly been captivated. Within the last three years the number of capital goods firms offering microelectronic controls in their sewing machines has increased from less than a half dozen to over twenty five. In addition, firms and individuals from outside the industry, who have extensive experience in electronics, have introduced the most radically new innovations in garment technology.

## INNOVATIONS IN GARMENT TECHNOLOGY

BELOW WE BRIEFLY DESCRIBE some of these innovations. There seems little doubt that these and subsequent innovations will eventually dramatically alter the structure and character of the industry in the advanced industrial economies. However, the rate at which this transformation will take place is still open to question given the deeply rooted and inbred nature of some of the structural and institutional problems referred to above.

Grading, laying out, marker making, and cutting have traditionally been separate and highly skilled manual tasks. Given the value of the cloth in the total costs of the finished product (often reaching 50 per cent), the phase of laying out patterns on the cloth is crucial if wastage is to be kept at a minimum. Likewise, cutting the cloth precisely according to the lay is equally important particularly since error at this stage can lead to defective garments at the assembly stage. The sequence of tasks from grading through to cutting have been the only operations for which automation with microelectronics has significantly bridged the gap between what were previously discrete activities. Technology is now available that combines computerised pattern grading with optimal pattern layout, marker duplication facilities and electronically controlled cutting. Several firms, notably Camsco and Gerber in the United States and Laser Lectric in France offer a mini-computer based system which allows operators already skilled in traditional techniques to increase the speed of grading and laying out from 2 to 6 times—grading which previously took 4 days now takes only 1 hour—while simultaneously reducing waste from between two and five per cent. When the volume of fabric usage runs as highly as $10 million as is the case in a number of medium sized firms even a 2 per cent fabric savings is substantial considering the low profit margin per unit. The outright purchase of these systems, however, costs between $300,000 and $600,000 and necessitates maintenance contracts in the range of

$1,800–$2,600 per month. Not surprisingly, the sizeable initial investment and the high fixed and running costs have led to the establishment of service bureaus which offer these services to users who cannot justify purchasing the systems themselves.

Cutting technology, having remained static until the early 1970's has moved from a manual technique using a hand held electrical or mechanical knife to completely mechanical cutting that is electronically controlled. A decade of experience has seen the virtual demise of automatic die cutting, as well as the use of lasers and high speed water-jets to perform cutting. These have largely been superceded technically by the Gerber computer controlled cutter which incorporates a self sharpening blade. This machine, which requires complete re-engineering of the cutting room to attain maximum efficiencies costs around $600,000–$1,000,000 and employs the programming facilities offered by the Camsco or Gerber marker makers described above. This has greatly reduced the time involved in these phases of the process from about one hour to a maximum of four minutes per suit. Skilled labour input is reduced dramatically with savings of up to 1000 per cent reported by some firms in their marking and cutting workforce.

Although technologically impressive, these innovations are associated with activities which give only limited increases to value added. And given the high level of investment required these can only be afforded by the larger firms.

While the value of fabric is a major proportion of final costs these costs are comparable for manufacturers worldwide. It is in the area of reducing labour costs in the assembly stage, accounting for about 80 per cent of all labour costs, where savings are required if any dramatic shifts in comparative advantage in favour of the developed countries are to occur. The technical changes which have occurred in the sewing room, while significant for certain sub-processes, are nowhere near the same magnitude as seen in earlier phases of the garment-manufacturing process.

The range of different sewing tasks that need to be carried out at the assembly stage is very wide and can involve complicated stitches requiring a very great deal of skill. In most cases, highly irregular shapes of different lengths are involved—all of which change as fashions vary. To compound the problem, many factories are required to make only a few units of the same size and shape at any one time, and often have to mix batches and orders.

The technical problems for mechanising and automating such a process are as obvious as they are formidable. The use of microelectronics has, however, allowed for two types of semi-automatic sewing which although they do not completely replace the operator, do increase productivity while at the same time reducing skill levels and training time. The first incorporates a dedicated microprocessor in specialised pieces of equipment for small parts assembly i.e. collars, cuffs, belt loops and pocket setting. These machines produced by a wide range of firms including Union Special, Pfaff, Durkoff, Neechi, Juki and Reece, cost in the range of $15–40,000, and allow a *high volume* producer to increase productivity by anywhere from 50 to 300 per cent by increasing the speed and combining a number of operations, e.g. in the case of collar attachment, these machines reduce the number of operations from eleven to three.

Long runs, infrequent style changes and more than one operating shift are necessary for these reliable, extremely efficient, but relatively inflexibile machines. The second category of sewing innovations is distinguished by the use of programmable memory chips. At five to eight thousand dollars these machines can be computer programmed and controlled to perform a large variety of stitches in either decorative or functional sewing tasks. When combined with photo-electric edge sensing equipment, these machines will monitor the edge of the material to be sewn and disengage the needle when required. One leading manufacturer quotes an independent evaluation of this system which claims that productivity is improved between 25 and 66 per cent for a variety of operations such as top stitching collars, setting shirt pockets and attaching collars.

With approximately 80 per cent of the average garment manufacturer's labour and capital costs associated with the assembly phase, the subsequent concentration of innovative activity in that area is justified. Given the current technical difficulties and potential savings in the sewing room it is also hardly surprising that the finishing stage has been relatively neglected. This is reinforced by the pervasive attitude among management that the function of pressing is primarily in making the product 'presentable' or 'saleable' rather than an essential component of value added.

This attitude is likely to change as modernisation in the industry fosters more sophisticated methods of cost justification. Two capital equipment manufacturers recognising that the 10 per cent of capital expenditure estimated for this stage also signifies a high degree of labour intensity, have now incorporated microelectronics into their pressing equipment.

Magpi/Sussman and Certus have introduced first generation pressing equipment for jeans, slacks and suits which incorporates microelectronics. In flat pressing jeans using this equipment, 6 pairs of jeans can be pressed per minute, or over 2,000 pairs per day, while suits can be pressed at the rate of 400 per day. While Certus' equipment goes some way in providing a more continuous pressing of separate parts of a garment (i.e. sleeves, armholes, shoulders etc.) and Sussman's allows the operator to choose from 12 pressing variables (i.e. steam pressure, temperature, etc.) neither machine, whose costs range from $30,000 to $200,000, has operated under production conditions long enough for an accurate assessment of their operating performance. What is currently evident is that for a limited number of products (jackets, jeans) these machines can be used to reduce both training times and operator skills, previously required to ensure quality.

## IMPLICATIONS FOR THIRD WORLD

WHILE THE FULLY automated garment production system is not at present a reality, the trends in technical change and the expectations in the industry are unmistakably oriented in the direction of increasingly comprehensive systems development—although it is likely that change will continue to be gradual rather than rapid—extending into medium to long term as opposed to the overnight changes predicted by some observers.

The structural and institutional problems discussed earlier which impeded the historical rate of technical change, will also slow the rate at which MBIs are introduced by manufacturers in the developed countries. The high capital costs of the systems and the advanced management and maintenance skills required to operate them efficiently means that the large manufacturers will be best placed to use the equipment. Interestingly, these large firms are also those most heavily involved in offshore manufacturing and producing under USA 807.00 import type clauses. Hence, they will be able to very accurately gauge the relative costs of producing garments offshore using cheap foreign labour as opposed to home based production using the advanced technology.

Although it has not happened yet, to any great extent, there is a feeling among the large producers that a large share of offshore production will be brought back. In those cases where firms have located manufacturing facilities in developing countries with large domestic markets, we would expect that the equipment would be introduced if competitive conditions required it. Both of these scenarios have significant implications for developing countries.

Locally owned Third World firms produce and export an extremely wide range of garments that vary enormously in terms of quality and cost. Traditionally, they have concentrated their efforts on standardised products of low to medium quality which have a wide demand in western countries—blue jeans, skirts, blouses, shirts, jackets, etc. More recently, however, they have been successfully moving 'up market' with their exports and are increasingly competing with developed country producers in high fashion sub-sectors such as suits and dresses, where profit rates are higher and there are fewer tariff barriers. Their comparative advantage, however, remains based on cheap labour rather than on improved technological capability.

Although Third World firms do have access to quite a large range of techniques on the international machinery market, they tend to rely extensively on conventional multi-purpose machines and cheap labour to manually perform the tasks which will become increasingly automated in the developed economies. This combination gives them the capacity to mass produce standardised products at low prices and enables them to respond quickly to ever changing fashion requirements and competitive conditions. These characteristics will allow Third World producers to resist the effects of microelectronic based technical change. But this ability to resist must gradually be eroded by the combined effects of rising domestic wages, higher and broader tariff barriers and innovations increasingly directed at precisely the activities where they now enjoy comparative advantage. As we have already mentioned the crucial variables in this process will be the rate at which these applications will be developed and their speed of diffusion within the advanced industrial countries. Across subsectors, this will be an irregular and discontinuous but nevertheless inexorable process that is already signposted by current developments.

Such a process will create problems for some Third World producers particularly those countries which are really only beginning to develop their apparel industries. The renewed strength of western manufacturers and the already entrenched position of the more advanced developing countries will work against any dramatic expansion of market opportunities for the least developed countries of the sort that fuelled industrialisation for the NICs. However the die is not yet cast and some time will elapse before these changes really begin to take effect. In the intervening period, these poorer developing countries will need to make a much clearer assessment of their strategies to develop the garment sector in light of the effects of microelectronics on the structure of the industry in western economies.

# New Technology and Employment: Extensions to the Case of Developing Countries

**_P. Stoneman_**
_University of Warwick, United Kingdom_

## 1. INTRODUCTION

THIS PAPER is in the nature of a report on research proceeding. Three earlier papers (Stoneman, 1980(a), 1980(b), 1981(a)), have discussed an approach to the analysis of the impact of new technology on employment which is reported further in this paper. The main purpose of this work is to attempt to get away from the microeconomic analysis of much of the work on the impact of new technology and to consider the macroeconomic questions. I am quite willing to accept that if a firm uses microelectronic technology then it may well produce a given level of output with less labour. However the impact on the demand for labour in the whole economy depends on many other factors and that is what is discussed here.

The main body of the work reported on here was designed for the analysis of a closed developed economy. In the Appendix below the details of the model itself are discussed. In section 2 we report on the employment paths that the model predicts will result from the introduction of new technology, and at that stage we will keep formal analysis to a minimum to save space and improve clarity. In section 3 we discuss the issues that will arise when the model is extended to an open economy. In section 4 the issues that are relevant in the developing economy context will be discussed. In section 5 conclusions are drawn. One point should be made at this point that is important to the whole analysis of this paper. In much of the literature in this area it seems to be implicitly assumed that new technology is introduced overnight. A basic principle underlying our analysis is that the change to new technology is a time intensive process. It can take 20 years for a new technology to be fully diffused. It is this period during which diffusion is occurring in which we are interested. The end result of the process is almost irrelevant given the rate at which new technologies appear in the modern world. Similarly it becomes very important as to how impacts will vary with the speed of diffusion. Thus the analysis of the impact of different diffusion speeds is another of our objectives.

The debate on the impact of technological change on employment has a long history. Every so often economics bursts into a flurry of activity on this question. The current concern with the topic has been prompted by the realisation of the potential of microelectronics. This realisation has generated a number of commentaries (e.g. Freeman, 1978, Barron and Curnow, 1979, Jenkins and Sherman, 1979), the major predictions of which are that the introduction of new technology will lead to unemployment. A rider to this is that the faster new technology is introduced (ignoring international competitive aspects) the higher will be the unemployment resulting. Unfortunately much of this literature is either devoid of theory, or if theory is included it is implicit rather than explicit. The outcome of this is that much of the literature ignores what Heertje (1977) calls compensation effects. In essence, if a labour saving technology is introduced in one sector of the economy there may well be automatic responses in the economy that lead to increased employment elsewhere. Compensation effects fall into three groups:

(1) Technological multiplier effects. These can be exemplified as the employment creating effects in industry *j* resulting from the introduction of a new technology in industry *i*, for a given final demand vector. Thus if new technology is wholly embodied the introduction of new technology will mean increased demand on the capital goods sector and thus employment effects in that sector.
(2) Income effects. If new technology leads to changes in income levels or distribution then the pattern of final demand may change with consequent effects on employment.
(3) Price effects. New process technology will probably mean lower costs and thus lower prices. New product technology should mean lower quality adjusted prices. The consequent demand increases should stimulate employment.

In this paper we consider all three compensation effects. Technological multiplier effects are introduced by using a two-sector framework, income effects are introduced by linking aggregate demands to wages and profits and price effects are introduced by linking prices to costs. Prices will not, however, be considered as perfectly flexible. It is well known (Neisser, 1942) that if prices react to notional excess demands and supplies in a Walrasian manner then technological change will not cause unemployment. In conformity with the Keynesian view of the world, prices will not be allowed to so react. Thus technological unemployment is a possibility.

## 2. TIME PATHS OF EMPLOYMENT

In the appendix (p. 270), the model is presented in some detail. In this section we present some results on the model's predictions on the time path of employment and the conditions governing these time paths. The model indicates that there are three important factors to consider (i) the wage path (ii) the diffusion speed (iii) the technical coefficients. In this section we will illustrate time paths of employment under two different wage regimes, discuss how the paths will depend on technical coefficients and then discuss how they change as the diffusion speed varies. We will not be proving the statements we make, these proofs if required can be found in Stoneman (1981a).

(*a*) *The fixed wage path.* Under this regime the wage remains fixed during the whole of the transition at the level prior to the introduction of the new technology. We then consider the behaviour of employment in two stages. We may then show that in Stage 1, the stage where both old and new capital goods are being built and used, that labour demand will increase relative to labour supply. In Stage 2 where only the new capital good is being produced, but new and old are being used, for higher values of $t$ (time) this increase will continue. However under certain conditions, the ratio of labour demand relative to labour supply ($\gamma_t$) may fall for a period. These conditions are neither simple to state nor clarify. In Figure 1 the two possibilities are illustrated (although it should be realised with all these graphs that curvatures may be incorrect for as yet, I have not been able to sign the second order derivatives involved).

The effect of an increase in the diffusion speed under this regime is to reduce, $T$, the length of Stage 1, but to increase employment for all $t < T$. It has not been possible to find the effects in Stage 2.

(*b*) *The changing-wage path.* In this scenario the wage is allowed to increase as new technology is introduced such as to keep the overall rate of profit in the economy constant. In Figures 2 and 3 we show two possible time paths for the employment ratio, $\gamma_t$, predicted from the model. In Figure 2, $b_1\alpha_1/b_2\alpha_2 > \beta_1/\beta_2$ and in Figure 3 $b_1\alpha_1/b_2\alpha_2 < \beta_1/\beta_2$.

The important condition determining the time path of employment is the ratio $b_1\alpha_1/b_2\alpha_2$ to $\beta_1/\beta_2$. If the new technology saves, relative to the old technology, more labour directly than indirectly i.e. it may be described as relatively labour saving rather than capital saving, then the employment ratio will initially fall (in Stage 1) and then rise (in Stage 2). If it is relatively capital saving the employment ratio will initially rise (in Stage 1) then fall then rise again (in Stage 2).

The effect of an increase in the diffusion speed is to always reduce $T$, the length of Stage 1. However in Figure 1, $\gamma_t$ will be higher for all $t < T$, in Figure 2 lower. In Stage 2 employment will be lower in Figure 1, and probably higher in Figure 2, but we have not been able to prove this latter point conclusively.

As one can see from this brief summary of results, there are no simple

**Figure 1. Employment Ratios: Fixed Wage Path**

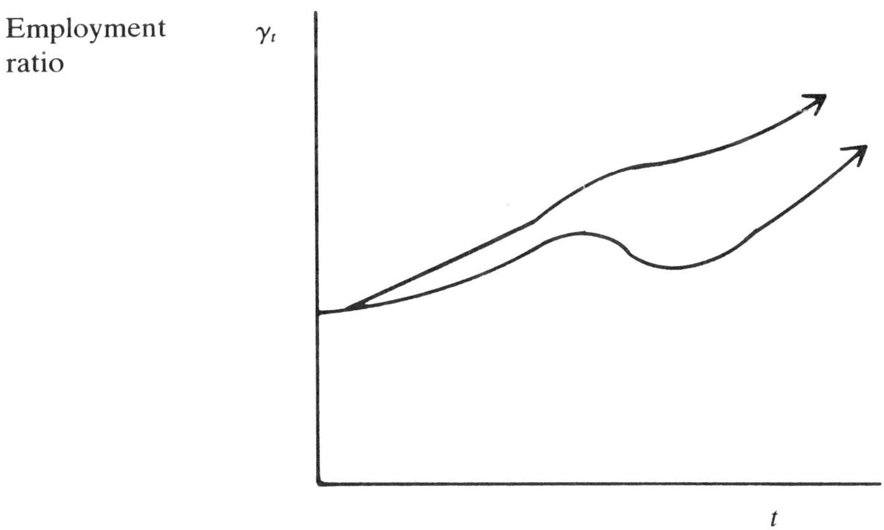


**Figure 2. Employment Ratio: Increasing Wages**

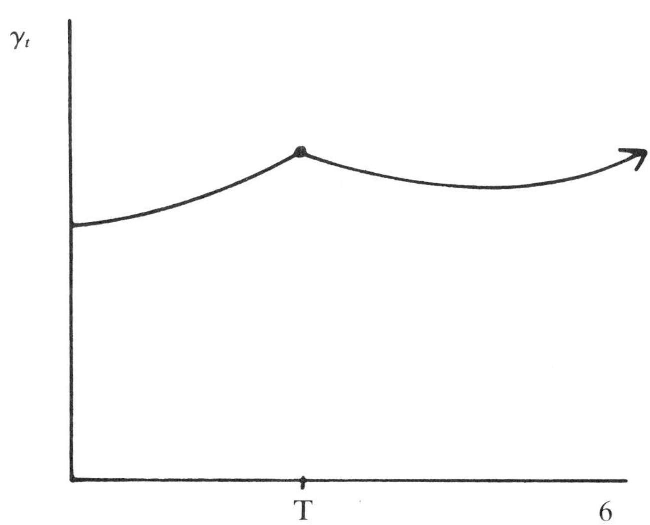

**Figure 3. Employment Ratio: Increasing Wages**

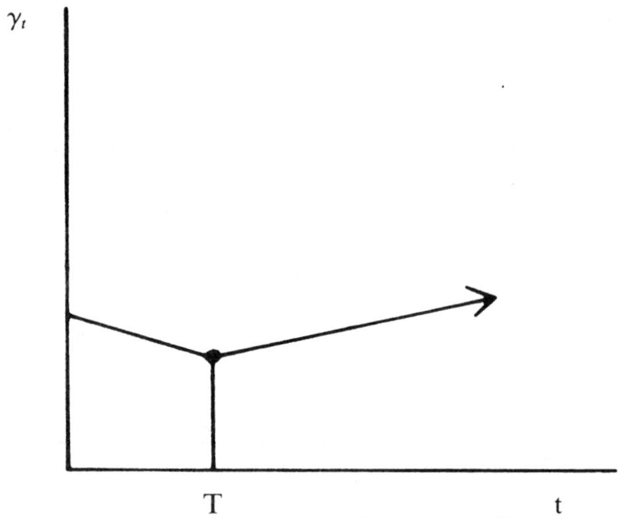


predictions on how employment will react to new technology. Even though we have allowed our technological change to be labour saving for a firm ($\beta_1 > \beta_2$) and also be capital saving ($b_1\alpha_1 > b_2\alpha_2$) we can still illustrate increases in labour demand. Such is the insight derived from macroeconomic analysis.

## 3. EXTENSIONS TO AN OPEN ECONOMY

THE RESULTS detailed above are derived from a model of a closed economy in which all technological change is of the process innovation type. When we open up the economy to trade we have to consider two major modifications.

(i) If we maintain the assumption of only process innovation occurring then in a world with trade we must consider how the economy will develop when it is possible that new type capital goods will be imported rather than produced at home. At first glance it appears that if new type capital goods are imported then the employment creating impact of the technology multiplier effect will no longer operate and thus technological unemployment becomes more likely. However a simple observation makes one query this reasoning. If new type capital goods are to be imported (whereas old type were produced at home) then the new type will have to be paid for. The only way that they can be paid for in the long run is by home production of other goods for export. Such production must mean the employment of labour. In principle therefore, although opening up the economy in this manner will affect our results, there is no presumption that unemployment is any more likely.

(ii) When we relax the assumption that innovation is only of the process type and also allow product innovation, a further set of complications arise in the open economy. I can see no reason why the model of the closed economy discussed above should be affected, it is only in the open economy context that the matter is relevant. Essentially if innovation is of the product improvement type and the home country does not take up the new technology as fast as other countries then home demand for consumer goods may be diverted towards imports, foreign demand may be diverted away from exports, and foreign buyers may be unwilling to accept home goods in exchange for foreign produced new type capital goods. Such impacts will tend to reduce the total compensation demand effects in the economy.

However the issue one must consider here is to what extent will these factors be offset by price changes. If the home producers fail to innovate with new products can the markets be maintained by price reductions? These price reductions could of course be the result of movements in a flexible exchange rate. If such price changes did occur then the impact on the home economy of not introducing new technology would be lower real wages rather than lower employment.

These comments indicate that work in this area has still got a long way to go, but one should not always start with pessimistic expectations on employment.

## 4. THE SPECIAL FACTORS INVOLVED IN ANALYSING DEVELOPING COUNTRIES

I BELIEVE THAT a fair way to start any analysis of the problems facing developing nations in a world of technological change is to characterise them as (a) skill deficient (b) labour intensive and short of capital (c) dependent on the developed nations as a market for their products and as suppliers of technology and (d) as followers rather than innovators. I will also find it useful to characterise them as having dual economies the parts of which I will call the rural and urban sectors. Such a view is a characterisation, developing countries do differ from one another. Thus for example it may be invalid to lump together the S.E. Asian economies such as S. Korea with its current bouyant electrical/electronic sector with Sri Lanka which does not have such a sector, but as long as this is held in mind we should not be too concerned.

In 1972 the ILO published the results of a round table discussion on the Manpower Problems Associated with the Introduction of Automation and Advanced Technology in Developing Countries (ILO, 1972). In many ways the issues discussed in that volume are as relevant today as they were in 1972. There are essentially three issues.

(i) What is the potential impact of new technology on a developing country?
(ii) What barriers exist to the introduction of new technology and can they be removed?
(iii) Should the introduction of new technology be encouraged or inhibited?

As far as the first issue is concerned I can see no good reason why the analysis developed above for a developed economy should not apply to the modern (or urban) sector of a developing economy i.e. that the diffusion speed, wage dynamics, trading environment and technological characteristics should determine the employment impact. It is only once these details have been specified that one can detail the employment path.

Of course, like all theoretical models the above only provides the core of our analysis, but it is worth mentioning that in the ILO Report (*op. cit.*) at least two authors are implicitly discussing issues in the context of similar two sector models. Although this model seems relevant to the urban sector we must still consider the impact on the rural sector and we shall mention that below.

The second issue concerns the factors that might encourage or limit the introduction of the new technology in a developing economy. Here the list of possible barriers is simple to generate: (a) capital shortages (b) lack of foreign exchange (c) skill and knowledge deficiencies (d) information deficiencies (e) labour opposition etc. There seems to be little advantage in expanding the list further here, it is surely discussed more fully in other papers. My own feeling however is that such factors will conspire to produce slow diffusion speeds in developing countries and thus we may expect the time paths of employment with slow diffusion speeds as detailed above to be the relevant ones.

Finally we come to the question of whether the introduction of new technology should be encouraged or inhibited. Here I want to consider three district areas

(a) Should private firms be encouraged or inhibited from using the new technology?
(b) Should the government encourage the use of microelectronics in the public sector?
(c) Should developing countries attempt to develop a microelectronics production capability?

In ILO (*op. cit.*) it is stated that developing countries should use new technology when "the productivity and profitability of ultra-modern equipment overcomes the properly weighted private and social cost of all inputs". One cannot disagree with this statement, but until one has detailed the properly weighted private and social costs and benefits one has not got very far. Moreover to provide such detail one needs knowledge of opportunity costs, which of course is no simple matter. However holding this view in mind let us discuss (c) above first of all. It would seem clear to me, looking from the point of view of a developed country that is currently attempting to develop a micro-electronic production capability, that even with resources of electronic and software expertise available this is a very expensive, capital intensive, risky enterprise. In the U.K. the opportunity cost of funds is probably much lower than in a developing country (consider for example relative returns to agricultural improvement) and yet even so such an investment is only marginally desirable. Given this I cannot see that it is a desirable path for a developing country to follow.

I will take it therefore that the major part of any microelectronics used in a developing country will be imported. Consider then whether the private firm should use microelectronics. The private decision will be made upon the basis of private costs and benefits. Essentially will the use of microelectronics be profitable given the prevailing domestic and world prices? If the answer is yes then the firm will use them. The welfare question is, should they be used and the answer to this centres on the social costs and benefits.

(i) What is the opportunity cost of the foreign exchange, could the foreign exchange be used for more socially beneficial objectives?
(ii) What will be the impact on income distribution? If the use of microelectronics creates a highly paid skilled class is it desirable?
(iii) What would be the effects of changes in urban employment and wages on the flow of labour from the rural sector and thus the overall unemployment in the economy and agricultural output?
(iv) What is the opportunity cost of the capital invested in the new machines, irrespective of the opportunity cost of foreign exchanges? Here, the relatively lower price of microelectronic devices relative to mechanical or electro-mechanical alternatives may well argue in the favour of microelectronics.

These answers are unfortunately framed as questions but I hesitate to be any more precise. Essentially, the decision on the desirability of the use of microelectronics is one of a country's social objectives. However, I might say, that if the developing country is trading in world markets then the product innovation aspects of microelectronics may well be so strong that no other investment opportunity has such low opportunity costs. The cost of non use may well exceed the cost of use.

When we turn to government decisions on the public sector the same questions arise again. However the main reason for considering this as a separate set of decisions is that one example nicely illustrates how microelectronics differs from earlier technological advances. That example is in telecommunications. A basic requirement of any developed nation is an efficient telecommunication system. Advances in microelectronics have generated systems that are both labour and capital saving with respect to old technology. Usually one is discussing capital using labour saving technologies. In a developing country building a telecommunications network, there can therefore be no reason (apart from a high shadow price of foreign exchange combined with domestic production of mechanical exchanges) not to use microelectronics. It is this capital saving nature of microelectronics that makes it so appealing to countries that suffer capital shortages (have a high opportunity cost of capital).

If one accepts therefore that there are definitely cases where developing economies should use microelectronics the next question that arises is at what speed should the acceptance take place? The thrust of the theoretical work above is that different diffusion speeds have very different implications as to impacts on the economy. I am, in this circumstance, going to avoid answering this question, despite its crucial importance. All I can say is that it is not necessarily the case that fastest or slowest is more desirable. Only once one gets much further into the theoretical analysis of the impact of different diffusion speeds will the answer become clear.

## 5. CONCLUSIONS

THE MAJOR CONCLUSIONS to be drawn from this paper are qualitative rather than quantitative. The analysis of the impact of new technology is not simple nor are conclusions drawn from such analysis easily interpreted. However, if nothing else, it is hoped that this paper shows that such analysis must be dynamic, and macro-orientated. However it also shows that factors such as wage paths, diffusion paths and technological characteristics are the keys to deriving any conclusions and thus any conclusions one draws are only as valid as the assumptions one makes on these key factors.

## APPENDIX—THE MODEL

AS ONE OF THE KEY FACTORS to be analysed is the technology multiplier effect, the model is a two-sector one with a consumption goods and a capital goods sector. We will consider only process innovations, and we will assume that all process innovations are wholly embodied. Thus to change technology a whole new set of capital goods is required. We will assume the economy is closed.

We consider two technologies, labelled 1 and 2, that are the old and new respectively. Both capital goods are made wholly by labour (although circulating capital could be allowed) but consumption goods are made by capital and labour. The technology of capital goods production is considered in this manner for it yields considerable simplification; however the omission of a capital goods input from capital goods production will bias any compensation effects we find. We represent the quantity relations of the two technologies as follows:

$$K_{it} = \alpha_i C_{it} \qquad i = 1, 2 \tag{1}$$

$$L_{it} = \beta_i C_{it} + b_i x_{it} \qquad i = 1, 2 \tag{2}$$

$$x_{it} - \frac{\mathrm{d}K_{it}}{\mathrm{d}t} \qquad i = 1, 2 \tag{3}$$

where $K_{it}$ = stock of capital good $i$ in time $t$
$C_{it}$ = output of consumption goods produced using capital good $i$ in time $t$
$L_{it}$ = labour employed producing consumption goods and capital good $i$ in time $t$
$x_{it}$ = output of capital good $i$ in time $t$.

We will consider that capital has an infinitely long physical life and there is thus no depreciation. Total labour supply will be given by (4).

$$L_t^s = L_0 e^{nt} \tag{4}$$

We will consider that at time $t$ the economy is in the process of making the transition from using technology 1 to using technology 2, and at any moment in time the two technologies are used in consumption goods production in the proportions $1 - \theta_t : \theta_t$. We specify that $\theta_t$ is the result of a diffusion process the time profile of which can be represented by a logistic curve (5).

$$\theta_t = \frac{1}{1 + \exp(-\eta - yt)} \tag{5}$$

This logistic diffusion process is assumed to be the result of forces outside the model. It summarises entrepreneurs' technique choice behaviour, subsuming within it their expectations and anticipations with respect to future prices and profitability of the new technology. The theoretical underpinnings of such a relationship can be found in, for example, Mansfield (1968), Davies (1979), Stoneman (1981b, 1981c). The apparent absence from this paper of expectations, profitability impacts on technique choice or explicit technique choice modelling, is thus something of a misapprehension for these are all presumed within (5). It is also further assumed that the policy makers can, if they wish, control $y$, the speed of diffusion. They are assumed to be unable to influence the appearance of new technology (represented by $\eta$).

We will assume that at any time $t$ the output of the relevant capital good is just sufficient to meet the demands of the consumer goods industry subject only to $x_i \geq 0$. Thus we have (6) for $i = 1, 2$

$$\begin{aligned} x_{it} &= \alpha_i \frac{\mathrm{d}C_{it}}{\mathrm{d}t} \quad \text{if} \quad \alpha_i \frac{\mathrm{d}C_{it}}{\mathrm{d}t} \geq 0 \\ &= 0 \qquad\quad\ \ \text{if} \quad \alpha_i \frac{\mathrm{d}C_{it}}{\mathrm{d}t} < 0 \end{aligned} \tag{6}$$

yielding (7) and (8) for $x_{it} \geq 0$,

$$x_{1t} = \alpha_1\left((1-\theta_t)\frac{\mathrm{d}C_t}{\mathrm{d}t} - C_t\frac{\mathrm{d}\theta_t}{\mathrm{d}t}\right) \tag{7}$$

$$x_{2t} = \alpha_2\left(\theta_t\frac{\mathrm{d}C_t}{\mathrm{d}t} + C_t\frac{\mathrm{d}\theta_t}{\mathrm{d}t}\right) \tag{8}$$

where $C_t \equiv C_{1t} + C_{2t}$. Defining $g_t \equiv \mathrm{d}\log C_t/\mathrm{d}t$ we may state from (7) and (8) that

$$x_{1t} > 0 \quad \text{if} \quad g_t > y\theta_t$$
$$x_{2t} > 0 \quad \text{if} \quad g_t > y(\theta_t - 1)$$

We will consider the transition process as the economy moves from using only technology 1 to only technology 2 in two stages. In Stage 1 both capital goods 1 and 2 are produced and used. In State 2 only capital good 2 is produced but both 1 and 2 are used. Thus we have

$$\text{Stage 1 } g_t > y\theta_t \text{ implying } x_1 > 0,\ x_2 > 0$$
$$\text{Stage 2 } g_t < y\theta_t \text{ implying } x_1 = 0$$

We will show that in Stage 2, $x_2 > 0$. We will also show that Stage 1 always precedes Stage 2, and that once the economy is in Stage 2 it never reverts back to Stage 1.

These quantity relations assume that there is only one type of labour, and we are thus assuming no skill differentials. To the extent that these exist we will assume that on the job training can remove them and such training is taken account of in the technical coefficients. In this part of the paper we are concerned with the aggregate demand for labour and not its skill breakdown and thus skills (and for that matter the geographical location of labour) are for all intents and purposes ignored.

Given the quantity relations we will now consider price relations. We assume that because there is no capital used in capital goods production, machines are priced at their labour cost. Let $w$ be the wage rate which we will assume to be the same for all workers no matter what they are producing or which machines they are using. Let $p_i$ be the price of machine $i$ and then we have (9)

$$p_i = wb_i \quad i = 1, 2 \tag{9}$$

Defining $r_i$ as the rate of profit in consumption goods production from the use of machine $i$, and letting the consumption good be the numeraire with a price of unity, then (10) holds.

$$1 = r_i p_i \alpha_i + w\beta_i \quad i = 1, 2 \tag{10}$$

For the new technology to be superior, and therefore for the entrepreneur's change of technique to be rational, technology 2 must be more profitable than technique 1. The superiority of technique 2 can be local or global, i.e. it can apply for one, some or all wage rates. We will assume global superiority to avoid switchbacks from new to old technologies with changing wage rates. For technology 2 to yield a higher profit rate ($r_2$) than technology 1 ($r_1$) for $w$ we require that

$$b_1\alpha_1 \geq b_2\alpha_2 \quad \text{and} \quad \beta_1 \geq \beta_2$$

with at least one strict inequality holding. We will be assuming, therefore, that technology 2 is absolutely labour saving relative to technology 1 (i.e. it requires the same or less labour in both direct and indirect use). This would seem to fit both the microelectronics case and be the most pessimistic assumption for unemployment implications. We have considered the behaviour of the model when technology 2 is only locally superior but the results add little.

Given the price and quantity relations we now close the model by specifying the demand environment. We assume that savings can be represented by a classical savings function with all wages consumed, all profits saved. We have experimented with a proportional savings function but as Hicks (1965) states, it sits much less happily in this framework. Defining $L_t = L_{1t} + L_{2t}$ we may then write (11)

$$C_t = w_t L_t \tag{11}$$

Define $\gamma_t$ as the ratio of the work force employed in time $t$

$$\gamma_t \equiv \frac{L_t}{L_t^s} \tag{12}$$

This paper concerns the time path of $\gamma_t$ as the economy makes the transition from using technology 1 to using technology 2. Sometimes it is useful to think of this as the actual level of employment relative to employment on a comparison path on which the new technology is not used. We may write (12) as (13)

$$\gamma_t = \frac{L_t}{\bar{C}_t(\beta_1 + b_1\alpha_1 n)} \tag{13}$$

where $\bar{C}_t$ is the level of consumption in time $t$ which will involve full employment using only technology 1, in which the natural interpretation of $\gamma_t$ becomes one of employment relative to that on a comparison path on which full employment growth using only technique 1 occurs. From (12) we may derive (14)

$$\frac{\mathrm{d}\log \gamma_t}{\mathrm{d}t} = \frac{\mathrm{d}\log L_t}{\mathrm{d}t} - n \tag{14}$$

From (11) we have (15) and thus (16)

$$g_t = \frac{\mathrm{d}\log L_t}{\mathrm{d}t} + \frac{\mathrm{d}w}{\mathrm{d}t} \cdot \frac{1}{w_t} \tag{15}$$

$$\frac{\mathrm{d}\log \gamma_t}{\mathrm{d}t} = g_t - \frac{\mathrm{d}w}{\mathrm{d}t} \cdot \frac{1}{w_t} - n \tag{16}$$

To derive $g_t$ we have from (2), (7), (8) and (11) that in Stage 1 (17) holds and in Stage 2 (18) holds

$$\frac{C_t}{w_t} = L_t = C_t\left[\beta_1(1-\theta_t) + \beta_2\theta_t + g_t(\alpha_1 b_1(1-\theta_t) + a_2 b_2\theta_t) + \frac{\mathrm{d}\theta_t}{\mathrm{d}t}(\alpha_1 b_2 - \alpha_2 b_1)\right] \tag{17}$$

$$\frac{C_t}{w_t} = L_t = C_t\left[\beta_1(1-\theta_t) + \beta_2\theta_t + g_t\theta_t\alpha_2 b_2 + \alpha_2 b_2\frac{\mathrm{d}\theta_t}{\mathrm{d}t}\right] \tag{18}$$

From which it is clear that in Stage 1 (19) holds and in Stage 2 (20) holds.

$$g_t = \frac{1 - w_t\beta_1(1-\theta_t) - w_t\beta_2\theta_t + w_t(\alpha_1 b_1 - \alpha_2 b_2)\frac{\mathrm{d}\theta_t}{\mathrm{d}t}}{w_t(\alpha_1 b_1(1-\theta_t) + \alpha_2 b_2\theta_t)} \tag{19}$$

$$g_t = \frac{1}{w_t\alpha_2 b_2\theta_t} - \frac{(\beta_1(1-\theta_t) + \beta_2\theta_t)}{\alpha_2 b_2\theta_t} - \frac{\mathrm{d}\theta_t}{\mathrm{d}t} \cdot \frac{1}{\theta_t} \tag{20}$$

From (16), (19), (20) we may thus state that the time path of employment depends on

(1) The time path of consumption which in turn depends on the time path of wages
(2) The time path of $\theta_t$, the diffusion process and
(3) The technical coefficients of the two technologies.

The time path of $\theta_t$ and the technical coefficients have already been specified. With wages we have two possible approaches (a) to model with wages determined by the excess supply of labour or (b) to consider wages determined exogenously. We feel that it is in the Keynesian (as opposed to the Walrasian) tradition to consider wages exogenously determined and thus we take this approach. In section 2 the time path of $\gamma_t$ is discussed firstly under a fixed wage assumption and secondly in a situation where wages are allowed to increase with productivity. We also investigate there the impact in the two regimes of an increase in the diffusion speed.

In section 2 we will assume that $t = 0$ represents the start of the transition process and at that date $\theta_0$ is given exogenously as the proportion of consumption goods output produced on the new technology. We assume that $\theta_0$ has been achieved without any impact on the economy. We will also allow that at time 0, $\gamma_0 = 1$ which from (13) can be interpreted as that prior to time 0 the economy has been growing on a full employment steady state growth path.

## References

Barron, I. and Curnow, R. (1979) *The Future with Microelectronics.* Frances Pinter Ltd., London.

Davies, S. (1979) *The Diffusion of Process Innovations.* Cambridge University Press, Cambridge.

Freeman, C. (1978) *Government Policies for Industrial Innovation.* The J.D. Bernal Lecture, delivered at Birkbeck College, London, May.

Heertje, A. (1977) *Economics and Technical Change.* Weidenfeld and Nicolson, London.

Hicks, Sir J.R. (1965) *Capital and Growth.* Oxford University Press, Oxford.

ILO (1972) *Automation in Developing Countries.* Geneva.

Jenkins, C. and Sherman, B. (1979) *The Collapse of Work.* Eyre Methuen, London.

Mansfield, E. (1968) *Industrial Research and Technological Innovation.* Norton, New York.

Neisser, H.P. (1942) *Permanent Technological Unemployment,* Am. Econ. Rev., March, pp. 50–71.

Stoneman, P. (1980a) *Technological Unemployment and the Speed of Diffusion,* paper presented to a conference on Methodologies to Assess the Impact of New Technology and Employment, CEI, Geneva, September.

Stoneman, P. (1980b) *New Technology, Demand and Employment,* paper presented to the European Industrial Production Study Group, Loughborough, September, to be published in D. Bosworth (ed.) *Technology and Employment,* Macmillan.

Stoneman, P. (1981a) *Technology, diffusion wages and employment,* mimeo University of Warwick, January.

Stoneman, P. (1981b) *Intra-firm diffusion, Bayesian learning and profitability.* Economic Journal, June.

Stoneman, P. (1981c) *The Rate of Imitation, Learning and Profitability.* Economics Letters. To be published.

SECTION 6

# Availability and Delivery of Information for Industrial Development: Services and Systems

Industrial development is the goal for which all countries aim. The developed countries are reasonably secure in the degree to which their industrial base has been built up, but with the less developed countries the task of establishing an industrial base is an increasingly urgent one, for a variety of reasons, social, economic and political. However, modern industrial development calls for a large and growing input of information. This section deals with a variety of ways in which national, regional and international organisations handle the problem of making information available to agencies and firms requiring it. Experience shows however that it is not sufficient to simply establish information systems and services and leave the rest to chance. A determined and sustained effort must be mounted to 'market' the information—again, a variety of means may be employed for this function.

# Information for Industrial Development

***Roch T. de Mautort***
*United Nations Industrial Development Organization (UNIDO)*

## INTRODUCTION

THE THEME OF THIS SECTION is the availability and delivery of information for industrial development, a field where information practice would seem to resist the onslaught of information science, and one where the resources of advanced informatics would find less convincing application than the field of research and development. This finds its expression in the distinction made between technological information flowing through systems towards users engaged in R & D, and information on technologies flowing through services to users engaged in industrial planning, investment and operations.

The former, in the language of acronyms favoured within the UN system, which it has in common with the professional community of information scientists, is usually referred to as STI (scientific and technological information). The latter, in UNIDO's usage derived from the operation of its Industrial and Technological Information Bank (INTIB), will be referred to as INTI. Like so many clear cut distinctions applied to many-faceted subject matters, vehicles and destinations, the one between STI and INTI may well be seen to be somewhat arbitrary, with zones of overlap and feedback, yet in twelve years of activity in the area of industrial information it makes a lot of sense to UNIDO. Relating this to the overall purpose of the conference, there will be great hopes on the part of UNIDO participants, at the time when rather long-time perspectives for their future work in this area are to be drawn up, as an outcome of the conference to be able to visualise for the not so distant future, what added dimension and value the resources for informatics may have in store also for INTIB.

## ROLE OF UNIDO

Within the United Nations system, the United Nations Industrial Development Organization (UNIDO) is the one with a specific mandate, to further industrialisation in developing countries, by a variety of means, towards the quantitative goal of the 25 per cent Lima Target* as well as the qualitative goal which is self reliance in technological ability to master the industrialisation process with all its social and economic implications. The means include the provision of information needed by all engaged in the industrialisation process in whatever capacity, and at whatever level of concern and responsibility.

## INFORMATION NEEDS FOR INDUSTRY

Information needs for industry in developing countries relate to every sector of industry, and, within each sector, to any level of complexity of the relevant technologies. Such technologies may be based on long 'proven' endeavours, or 'advanced' and freshly developed ones. They may be freely accessible 'text-book' technologies or proprietory technologies patented or not, licenced or not, and difficult to access and implement. Needs, furthermore, relate to all cottage, small, medium or heavy scale industry, in the public and private sectors, whether for countries well on the way to industrial development or among the LDC's. There is a corresponding diversity of technological capabilities, industrial environment, and linguistic versatility of end-users. Their responsibilities range from broad planning through advisory, investment decision-making, managerial, engineering or marketing functions.

Information needs in industry relate to the existence and availability of technologies, the state of the art within technological alternatives, and the terms and conditions of access to industrial techniques for pre-investment assessment. They also bear upon the very contents of knowledge and know-how of such technologies (their software in the language of informatics). There are needs for data on industrial R & D, where telecommunications between inter-active data banks and terminals are already helping in the communication of STI to scientists and technologists. They will be for processed information on industry proper, including information on the output of R & D that may be ripe for full scale industrial exploitation. Investment decisions must be based on clearly delineated alternatives, and industries developed from adequately packaged information material.

* Lima Declaration and Plan of Action on Industrial Development and Co-operation, March 1975, para 28: "The share of DC's in total world industrial production . . . should be increased . . . to at least 25 per cent of world industrial production by the year 2000".

## INFORMATION RELATED TO INDUSTRIALISATION

THERE IS LITTLE HOPE of comprehensively categorising information needs for every single sector of industry, such as the twenty sectors in which INTIB operates. The diversity of users and of information needs defy the imagination on how all this can be accommodated in a single system apparatus however much 'informatised'. Nevertheless industrial inquiries in their thousands have been received and processed by the Industrial Inquiry Service of UNIDO over the years. The answers provided and retained on file for repeated use for similar questions, make up an important documentary base relating to a wide diversity of problems in industry, classified and recorded under subject files, and are dealt with on a formalised data bank (LINK) which also accommodates listings of institutions active in the area of technology and industrial information.

At the time when world-wide subject-centred data banks were being set up within the UN system—INIS by the IAEA, AGRIS by FAO, etc.—some thought was given to providing bibliographies, and also to obtaining industry-related references or abstracts from central input points in each country. An acronym had even been produced, INDIS for Industrial Development Information System. What eventually resulted was a data bank of abstracts of UNIDO's own reports, studies and publications, 10,000 to date, retrievable with reference to a thesaurus of industrial terms developed by UNIDO, kept on tape and published as a series. Copies of the tapes are available to centres in developing countries with adequate computer equipment. Requests for short lists are processed by UNIDO and original documents from these lists are made available. There is no intention now to emulate INIS or AGRIS. The acronym now designates the full range of classical industrial information services offered by UNIDO, in contrast to INTIB, the Industrial and Technological Information Bank.

Among those long ongoing services is the circulation, to some 30,000 readers, of UNIDO's *Newsletter* in four languages, all listed on a computerised mailing list allowing for the extraction of short lists by sectors and special interest for the mailing of documents of corresponding relevance. Likewise on tape and retrievable by sectoral categories is a roster of industrial consultants, including consultant firms from developing countries, the numbers of which are increasing. A similar roster of individual experts willing to offer direct services to developing countries is being built up.

## THE ROLE OF NATIONAL INFORMATION STRUCTURES

WHATEVER THE NUMBER of inquiries processed at UNIDO, their absolute number will, of course, only represent a small percentage of those questions that either remain unanswered, or are not perceived as problem areas in developing countries, or that find their answers from local services. Strengthening the latter is part of UNIDO's task, and over the years numerous industrial information facilities, oriented towards industrial planners and operators, i.e. with end-users of information rather than towards their documentation function, have been set up or adapted to industrial requirements with technical assistance from UNIDO, with expert advice on site and training of industrial information officers. In fact, there is no substitute for the establishment of national information structures with a growth function.

In many developing countries, successful industrial information structures, and even regional networks, like the Singapore-based TECHNONET, are in operation. However, these mainly service the small scale sector of industry, rather than medium scale, which is often the most technologically sophisticated, and the heavier sector. These sectors need to be served by groups of industrial information officers with polytechnical training coupled with a grounding in the economics of industrial enterprise; efforts will be made to develop a consciousness of this necessity at the governmental level in developing countries. Much has been said and written on the necessity for developing countries to develop information policies, as part of the responsibilities of government in education, science and culture. It seems that more will have to be done about stressing the information component of industrial, agricultural and other sectoral policies, and, as far as UNIDO is concerned, to persuade ministers of industry that information for industry is a necessary 'commodity', as pointed out by the International Federation of Documentation.

## UNIDO WORK ON INFORMATION FOR INDUSTRY

THE LIMA PLAN OF ACTION remains the charter of UNIDO's work in the area of information for industry under whatever acronyms it has been described in the Directory of UN Information Systems. This has been updated and complemented by a subsequent decision of the Industrial Development Board and the UN General Assembly and is bound to be marked by the establishment of a global network for the exchange of scientific and technological information as an outcome of the Vienna Programme of Science and Technology for Development.

The Lima Plan of Action which set the 25 per cent target, endorsed the long range strategy of UNIDO that the industrial information clearing house function of UNIDO should be given priority and that an Industrial and Technological Information Bank should be considered.* Because of the great diversity of user

* Lima Declaration and Plan of Action on Industrial Development and Co-operation, March 1975, chapter III: Co-operation between developing and developed countries: (j) "The DC's should be granted access to technological know-how and advanced technology . . . (k) Appropriate measures, including consideration of the establishment of an industrial and technological information bank, should be taken to make available a greater flow to the developing countries of information permitting the proper selection of advanced technologies".

needs, the original INTIB terms of reference have been extended to relate to co-operation between developed countries, and also not only to necessarily advanced but to 'proven' technologies (the degree of appropriateness being perceived by the developing countries); also to relate to individual technologies rather than to comparable generalities on those available.

An important outcome of the New Delhi Forum on Appropriate Technology, November 1978 has been the introduction into INTIB of not-so-advanced technologies. At present there are also plans for holding a Forum devoted to technological advances and their implication for industrialisation. UNIDO's interest in this present Conference is part of this trend, as has been the holding of other recent expert meetings. The forthcoming (August 1981) UN conference on new and renewable sources of energy will draw attention to this complex sector, which has been dealt with by INTIB for a long time, and has also been the subject of a new venture in UNESCO's programme. Inter-agency co-operation, particularly active between UNIDO and WIPO, making the most of the information content of patent descriptions and furthering co-operation in developing countries between patent offices and information services will continue.

## INTIB – INDUSTRIAL AND TECHNOLOGICAL INFORMATION BANK

In an ideal situation individual countries would be able to do without UNIDO and INTIB and to develop their own external development relations. The tools for this are manifold, UNIDO provides *Guides to Sources of Information* in individual sectors of industry the world over, with emphasis on sources in developing countries. UNIDO, with the operation of INDIS and INTIB, draws on such sources, some of which are part of its "Network of Correspondents". A *Directory of Industrial Information Facilities in Developing Countries* will shortly be issued and disseminated throughout developing countries to facilitate direct communications and information exchange.

A brochure describing the working of INTIB and listing the 20 sectors in which it operates is available from UNIDO. Designated as computer supported services, they do not yet contribute much to the furthering of a world of fingertip access to blue-prints and technological know-how that the resources of advanced information science and "telematics" (to go one semantic step beyond informatics) will bring about. But in UNIDO there is clearly perceived a challenge that might force reappraisals upon us, new formulation of future programmes, and if light can be cast by the present Conference on this challenge, we shall be grateful to it and have one more reason to co-sponsor it than to publicise our Industrial and Technological Information Bank.

# UN University's Role in the Dissemination of Knowledge

*I. Wesley-Tanasković*
*United Nations University, Toyko*

## INTRODUCTION

The charter of the United Nations University was adopted by the General Assembly of the United Nations on 6 December 1973 (resolution 3081/XXVIII). The following articles are considered to be important in the context of dissemination of information and the relief of intellectual isolation (key words are shown in italics):

1. The United Nations University shall be an international community of scholars, engaged in research, post-graduate training and *dissemination of knowledge*... it shall function... through the central programming and co-ordinating body and a *network* of research and post-graduate training centres and programmes located in the developed and developing countries.

4. The University shall disseminate the knowledge gained in its activities to the United Nations and its agencies, to scholars and to the public, in order to increase *dynamic interaction* in the world-wide community of learning and research.

6. The University shall have, as a central objective of its research and training centres and programmes, the continuing growth of vigorous academic and scientific communities everywhere and particularly in the developing countries... It shall endeavour to alleviate the *intellectual isolation* of persons in such communities in the developing countries....

To summarise, the mandate of the UN University is, among other, 'disseminating knowledge' through 'networks' which would by 'dynamic interaction' contribute to the growth of vigorous academic communities in the Third World.

In its first five years from 1975 to 1980 the UN University has been organised into an operational institution with an extensive, international system of 19 problem-oriented networks of scholars and institutions engaged in research, advanced training and dissemination of knowledge, as stated in the UNU Council Report for 1979–1980.

The group of experts who met at UN University headquarters in early 1977 to consider the University's appropriate role in the dissemination of knowledge characterised the problems it faced:

> "Surveying the field of scholarship as a whole, we see groups and individuals scattered all over the globe. They usually know little of what each other is doing, outside of their particular country, language and discipline. Their grasp of sources of information is often seriously incomplete. Their access to policy-makers is more or less distant. They tend to write for a few colleagues in a language too esoteric for the general public or even scientists in other disciplines. As the generation of knowledge accelerates . . . it takes longer and longer for new knowledge to be diffused and applied".
>
> "It is often assumed that once new knowledge is available it naturally flows to useful outlets: in fact, the reverse is more frequently true—much of the world's knowledge lies unused behind dams of ignorance, indifference and inefficiency. A particularly crucial problem is the imbalance in the sources and distribution of knowledge, with far too little coming from the Third World and reaching other developing as well as industrialised countries."

## DISSEMINATION ACTIVITIES OF UNU

THE UNIVERSITY'S DISSEMINATION of knowledge activities have thus far responded to the concerns expressed in 1977 mainly in two more or less traditional ways: publications and meetings (workshops, seminars and symposia). The University now publishes two regular periodicals, both designed to answer specific needs in their fields. One, which may be interesting in the context of this Conference, is called *ASSET*, which stands for "Abstracts of Selected Solar Energy Technology". This is a secondary publication produced monthly in cooperation with the Japan Information Center for Science and Technology, the Technical Information Center (DOE), USA, and the Technology Application Center (NASA), providing computer files and/or documents.

Intended mainly for Third World scientists, it abstracts publications on solar, wind and other forms of renewable energy which are of special relevance to rural communities in developing countries. The wealth of information collected in *ASSET* could be processed for wider dissemination.

It is issued to 550 scientists and engineers in the field of solar energy technology in 80 developing countries, and provides up-to-date information on renewable energy sources applicable to their local situations. In turn, they can send their own papers to the *ASSET* office for abstracting, thereby creating a self-generating information network.

The University's knowledge dissemination efforts are not limited to publications. Much valuable information is also generated and disseminated in the many workshops and other meetings which the University organises throughout the world on specific topics of timely importance, such as alternative energy strategies for the future or appropriate transformation of technology for development. These meetings bring together scientists, engineers, policy-makers and social workers from many regions for an open exchange of views. During the past year, the University sponsored some 78 meetings, in 25 countries, with some 1500 participants. The University's series of technical publications include the proceedings of workshops, seminars and symposia in addition to the reports of research projects.

As may be seen, these are traditional information transfer activities even when based where available, like *ASSET*, on computerised information retrieval services. Thus, informatics, meaning 'the rational and systematic application of information technology' to social problems, has not yet been implemented by the University in its world-wide sharing of knowledge. However, one has to recognise that several attempts in this direction have been made, but they failed so far.

## LINKAGES BETWEEN UNU AND ASSOCIATED INSTITUTIONS AND SCHOLARS

THE LINKAGES ESTABLISHED by the University with associated institutions and individual scholars through its 19 operating networks, provide a means of communicating knowledge far more widely and rapidly than the traditional route of publication and seminars. The networks should provide opportunities for scholars to be in direct touch with colleagues in the same field elsewhere in the world, sharing information as research progresses. The potential of these networks for research communication cannot be fully exploited unless much use is made of informatics and 'telematics'. In the future interconnected computer networks will enable efficient research cooperation and scientific information exchange sponsored by the UN University.

Possibilities for doing this with the help of digital networks should be assessed by the University through feasibility studies or surveys of existing projects in the field of computer-assisted remote team research and information exchange. This is crucial for the survival of the traditional networks established by the University.

## FUTURE UNU DEVELOPMENTS

MUCH REMAINS to be done by the University in the dissemination of knowledge, but on a potentially larger dimension of relevance to this Conference, I would wish to outline some on-going processes in the UN University.

From 1979, a review of the initial programmes of the University has revealed the strengths and weaknesses of the programmes and their potential for further development.

Out of these evaluations and prospective exercises, one of the major themes emerging is the specific role of the UN University in raising the level and comprehensiveness of the world's review of knowledge and information on science and technology and examining their practical implications and their socio-economic and cultural impact on different kinds of societies.

In beginning to study this major theme, it is tentatively envisaged that the UNU will organise reviews on the state-of-the-art and the state-of-research in a few selected scientific areas where breakthroughs are taking place, are imminent, or are likely to occur in the next decade or so. Examples that come to mind are energy, communications and molecular biology.

To undertake the state-of-the-art and state-of-research assessments, interdisciplinary panels would be needed in which natural scientists and computer specialists will be joined by social scientists and humanists. Specific papers in sub-areas would be commissioned and the resulting reports could serve as a guide to researchers around the world and become a basis for the next step of examining their implications and their international patterns of relationships.

Alternatively, the UNU could organise research with a prospective view on selected global issues, such as the issues of the computer age or of the nuclear age. Among such issues are those concerning the access of all countries to modern information processing, the patterns of centralisation and dependency consequent on computerisation and integrated energy systems development.

# The Role of Data Bases in Industrial Development

***André J. Michel***
*Intergovernmental Bureau for Informatics, Rome*

## THE CONCEPT OF DATA BASES AND ITS RELATION TO DEVELOPMENT

ALTHOUGH THE PHRASE 'data base' is no more than twenty years old, (origin: the military systems developed for processing great amounts of data together with operations research), the concept in itself is much more ancient.

Indeed, all through the history of Art and Science many scientists and philosophers have been tempted to sum up, at one time or another, all accumulated knowledge in a reduced and organised format (encyclopedia) to make it available and retrievable at their fingertips.

In fact, the concept of data bases has implicitly inspired the computer manufacturers from the very beginning: i.e. the production of magnetic drums, disc-packs, the famous IBM 'data-cells', etc.

Data base management is currently the object of very sophisticated studies giving rise to a great amount of literature. This situation could be compared to the mathematician's study of abstract algebraic and topologic structures geared to help the physician and the engineer to solve their problems.

The data base is in fact a data-structure and therefore an information (knowledge) structure. There is no possible economic growth without knowledge and no economic development (industrial development) without economic growth.

## THE INFORMATION INDUSTRY

The technological convergence of computers and telecommunications makes information travel like merchandise. Well packaged information can be carried, distributed and sold. This sector of activity is in full expansion and the description of the corresponding economic agents is a classic one: the producer, the server, the carrier and the end-user.

There is a strong analogy between information as a resource, and energy resources: geographical localisation—extraction—packaging (data base)—transport—distribution—marketing.

The only differences between the two kinds of resources are the following: information is not exhausted by duplication and energy resources are, for a great part, not renewable.

## INDUSTRIAL DEVELOPMENT

It may be recalled that:

(1) The success of the industrialisation of developing countries is based on:
 the acquisition and mastering of the new technological process;
 the capacity to adapt and innovate technical and organisational changes with a view to increase productivity;
 the possibility of having access to engineering consultancy.

(2) Technology transfer to developing countries is channelled by:
 the importation of manufactured goods
 direct investment by transnational companies
 consultancy
 training
 turn-key projects
 licensing, etc.

(3) Industrialisation is not only the importation of technology but also the export of manufactured goods and/or services in order to get a foot in world markets.

Whatever facet of industrial development is considered among the three quoted above, information appears to be the primary fuel for the decision-making process. Numerous data bases exist for each sector.

## ONE MORE STEP AHEAD

THE MOST SOPHISTICATED data bases provide more than simple bibliographical or factual information. They are, in fact, interactive models based on computerised algorithms answering, through an on-line terminal, like an intelligent consultant would do, some complex questions of industrial applications.

For instance:

mapping the supply and demand of technology at world level;

displaying the best license for an industrial plant in given geopolitical conditions;

estimating the PERT-cost of an industrial project in any part of the world.

Industrial consultancy is becoming more and more a type of 'artificial intelligence' thanks to advances in data base organisation.

## INDIGENOUS DATA BASE

THE TECHNIQUES OF data base management systems must be transferred to the professionals in developing countries.

Indeed, in building national data bases, the professional in developing countries will create essential tools for:

raw data collecting

development planning and forecasting

evaluation of a strategy for development

in all sectors of the economy.

## CONCLUSION: A NEW INTERNATIONAL ORDER FOR INFORMATICS

THE UNBALANCED SHARE of information resource at world level can be compared to the unbalanced share of goods and energy resources. The inescapable conclusion must be that developing countries, with the aid of developed countries and international agencies must work towards better access to the existing networks of data bases.

# Delivery Mechanisms for Industrial and Technological Information in the South-East Asian Region

*Leon V. Chico*
*Technonet Asia, Singapore*

## A REGION OF GROWTH AND DEVELOPMENT

SOUTH-EAST ASIA is the fastest developing region in the world today. In terms of economic growth, the combined GNP of the countries in the region has increased annually by over 6 per cent in real terms during the last ten years with some countries recording double figures. This growth has been remarkable and in spite of various world crises. The future of the region is one of great promise and high aspirations.

In spite of this rapid development, much of the population, particularly the rural communities remains poor and, except for a few countries, basic needs for food, clothing and shelter are not adequately met. The great majority of the people live in the countrysides where life for vast numbers is a question of survival. Many migrate to cities in search of jobs and a better life, but there is not enough work for all of them. The problem of unemployment continues to be one of the major problems in most of South-East Asia today. This is simply because over 50 per cent of the increase of the world's labour force during the 1970s has occurred in Asia.

Mostly dependent on agriculture and their mineral and other natural resources, many South-East Asian countries have opted for a blanced agro-industrial development. Industrial development has been thought of as a means to employment and eventual prosperity. Many are seeking the establishment of industries that will be labour-intensive, that will produce substitutes for imports, or that will produce goods to meet the quality standards of international trade and can compete in export markets. Most of these developing countries, with rapidly growing population and inadequate capital resources, have given high priority to the promotion and development of small and medium scale industries—to complement the capital-saving technologies that are brought in by large enterprises, mostly multinational companies.

But small industries are often ineffective because of lack of technological know-how. They are often saddled with excessive costs—particularly through wastage or improper use of raw materials—or because their products are of low quality due to poor methods of production. To overcome these problems and meet the objectives, industry needs technological advice and information. Quite often, such advice or information needed is at a very basic level—it is not a question of providing advanced technology, but of having an experienced engineer or technician look at a plant and make suggestions that will improve the processes or the products. Of course, these engineers or technicians must have access to modern industrial and technological information—on a national, regional and international level—in order to upgrade their expertise and be effective.

A 'total' approach to promoting the establishment of small business requires that modern technology be accessible to them. Generally lacking in technical expertise to solve their problems, small enterprises require assistance from information centres and industrial extension services. Governments can also play an important role in promoting the types of technologies consistent with development goals and stimulate their transfer by providing incentive schemes. The sharing of technology—of processes and expertise—among developing countries could be effective in upgrading the capabilities of small enterprises. This is not to say that technologies from industrialised countries cannot be adapted to suit the needs of small business in developing countries.

The above premise was a major rationale for the development of various mechanisms and organisations devoted to industrial and technological information, in particular, for industry and, in general, for other sectors in the development process. The following section described some of these mechanisms.

## REGIONAL ORGANISATIONS AND MECHANISMS OF ENGINEERING AND TECHNOLOGICAL INFORMATION

*TECHNONET ASIA* (*Asian Network for Industrial Technology Information and Extension*): Often referred to as an 'experiment in cooperation' and in the sharing of technology, TECHNONET ASIA is a cooperative grouping ('network') of thirteen Participating Organisations in ten Asian-Pacific countries, which aims at improving the quality and efficiency of production in those countries' small and medium scale industrial enterprises. Particular emphasis is given to the application of knowledge concerning known processes, methods, techniques, equipment, modifications and approaches to existing operations, effected by the transfer of technical information, provision of industrial extension services, the promotion and development of indigenous entrepreneurs and new enterprises.

The countries and organisations involved are the following:

| | |
|---|---|
| Bangladesh | *Bangladesh Small and Cottage Industries Corportation* (BSCIC) |
| Fiji | *Fiji National Training Council* (FNTC) |
| Hong Kong | *The Hong Kong Productivity Centre* (HKPC) |
| Indonesia | *Direktorat Jenderal Industri Kecil, Departemen Perindustrian* (*Directorate General for Small Industries, Ministry of Industry*) (DP/DJIK) |
| Korea | *Korea Scientific and Technological Information Centre* (KORSTIC) |
| | *Small and Medium Industry Promotion Corporation* (SMC) |
| Malaysia | *Standards and Industrial Research Institute of Malaysia* (SIRIM) |
| | *Majlis Amanah ra'ayat* (MARA) (*Council of Trust for Indigenous People*) |
| Philippines | *Institute for Small-Scale Industries, University of the Philippines* (UP ISSI) |
| | *Economic Development Foundation* (EDF) |
| Singapore | *Singapore Institute of Standards and Industrial Research* (SISIR) |
| Sri Lanka | *Industrial Development Board* (IDB) |
| Thailand | *Department of Industrial Promotion, Ministry of Industry* (DIP) |

The network started in 1973 when the International Development Research Centre (IDRC) of Canada agreed to provide support for a period of five years. It brought together eight organisations in ten Asian countries—later expanded to the present thirteen organisations in ten Asian-Pacific countries—into a network for industrial technology information and extension services. Spurred by the success of its programs, the organisation was registered as an official and legal entity in January 1980. Beginning in 1980, the Canadian International Development Agency (CIDA), through its Industrial Cooperation Division, has assumed a major role in supporting TECHNONET ASIA. IDRC, however, continues to support a substantial portion of the organisation's 1980–1983 budget.

In addition, Participating Organisations make annual contributions to TECHNONET and provide counterpart support in its activities. The assistance of other donor agencies and international organisations has also been sought on specific projects, notably that of the Japan International Cooperation Agency (JICA), which has provided funds and expertise for joint research on the small and medium scale metalworking industries and other related activities. Several joint activities have also been undertaken with the United Nations Industrial Development Organization (UNIDO) and its Industrial and Technological Information Bank (INTIB). TECHNONET also has links with some sixty other cooperating institutions worldwide.

TECHNONET Centre, located in Singapore, acts as the focal point for the network. It is headed by an Executive Director selected from one of the Participating Organisations. The heads of the participating organisations, together with the Executive Director, comprise the Council of the Centre. The main activities include:

training of industrial development officers;
strengthening industrial (technical) information through 'direct networking' among participating organisations;
promoting exchange of technical enquiries among participating organisations;
establishing links with specialised technical information sources;
facilitating interchange of extension and information personnel;
providing current awareness services, state-of-the-art reviews, appropriate pamphlets, films etc.;
facilitating transfer and sharing of appropriate technologies;
assisting in promoting entrepreneurship, establishment of new industries, and joint ventures.

Since its inception in 1973, much improved technical information services for small and medium scale industry have been established in the TECHNONET network which now has computerised services in seven participating organisations. Further, the network is seeking more effective ways of facilitating the flow of technical information for industry, of Asian origin, between participating organisations and other interested regional bodies. The Technical Information Service of the National Research Council of Canada (NRC/TIS) backs up TECHNONET's technical information needs when the information it has access to is pertinent to Asia. Linkages with other international sources of information have also been effected.

TECHNONET relies heavily on the 'human element' in the delivery of technological information. These are a group of 1500 industrial extension and information officers, who are constantly involved in factory visits to solve problems on the shop floor. They also act as 'processors' of technological information, interpreting and adapting it to the needs of small factories. A quarterly TECHNONET Digest keeps participating organisations informed on the latest technological developments in countries within and outside the network. Over 30,000 cases are now being handled each year within the network.

In 1975, the Asian Industrial Extension Officers' (ASINDEX) Forum, under the aegis of TECHNONET, was established to give added impetus to this emerging profession. Local chapters have been formed in countries within the network. Annual awards are given to outstanding industrial extension (development) officers within the network.

*APPROTECH ASIA* (*Asian Alliance of Appropriate Technology Practitioners*): This is a new mechanism set up early in 1980 during a meeting of the Appropriate Technology Delivery Group which was held in Bangkok, Thailand. TECHNONET ASIA acts as its Secretariat and initial support has been granted by A T International (Washington, D.C.).

APPROTECH ASIA has been formed in recognition of the significant work underway in Asia to increase the access of the opportunity-poor to technologies and processes appropriate to their needs and expanding capacities. Sustainable self-development through widened access to opportunity is the common objective of member organisations. The organisation is a regional appropriate technology service mechanism composed of institutions, formal and non-formal groups, and individuals experienced in appropriate technology issues and incorporating appropriate technologies in successful development activities. On a regional level, APPROTECH is intended to:

promote exchange of processes, technologies, and practitioners amongst active groups;

promote the discussion of appropriate technology issues, and advocate government and institutional policies in support of appropriate technology activities;

become a focus for evaluating, screening and disseminating appropriate technology-related information;

link appropriate technology groups with other professional organisations seeking to advance appropriate technology interests;

encourage development of national alliances of appropriate technology practitioners;

On the global level, APPROTECH is intended to:

aggregate useful regional appropriate technology experience and expertise for wider use and application;

serve as the Asian regional affiliate of other development organisations as may be requested;

seek access to appropriate technology experience in other regions of the world while providing those regions with access to relevant Asian experiences.

The initial countries and organisations involved are:

| | |
|---|---|
| Bangladesh | *Gonoshasthaya Kendra* |
| | *Village Education Resource Center* |
| | *Bangladesh Rural Advancement Committee* |
| India | *Centre for Science and Environment* |
| | *The Bhagavatula Charitable Trust* |
| Indonesia | *Yayasan Dian Desa* |
| | *Yayasan Indonesia Sejahtera* |
| Malaysia | *Institute Masyarakat Berhad (People's Institute)* |
| Philippines | *Philippine Center for Appropriate Training and Technology* |
| | *Philippine Business for Social Progress* |
| | *Philippine Women's University, Barangay Technology Center* |
| | *Manila Community Services, Inc.* |
| Sri Lanka | *Lanka Jatika Sarvodaya* |
| | *Community Development Services* |
| Thailand | *Community Based Appropriate Technology and Development Services, Population and Community Development Association.* |

APPROTECH proposes to undertake the following information activities:

Finance distribution of member publications, periodicals, etc. to other mem-

bers, and seek out non-member publications of special interest to members.
Compile reports on appropriate technology as a means of learning about effective technology transfer efforts.
Work with TECHNONET participating organisations to expand their services to include topics such as technical assistance, appropriate technology information, etc.
Provide technical information services normally available to TECHNONET participating organisations.
Assist in developing regional and country strategies for enhancing appropriate technology activities and promoting appropriate technology concerns.
Bring together regional experiences on: overcoming obstacles to appropriate technology, improving the intermediary function of development organisations, and passing on knowledge at grassroot level.

It is too early to assess the impact of this new mechanism. The record of individual members in delivering appropriate technologies, particularly to the rural poor, speaks highly of the potential of the new grouping.

*APO* (*Asian Productivity Organisation*): This is an inter-governmental organization established by convention in 1961 by several Asian governments to hasten their respective economic development. It is non-political, non-profit making and non-discriminating. APO aims to increase productivity and accelerate economic development, in the Asian region, by mutual cooperation. It seeks to realize its objective by:
propagating productivity consciousness
disseminating modern productivity knowledge, techniques and experience in agriculture, industry and service sectors.

Membership is open to all Asian governments that are members of the UN Economic and Social Commission for Asia and the Pacific (ESCAP). Governments outside the Asian region may become Associate Members. National Productivity Organisations (NPOs) in member countries deal with national productivity activities. They also act as implementing agencies for APO projects and participate in APO's multi-country projects. The Heads of NPOs meet once a year to exchange experiences and to review and work out project and implementation details of APO's programme of activities.

The Governing Body of APO is composed of Directors appointed by member governments and meets once a year to decide on policy matters. The Secretariat, headed by a Secretary-General, is the executive arm of the Governing Body. The Secretariat is drawn from qualified persons amongst member countries and is located in Tokyo, Japan. The information programme of APO includes:
Propagation of productivity consciousness and projection of the APO image, mainly through distribution of publications (including the monthly newsletter, *APO News*) among member countries for dissemination and use in their own programmes;
Dissemination of productivity knowledge by means of technical publications (in management and technology fields) through commercial booksellers;

Compilation of a catalogue of audio-visual materials possessed by National Productivity Organisations (NPOs), and of training manuals developed by them;

Periodic meetings of NPO information officers; these guarantee co-ordination between APOs and NPOs and provide a forum for reviewing and strengthening the information programme;

Provision of technical expert services in the development, production, and effective utilisation of audio-visual aids;

Referral of technical inquiries to UNIDO or to appropriate member countries for assistance.

APO is considered to be a major instrument in bringing about productivity consciousness in the Asian region and has effectively prompted mutual cooperation among member countries and with other organisations.

*RCTT* (*Regional Centre for Technology Transfer*): This organisation which was established in Bangalore in July 1977, views technology transfer as a comprehensive process involving the transition of technology from its development and generation through effective diffusion, and appropriate adaptation to practical applications. 'Technology transfer' denotes the whole spectrum of building up technological capabilities through development, adaptation and transfer of technology. The Centre is based on the 'network' concept, i.e. that it would not become the location of activities for transfer and development of technology but would assist in strengthening national capabilities. Activities to be undertaken by the Centre will provide support to the national efforts or will be of catalytic nature. The RCTT will also act as a regional instrument to pursue major concerns of the organisations of the UN system working in the field of technology such as UNIDO and UNCTAD.

An adequate information system is essential for such an approach. The *RCTT Newletter*, published quarterly, seeks to provide a channel of information for use by the institutions participating in the network activities as well as the other institutions and individuals interested in its activities. It reports on programmes and activities of RCTT and national focal points, developments in technology policies and other events of interest.

In deciding upon its work programme, RCTT adopted the approach of ascertaining the needs of the participating countries and of formulating programmes to fulfill them. A workshop in Bangalore in April 1978 attended by directors/officers-in-charge of national focal points, and representatives of the UN agencies identified the following fields of concern for the RCTT:

Food and agriculture e.g. industrial utilisation of agro-wastes, processing of essential oils and medicinal plants; post-harvest technologies; packaging technologies;

Industry e.g. machine tools, metal processing, textiles, leather, ceramics, handlooms and low-cost automation;

Housing e.g. cement production;

Energy e.g. mini-hydro plants, biogas, wind, solar and coal;

Natural resources e.g. mineral processing;

Health e.g. food fortification and family planning.

RCTT has since been engaged in developing specific activities in the above areas. In the field of technological information, RCTT has taken the following steps:

Building up the library of the Centre, the nucleus of which was already in existence;

An information service on a modest scale now disseminates information regarding important technological developments in the region and answers enquires from institutions in participating countries regarding sources of technological information. This will be systematised and expanded. Experts from the USSR and Israel assisted in establishing the information network; the main feature that will distinguish the RCTT information service from global information services in operation will be that it will be specific to the requirements of participating countries;

A periodical newsletter has been started;

Two volumes of national papers prepared by the ESCAP member countries in preparation for the United Nations Conference on Science and Technology for Development (August 1979) have been published.

RCTT activities to stimulate the flow of technological information within the region are being pursued with vigour. Its library has achieved steady growth. Since its inception RCTT has undertaken projects to attain its objectives and more are planned. RCTT is expected to play a more important role in future in the flow of technological information within the region.

*AIT* (*Asian Institute of Technology*): founded in 1959 as the SEATO Graduate School of Engineering to help meet the growing need for advanced engineering education in Asia, it became fully independent in 1967 as AIT. Its Royal Thai charter accords it the status of an autonomous international institution empowered to award degrees and diplomas.

It provides advanced education in engineering, science and allied fields, its academic programmes being related closely to the needs of Asia.

The following AIT units specifically deal with information:

*Regional Documentation Centre*: established in 1977 to co-ordinate present and future information activities being developed to meet the information needs of both AIT's Academic Divisions and the Asian countries. The RDC has four operating specialised information services.

*International Ferrocement Information Centre* (*IFIC*): established in 1977 as a joint project with the Division of Structural Engineering and Construction, the Centre is sponsored by the Government of New Zealand, the US Agency for International Development and the International Development Research Centre of Canada (IDRC). It is engaged in collecting, processing and disseminating information on ferrocement and related materials. *Journal of Ferrocement* (quarterly) is published in collaboration with the NZ Ferrocement Marine Association.

*Renewable Energy Resources Information Centre* (*RERIC*): officially established at AIT in 1978 with financial support from the Government of France and UNESCO, this covers solar energy, wind, biogas, small scale hydropower with an 'appropriate technology' approach of interest to tropical developing countries. Publications include *RERIC NEWS* (newsletter), *Renewable Energy Review Journal*, and occasional bibliographies and booklets.

*Environmental Sanitation Information Centre* (*ENSIC*): begun in 1978 with financial assistance from the International Development Research Centre (Canada) it aims to provide developing countries with information on the disposal and recycling of wastes. It publishes regularly *ENFO Newsletter, Environmental Sanitation Abstracts—Low Cost Options* and the *Annual Technology Review*, and occasional monographs and booklets.

*Asian Information Centre for Geotechnical Engineering* (*AGE*): established in 1973 under joint sponsorship of the Division of Geotechnical and Transportation Engineering and the Library and Information Centre with financial support from the IDRC. Serves as a clearing house for information on geotechnical engineering projects, i.e. data on design, construction and research projects are analysed, abstracted and indexed for effective storage, retrieval and dissemination. Regular publications include *Asian Geotechnical Engineering Abstracts, AGE Current Awareness Service, AGE Journal Holdings List, AGE Conference Proceedings List, AGE Research Reports Holdings List, AGE News*, and *AGE Digest*.

General: The foregoing mechanisms have all played a major role in providing engineering and technological information in the South-East Asian region. There are, of course, other mechanisms which deal with more specialised and specific fields. Inter-governmental arrangements such as ASEAN (Association of South-East Asian Nations) and UN ESCAP (United Nations Economic and Social Commission for Asia and the Pacific), the educational systems and their exchange programmes, professional associations, industry federations, and other bodies and mechanisms all play important roles to complete the total 'delivery' system.

## PROBLEMS OF INFORMATION TRANSFER FOR TECHNOLOGICAL DEVELOPMENT

WHILE NO ONE DENIES that the developed countries and the international sources of information could be more effective in stimulating the diffusion of technology, there is a growing realisation that the greater part of man's technological know-how is freely available. The biggest problem is that developing countries are ill-equipped to find, evaluate and apply it. Often, information needs are not apparent or properly identified, hence, the need of strengthening the capabilities of Third World institutions.

Information seekers are also faced with both extremes in their search for technological information. On one hand, it can be very limited because they have no ready access to sources of technological information, and even if such information is available, it may not be in an appropriate form. On the other hand, the range of technological information can be very wide and the problem becomes that of choosing appropriate technology or the adaptation of technology to suit a specific situation. In addition, there is the oft-repeated criticism that information obtained from developed country sources is not relevant to the needs of developing countries. But who should really decide what is relevant? Again, the role of information 'processors' or development agents is extremely important in developing countries.

Developing countries have much to share in terms of relevant technological information, processes and expertise, but the stimulus or the capability is not always there. Most developing countries are competitors in international trade. There appears to be inadequate awareness that the transfer of technological information is a two-way affair—between developing countries and even, to a lesser degree, between developed and developing countries.

The physical infrastructures for effective information transfer are not always present in developing countries. Countrysides are sometimes inaccessible, communication links inadequate. Hence, the problem of delivering information at the right time, at the right place, in the right form.

All these problems are not insurmountable; much progress has been made. Mechanisms that have been set up have made their contributions to minimising some of them and have had their impact on technological development. Some problems require political will, others time and funds. Most important, perhaps, is that *cooperation* and *sharing* among countries in the field of technological information do not become mere slogans. They can be made to work.

# Informatic Technologies and Information Transfer

*John R.U. Page*
*IIASA, Laxenburg, Austria*

## INTRODUCTION

THE APPLICATION of informatic technologies has had a major impact on information transfer and the process seems likely to continue in the foreseeable future. The impact, however, is not simply a matter of a straightforward application of a technological innovation to solve known problems at an acceptable cost, but appears to involve a complexity of factors including cross-fertilisation between two previously independent areas of development (computers and telecommunications), the explosive growth of general electronic data processing in the industrialised countries, market forces, and user reaction. An historical perspective may therefore help to establish the dynamics of these continuing interactions between developments in informatics and the concomitant rapidly occurring changes in information transfer possibilities and methods.

During the 1960s developments in random-access memories, retrieval software and computer time sharing were associated with the use of machine-readable material prepared as a part of the production process for secondary abstracting and indexing journals to produce experimental versions of today's interactive online information systems which began to appear late in that decade. It may be noted, however, that the real impetus for the explosive growth of information transfer systems of this kind came from the marriage of computers and telecommunications technologies in the form of time sharing and networking, and these were driven almost entirely by the possibilities opening up for the business community: scientific and technical information transfer in the widest sense of that term then, as now, rode on the backs of other applications which exerted greater market pressure and therefore determined the path of technological development.

At all events, growth of online information retrieval during the latter half of the 70s reached the point at which probably well over two million searches per year were being carried out in the USA by the end of the decade. Important changes had also occurred within the data base industry itself: whereas at the start of the decade data bases available online numbered perhaps about 20, these being institutionally-produced bibliographic data bases in the main disciplines of pure science and technology, today's user in the industrialised countries will probably have three or four hundred data bases available to him, many of them being fact or source data banks. These latter have shown consistently high growth rates, particularly in the fields of finance, trade, statistics, etc. Much of current data base and data bank production is now in the hands of private, for-profit industry rather than professional institutions and government agencies.

With regard to the users themselves, it is probably true that the early expectations that the technology would favour end-users searching for themselves have not yet materialised: today's users of online systems are probably librarians, information brokers and information personnel generally. The growing emphasis on fact data banks and on subject fields well removed from traditional pure and applied science would, however, suggest that online information transfer is beginning to move out of the scientific area into business, finance, planning and management. If this is so, then conventional online technology may be inadequate to promote a further round of explosive growth in utilisation.

## INFORMATION CAPTURING AND ORDERING

### Present Position

Most of the formal information resource is still captured by conventional means, i.e., authors' draft papers which are then reviewed and eventually edited and published as articles in a journal or as monographs. While the printing industry is now becoming computerised, this has had little impact on the timeliness and effectiveness of the process as a whole, viewed as a means of rapid information transfer within a relatively small and rather homogeneous professional community. In this context, 'ordering' of information means the process by which the stock of information is systematically arranged so that items can be identified, located and retrieved according to relevance to a subject or a problem area, etc. The present system still leans heavily on the classical procedures of librarianship, i.e., cataloguing and indexing, even though the means for applying these procedures are computerised in the form of inverted-file data bases and sophisticated retrieval software. There is, however, no doubt that the power of these mechanised components, by which a virtually complete set of valid references can be retrieved in a half hour search across data bases containing many millions of references, has brought into sharp relief the inadequacies of the other components of the system as a whole. It is rare that the editorial and publication process is completed in less than a year after the first draft of a manuscript, and the heavy professional and clerical load required by this and the secondary publication process, which involves highly skilled abstracters and indexers, builds up to a very heavy overhead per reference retrieved.

## Areas for New Information Technology Applications

The whole area of improving timeliness in information capture and ordering deserves attention, especially in those fast growing areas of applied science and technology having important economic implications, and in transdisciplinary studies having impacts outside the pure research and development field. Electronic publishing could be a partial answer, since the production of a machine-readable text as the primary product of the publishing process provides a ready vehicle for on-demand publishing, personalised journals and other forms of information which can be rapidly distributed by electronic or other means to those requiring it. Electronic publishing, while improving the situation would not solve all the problems of timeliness: much of the delay is taken up by the drafting, submission, review and redrafting of the original text required under present procedures. All these processes could be simplified and speeded up by using electronic means of text processing and communication, but this would perhaps not be consistent with other requirements in scientific publishing (one possible scheme has been described in Page, 1980a).

The proliferation of data bases has increased the difficulties inherent in the where-to-look problem: a highly skilled intermediary cannot know the contents of all possible data bases which might be used in a search: there is an urgent need for automatic referral systems and constantly updated in-depth data base directories online which will guide an intermediary or an end-user through the possibilities.

## Application in Developing Countries

While improvements in timeliness and in information location will be of value to all users, there are specific problems in information management which presently constitute an obstacle for the full exploitation of locally produced information in developing countries. As Woodward (1980) has pointed out, the organisation and retrieval of locally produced information are significant factors in development. Part of the problem may be organisational, but relatively simple informatics applications in the shape of simple-to-use input procedures for local data bases and the use of mini- or even micro-computers as stand-alone devices for data base creation and information retrieval offer useful possibilities. With regard to organisational aspects, Tell (1980) points out that where responsibilities for information management are reasonably centralised in an executive ministry, and this ministry possesses a satisfactory computer, locally produced and other information may be aggregated into what amounts to powerful local data bases, tailored to the needs of the decision maker concerned. Such solutions, however, presuppose a certain level of infrastructure and technical expertise which may not be available in a very large number of cases.

## INFORMATION RETRIEVAL—INFORMATICS AND THE USER

ONLINE RETRIEVAL is no longer a cottage industry in industrialised countries: Cuadra (1980) has identified over 60 organisations offering online retrieval services internationally, and some authorities believe that the current US market is about $40,000,000 per year in size, with Japan and Europe each at about the $10,000,000 per annum level. While a number of online searches are being made by countries outside these regions, these are "imported", mainly from the USA: there appears to be little activity as yet in supplying online services from other regions.

### Retrieval Software—the User/System Interface

Although there are now a large number of retrieval packages available off-the-shelf, they are, by and large, far removed from the ideal of natural language systems. One must learn the commands appropriate to each, their abbreviations and combinations which may be represented by punctuation marks of one kind or another. In this respect, they are insufficiently user-friendly, although there have been improvements here in recent years. An instance is the development of the common command language by the Commission of the EEC which, when implemented on a host computer, will automatically translate commands from the user in CCL into the host's own retrieval language.

Seen from the user's stand point, the retrieval system and the structure of the data base (indexing and abstracting) form a black box with which he must interact to obtain relevant answers. This can be a frustrating experience, not only because of the intricacies of the command language, but also because the system gives him little if any guide to the indexing philosophy used in a particular data base. While really major improvements in this respect are not yet in sight, as they will require the further development of artificial intelligence and associative memory concepts, certain palliatives could be investigated, such as more and better tutorial systems, both online and offline, and easier-to-handle information on data base structure; for example, online and offline synonym and related terms structural diagrams. In respect of data base-specific and retrieval system-specific training, intermediaries in developing countries are worse off than their colleagues in industrialised countries, since intercontinental journeys may be required for attendance at training seminars, and it will be similarly difficult to reach the advisory service of the data base or system concerned. Therefore, well-organised and in-depth training and explanatory material, both on and offline, is an important area for improvement. Another area for development relevant to the needs of developing countries is that of multilingualism: the great majority of data bases now available online are in the English language, as are almost all retrieval languages. This is not a situation which can be changed overnight, but short of full automatic translation there are possibilities in the shape of multilingual online thesauri which could assist users to search in another language even though the search output would be in English.

## ACCESS TO ACTUAL INFORMATION

A BIBLIOGRAPHIC SEARCH is of minimal value to the user unless he can subsequently obtain, at a reasonable price and with an acceptable delay, copies of the original document cited. This is a problem common to users in both industrialised and developing countries, although the latter's situation is worse in that local library stocks are probably inadequate: securing copies from one of the major supply centres overseas will require the expenditure of scarce hard currency, probably requiring prior approval. A further problem affecting users in developing countries more severely than their colleagues in industrialised countries is the heavy cost of telecommunications to reach online vendors or host organisations operating the data bases they need to consult.

### Document Delivery

The Commission of the EEC has recently been active in studying the possibilities of electronic document delivery and its relation to electronic publishing (see Norman, 1981, and Page, 1980b). These indicate that electronic document delivery is feasible and technically within the state-of-the-art, although more work needs to be done on systems integration and in particular, on developing terminals for receiving electronic copies in both facsimile and coded character formats, with the ability to switch between each. Open questions at the moment are user reactions and costs per item delivered in operational systems: we do not know, for example, what kinds of users want what types of documents with ultra rapid delivery (e.g. overnight), and what they would be prepared to pay for such a service. Cost components are dominated by the cost of transmission, which, using typical European packet-switched network tariffs and the large number of bits associated with facsimile, could be well over one dollar per page. Full consideration has not yet been given to the application of these techniques to the problem of document delivery to developing countries, but at present this will require expensive leased circuits, and could make economic sense if there were a sufficient traffic volume to spread the costs over a large number of items: in this respect it should be noted that modern page imaging equipment for the production of data-compressed digital facsimile pages can operate at speeds of one to two seconds per page.

There is, however, another possibility, just becoming state-of-the-art, and this is the use of videodiscs to store electronic versions of documents. At least one current information retrieval system, Pergamon's PATSEARCH, offers the possibility of full-text document retrieval on a stand-alone videodisc-videoplayer terminal at the user's desk. Other developments now in the prototype stage such as the Philips MEGADOC system provides for full-text storage on ultra high density read-only digital optical (laser) discs with very fast retrieval and the possibility of associating banks of such discs in a jukebox arrangement. Such a system could conceivably offer a locally operated full-text information system with a minicomputer-based retrieval capability. Under certain circumstances this might be more cost-effective than paying intercontinental telecommunications charges for data base searching plus charges for delivery of relevant document copies from a remote centre.

## Telecommunications as a Means of Access to Information Systems

Aside from the possibilities of total decentralisation to local information centres, or even to groups of end-users, offered by advanced videodisc systems, new information and telecommunications technologies offer alternatives to the present use of networks derived from the national and international telephone systems. It may be noted that, in practice, present day packet-switched networks and data circuits are in many ways limited by the characteristics of the circuits they use, which were primarily designed for voice telephony. Although all digital-networks are being gradually introduced in the industrialised countries, these limitations are likely to remain for some long time to come. The basic limitation is that in terms of bandwidth, which in practice may be equated with the number of bits of data which may be transmitted per second. In general, the maximum transmission speed which can be dealt by a telephone circuit is insufficient to carry bit streams representing pictorial information or facsimile data at a speed at which modern page imaging devices can operate. To match the high bit-rates needed for these applications requires the grouping of several telephone circuits together in parallel or the use of special coaxial cables, such as in cable television systems.

Telecommunication satellite systems are not subject to the same bandwidth limitations and channels within the satellite are typically of megabit speeds. Hitherto, satellites used for international telecommunications have required large earth stations to and from which traffic must be fed by the terrestrial telephone, etc., networks: thus, from the users point of view, the basic bandwidth limitations of the terrestrial networks remain unchanged in spite of the satellite link. Moreover, the telecommunications administrations concerned will charge the user similar tariffs whether the traffic goes by satellite or by under sea cable, etc.

With the advent of regional communications satellites, and technical innovation which allow the use of small (three metres) disc antennas in place of the large ground stations, a new range of all-satellite data services become possible, such as direct TV broadcasting, high capacity text delivery systems, either broadcast or point-to-point, massive computer file transfer, video teleconferencing, etc. Many such facilities are now being offered by the Satellite Business Systems Cooperation in the USA, and are planned for the European Communication Satellite system (ECS), starting operational life in 1982/3. The degree to which these developments can significantly impact information transfer, particularly to and within developing regions of the world, will depend on tariffs, but international bodies are now becoming aware of the importance of low international telecommunications tariffs for information transfer in developing regions. It is to be hoped that initiatives in information transfer, such as the PADIS proposals by the UN Economic Commission for Africa, in which, *inter alia*, an African regional communications satellite for information transfer is proposed, will result in a more favourable situation for users in developing countries than is given by present international tariff structures.

## References


Cuadra, C. (1980) *Surviving the Eighties: New Roles for Publishers, Information Service Organizations and Users.* NFAIS Newsletter, **22(2)**.

Norman, A. (1981) *Electronic Document Delivery. The ARTEMIS Concept for Document Digitalisation and Teletransmission.* Arthur D. Little. Learned Information, Oxford, UK.

Page, J.R.U. (1980a) *Using Text Processing, Computer Networking and Satellite Telecommunications Technologies to Publish Primary Scientific and Technical Information* in Impact of New Technologies on Publishing, K.G. Saur.

Page, J.R.U. (1980b) *Proceedings of the Exhibition and Workshop on Electronic Delivery of Primary Documents.* Learned Information, Oxford, UK.

Tell, B.V. (1980) *The Awakening Information Needs of the Developing Countries*, J. Information Sci., Principles and Practice **1(5)**, 285–289.

Woodward, A.M. (1980) *Future Information Requirements of the Third World*, J. Information Sci. **1**, 259–265.

# Information Flow in a National System for Research and Development

*Konrad Fiałkowski, Henryk Rybiński and Bolesław Szymański*
*SINTO, Poland*

## INTRODUCTION

INFORMATION FLOW for research and development in Poland is organised within the framework of the National System for Scientific, Technical and Economic Information (SINTO). The aim of the SINTO system is to perform two basic tasks, namely:

(1) to supply all individual users (chiefly scientists, teachers, engineers etc.) with full and relevant information;

(2) to supply all institutional users (chiefly Government officers) with aggregate information and statistics concerning scientific research activities, initiations of new technologies, etc., within the country.

Such an information system is a substantial step towards improving the effectiveness of research projects and technology development. Wide cooperation among information subsystems of SINTO is assumed. Moreover a considerable exchange of information between SINTO and international information services is planned. Therefore the organisation of information flow inside SINTO is of great importance for SINTO functioning. The aim of this paper is to discuss some solutions to information flow organisation in the SINTO system.

## NATIONAL SYSTEM FOR SCIENTIFIC, TECHNICAL AND ECONOMIC INFORMATION (SINTO)

THE PRELIMINARY PLAN for the national system of scientific, technical and economic information (SINTO) was formally approved in 1976 (Nitocki, 1979). The system is currently in the process of implementation. Its aim is to facilitate the identification of source material, assist in the retrieval of the requested material, and classify, index, and disseminate the abstracts of the documents to all participating users. There are three types of institutions which are directly involved: information centres, libraries and archives.

The main purpose of SINTO is:

(1) to integrate activities of all three types of institutions without changing their organisational structure and dependencies;

(2) to enable a more exhaustive inter-field of primary and secondary information.

In that sense SINTO can be regarded as a system which provides functional co-ordination of the institutions which operate in the information area.

Three basic types of automated information systems operate within SINTO. They are specialised branch and regional. The systems differ in their territorial range and processed information scope.

The specialised systems maintain and co-ordinate all information activities concerning special documents such as patents, research reports, etc. In Poland the main specialised systems are:

SYNABA (information system for research reports) (Dobosj and Wojewoda, 1980)

SAZAPS (information system on Polish international scientific cooperation) (ISTEI, 1980)

PATENT (information system for patent documents) (Brajczewski, 1979)

TRANSLATIONS (information system for translations) (IINTE, 1980).

The branch systems maintain information activity for specific fields of science, technology or industry.

At present the Agriculture Information System (Hancko, 1978), Mining Industry Information System (Bartosz) and the Chemical Information System (IEPCH, 1979) are examples of advanced branch systems which are being developed within the framework of SINTO.

The range of specialised and branch system activity is national.

In addition, territorial systems operate in particular regions of the country. They maintain and co-ordinate all information activities in their region. Usually they are implemented by large university and research libraries. The system KRAKUS which is operating in the Kraków region is a good example of such a system (Czerni, 1979).

These three types of systems also differ in their roles within SINTO. The specialised systems are mainly responsible for preparing and preprocessing rele-

**Figure 1. The general structure of SINTO**

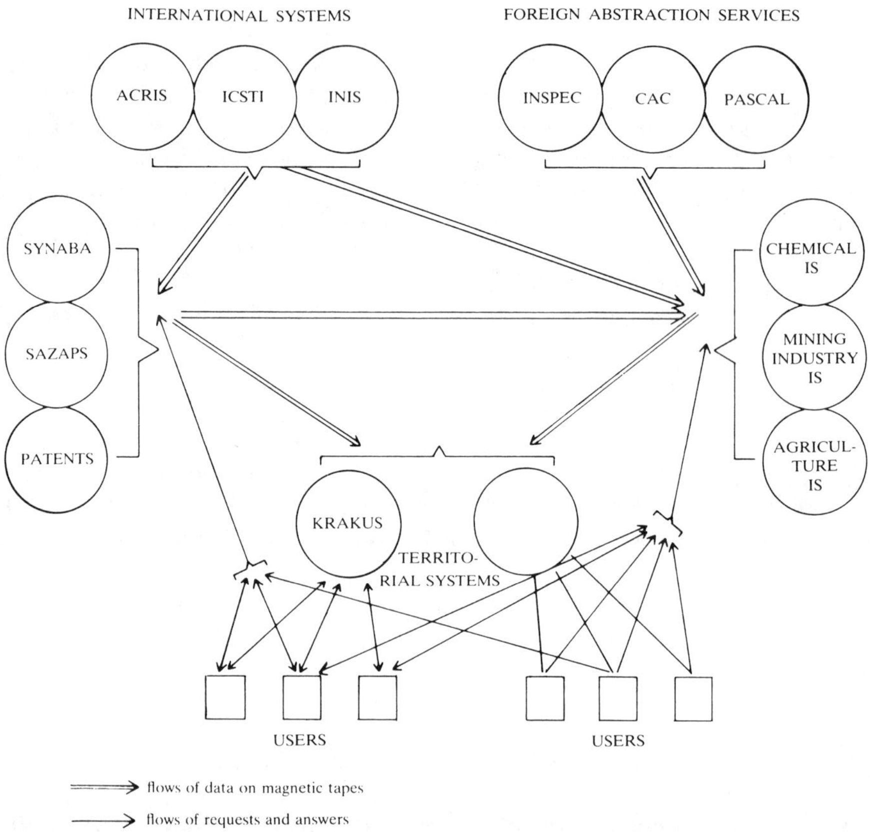


vant information material. The data bases thus created are broadly divided by topic and disseminated to appropriate branch and territorial systems. Those data bases are also used by individual users. In particular they are used to obtain a variety of statistical and planning information for management and administration purposes.

The branch information systems are responsible for servicing users interested in specific fields or branches of science, technology or industry. Their data bases are also augmented by information from

(a) specialised systems, relating to information created in Poland,
(b) foreign abstracting services on magnetic tapes relating to world-wide bibliographic data,

(c) international information systems, such as ICSTI (Moscow), AGRIS, INIS, etc.

On the other hand the branch systems produce magnetic tapes for international exchange and for territorial systems under various agreements.

The territorial systems are augmented by other systems of SINTO. They are responsible for direct cooperation with individual users from given regions.

## PRINCIPLES OF SINTO IMPLEMENTATION

SINTO INFORMATION SYSTEMS integrate the functions of both scientific information systems and management information systems. An analysis of CDS ISIS facilities has shown that, in general, the CDS ISIS package (Modrzjwska, 1980) can be successfully used as a standard software for SINTO. However, because of the special needs of SINTO several extensions of CDS ISIS have had to be made. These extensions permit the creation of the so-called Multilevel Information System (Rybinski, in press), which allows ISIS databases to operate in a more flexible way.

The main idea of that approach is to provide the users with additional levels of access to data in order to meet their specific needs. The new levels of access are obtained by means of DML enabling the production of application programs, and the generation of statistical and aggregated information.

The extensions mentioned above have been widely applied within specialist systems, especially SAZAPS, and SYNABA. They are also applied within some branch and territorial systems.

## CONCLUSION

THE NATIONAL SYSTEM for Scientific, Technical and Economic Information (SINTO) is being designed in Poland to support development of science and technology. To provide SINTO users with accurate and relevant information co-operation among SINTO subsystems services and an exchange of information between SINTO and international information services is required. Therefore a rich information flow within the framework of SINTO is being organised. To facilitate such flow uniform software has been proposed to implement SINTO subsystems.

## References


Bartosz, J., Private communication.

Brajczewski, H., *et al.* (1979) *Projekt zautomatyzowanego podsystemu informacji patentowej.* UP PRL, Interim Report.

Czerni, J. (1979) *Miejsce Krakusa w systemie SINTO, 1979*, P. Krakowska, Interim Report (in Polish).

Dobosz, J., and Wojewoda, A. (1980) *Problemy wdrazania pakietu ISIS do obsługi wybranych funkcji systemu SYNABA.* ISTEI Reports, no. 34, Warsaw (in Polish).

Hancko, R., *et al.* (1978) *System informacji o gospodarce zywnościowej—Projekt techniczny.* CBR, Interim Report.

IEPCH (1979) *System informacji chemicznej, Projekt wstępny. Część ogólna.* Interim Report (in Polish).

IINTE (1980) *Załozenia ogólnokrajowego systemu informacji o tłumaczeniach, Praca zb.* Interim Report (in Polish).

ISTEI (1980) *Wstepny projekt techniczny ogólnopolskiego systemu SAZAPS.* Interim Report, Warsaw (in Polish).

Modrzjwska, A., and Rolecki, J. (1980) *Adaptacja wersji wsadowej systemu CDS ISIS na maszyny jednolitego systemu.* Prace jubileuszowe INTE.

Nitocki, J.Z. (1979) *National Network of Information in Poland.* J. Amer. Soc. Infor. Sci. **30** (5).

Rybiński, H., and Szymański, B., *Multilevel information system towards more flexible information retrieval systems.* Info. Process. and Manag. (in press).

# Information for Progress: Revising Our Priorities

*Sean Cooney*
*Agricultural Institute, Dublin*

## INTRODUCTION

DURING THE PAST TWO DECADES increasing attention has been paid to the problems confronting economic development and social progress around the world, and especially in LDCs. The role played by science and technology in such development has rapidly assumed an important, even a dominant, position in these debates. During the 1970s, in particular, the view has become widely accepted that the intersection between science and development occurs in the process of information transfer. Science, according to this view, affects both economic growth and social progress through the dissemination and use of new technical knowledge. Conversely, the type and amount of research, and hence the pace and direction of the growth of our stock of knowledge, is strongly influenced by demands for information to solve problems connected with promoting growth and progress (Allen, 1966; Anderla, 1973; Robertson *et al.*, 1972; Allen *et al.*, 1971; OECD, 1971).

Furthermore, it has been widely believed (until recently at least) that more effective information transfer mechanisms would simultaneously promote both economic growth and social progress. This belief is shown by the rapid increase in the level of funding not only for research, but for information transfer mechanisms, and the growing attention given to information policy in the economic development strategies of governments and other agencies, in the Third World as elsewhere (NAS 1972, OECD 1971). Informatics, in other words, has moved to a central position in studies of development policy. According to the programme for this conference "Informatics is the rational and systematic application of information technology to economic, social and political problems". It therefore includes the study of how the expansion of information through research, and the dissemination of research results, interact with economic and social change, and of how to improve the effectiveness of information systems in promoting both economic growth and progress.

Developing countries have a special need for effective access to information to promote growth and progress. Their resources are rarely adequate to meet the costs of meeting this need by the mechanisms used in richer countries. The problem is made more difficult by the limited general infrastructure of LDCs. Poor roads, phone systems, postal services and other similar factors have a serious effect on information transfer. Scientists often work at centres remote from the urban locations where information is available and have special difficulties in travelling to or communicating with, these centres.

Here are some characteristics which a good information service should have (Cooney, 1974b): it should (a) be accessible and easy to use (Zipf, 1949); (b) deliver a high rate of messages for a given time and effort by the user (Allen, 1966); (c) make available a wide range of information not usually available in or near the scientist's own work place; (d) maintain a high level of quality in the information content supplied; (e) in a less developed region more than elsewhere, a service must also be cost effective.

## CHANGED PRIORITIES

THE FOLLOWING SCALE of priorities would appear to offer the best hope of providing these requirements quickly and effectively.

The first priority should be given to promoting meetings between scientists, that is, developing the "technological gatekeeper" function (Allen, 1966, Allen *et al.*, 1971). It has been shown that scientists and technologists get most of the information which benefits their work from talking to colleagues. The colleagues who are most relied on are termed "gatekeepers". Research carried out in Ireland (Allen *et al.*, 1971) identified the "international technological gatekeeper". It was shown that a nation, like an organisation, depends on its gatekeepers to bring technical knowledge into the country to aid development. Since no country can hope to develop rapidly without knowledge from abroad, this finding has important implications for national policies on travel by scientists.

In most countries the official attitude to experts who seek funds for travel, especially abroad, has often been "why should we spend money sending these people on glorified holidays?". In LDCs this attitude is reinforced by the severe shortage of foreign currency for any purpose. However, it is important to remember that knowledge goes out of date very quickly and must be replaced by new knowledge. The best way is to talk with someone who knows and such people are often in another country.

A high priority should therefore be given to enabling trained scientists from developing countries to keep up to date by travelling and meeting colleagues abroad (Cooney and Allen, 1974). Budgets for attending international meetings should be made more generous. The total bans on foreign travel, which seem to apply in some countries, should be relaxed. Aid programmes should lay greater emphasis on helping this type of information function, as recommended in a report on US AID some years ago (NAS, 1972).

## LITERATURE SERVICES

THE SECOND PRIORITY AREA is to make primary current literature, mainly in scientific journals, more easily and quickly available to users throughout developing countries. If there are one or two good, large collections of journals in a country, it is appropriate to use photocopying technology to spread the material around to those who need it. The literature services established in some countries in East Africa and Asia go a long way to achieving this (Cooney, 1968, 1974a, 1978b; Munn, 1973; Schlie, 1978).

The service operates as follows. Working from a good library with a heavy duty photocopier, it supplies users at locations around the country with copies of contents pages of every issue of any available journals they ask for. Copies of articles can then be requested after the user, who may be hundreds of miles away, has studied the contents pages. Users may ask for articles from current journals, or look back over contents pages of older issues for earlier material. This type of service is relatively inexpensive, and has proved extremely popular wherever it has been installed. Among its advantages are these: (a) It operates in parallel, not sequentially, i.e., all users who require a particular article receive it simultaneously (thus eliminating delays inherent in circulation systems). (b) It provides full texts, not a summary, of the original documents. Circulation systems do this, but by no means as quickly. (c) Operating it calls for the minimum of high level bibliographic skills and no computer skills at all. (d) It accommodates a large number of users as easily as a small number (whereas a circulation system tends to break down when the numbers served grow large). (e) It copies all the material required from current journals at one time, so reducing the time involved. (f) It provides a convenient mechanism (the contents page) enabling the user to select material of interest and to ignore the rest. (g) It takes the initiative in reminding the scientist of the flow of new information.

The first literature service of this kind was established in East Africa in 1967 to serve over 100 centres in Kenya, Uganda and Tanzania. The method described was used from the outset and has remained unchanged. At the start there were some 900 journals, and during 1968 nearly 200,000 pages were copied. An evaluation was carried out in 1969 (EAAFRO 1969), following which the East African Community took over the service from the Rockefeller Foundation. It continued until 1976, when for political reasons the Community ceased to exist. Since then the service has been confined to Kenya, where it operates from the library of the Kenya Agricultural Research Institute. In 1980 the service supplied users at 77 centres in Kenya with 17,000 contents pages and 11,000 articles. Over 20 new centres were seeking enrollment (Njorage, 1981).

In 1974, I was invited to help set up a similar service for the Philippine Council for Agricultural Research and Resources. The service was, by the end of that year, servicing 52 research and education centres involving about 2,000 scientists, using a collection of over 1,100 journals at the agricultural library of the University of the Philippines at Los Banos. Interest was expressed in developing similar services in Malaysia, Thailand and Indonesia.

An agreement between the Irish and Tanzanian governments led to a new service at the University of Dar es Salaam agricultural library in Morogoro (Cooney, 1978b).

It would seem that many other countries in the Third World would benefit from such services, which have not, however, received the attention from aid agencies that their proven popularity and effectiveness deserves.

## COMPUTER BASED SYSTEMS

THE DEVELOPMENT of gatekeeping networks, and literature services, should, it is suggested, rate a higher priority in developing countries than computerized information systems. The great advantage of computerised systems using data bases is that they cover a very large proportion of the world's literature, and they can be searched quickly through an online interactive terminal.

However, online services are expensive to use, especially in developing countries. They are also heavy users of foreign currency. They provide only references and in some cases abstracts: obtaining the original articles requires further effort and expense, mostly involving foreign currency. And there are technical problems to be overcome as well.

The main technical problems affecting computerised services in the Third World are concerned with electricity and telecommunications. Computing equipment requires reliable supplies of electricity: blackouts and severe voltage fluctuations make the equipment unusable. The quality of telecommunication lines needed for on-line transmission is higher than for normal speech. Dialling up on an international network is almost indispensable. These facilities are rarely available in Third World countries at present, especially in the case of centres remote from main urban areas (Cooney, 1978a).

For these reasons, it is suggested, the priority to be given to computer information services in these countries should be considerably lower than for gatekeeping and photocopying services.

## ECONOMIC GROWTH VERSUS PROGRESS

I NOW TURN to consider briefly the second area where re-assessment of the role of information in development is needed. This arises from the conclusion that progress is not the same as economic growth, and that these two processes are affected in different ways by the flow of new information. This question can only be touched on here (it is the subject of a current research project under a contract from the European Community) (Cooney, 1981).

It is widely believed that technical innovation, based on new information, promotes increased "productivity" and economic development, and that this in turn promotes social progress by generating increased monetary wealth, more jobs, and more government revenue for social and development policies.

Much recent evidence throws doubt on this model, including for instance the recent slowdown in monetary growth in rich countries, together with rising unemployment and damage to the environment. These trends seem likely to affect the Third World both directly and indirectly in the near future. It seems highly likely that the increasingly rapid flow of new technical information, and of innovation, plays an important but until now little recognised role in these unexpected and unwelcome trends (as well as in the welcome and progressive ones such as economic growth). If this is true, the prevailing development policies in Third World countries, based on the western model as a rule, and current aid programmes, can no longer be regarded with confidence as the right road for these countries to take (Cooney, 1980).

The Green Revolution has been, up to recently, an example of how changing technology and the money economy combine to frustrate some of the objectives of aid for development. The Green Revolution was a great scientific advance and it fed, and continues to feed, millions who would otherwise starve. But it also causes dislocations in the economy and in society which social scientists did not anticipate. The new crops cannot be raised successfully without costly chemicals, money is borrowed to buy these, the bumper crops cause prices to fall and not enough money is earned to repay the bank loans. Economically people are worse off than before because they have become victims of the money economy. Currently, efforts are being made to overcome these problems by developing crop varieties which have high yields but do not require investment in chemicals and machinery—a return to self sufficiency at a higher level of physical productivity.

By leading LDCs along the road towards involvement in the money economy, richer nations are confronting them with problems. Currently the most basic of these is the inexorable process of technical changes, forced by competition which is at the same time undermining the economic basis of our own rich economies by replacing workers with machines.

## References

Allen, T.J. (1966) *Managing the flow of scientific and technological information.* Ph.D. dissertation, Sloan School of Management, Massachusetts Institute of Technology, 1966, R & D Management **5**.

Allen, T.J., Piepmeier, J.M. and Cooney, S. (1971) *The international technological gatekeeper.* Technol. Rev. **17**, 37–44.

Anderla, G. (1973) *Information in 1985: A forecasting study.* Paris. OECD.

Cooney, S. (1968) *The East African Scientific Literature Service.* Dublin, An Foras Taluntais.

Cooney, S. and Allen, T.J. (1974a) *The technological gatekeeper and policies for national and international transfer of information.* R & D Management **5**: 1.

Cooney, S. (1974b) *Criteria for a user-oriented cost-effective information service. Special Libraries,* Vol. 65, No. 12.

Cooney, S. (1978a) *Potential role of online information services in developing countries.* Lecture to SDP students, Trinity College, Dublin.

Cooney, S. (1978b) *Proposals for a Tanzanian Literature Service.* Report to the Irish and Tanzania governments.

Cooney, S. (1980) *Aid—Motivation and objectives.* Comhlamh, Dublin.

Cooney, S. (1981) *Productivity and progress—a review.* EEC/FAST Programme.

EAAFRO (1969) *Report on the East African Literature Service.* Muguga, Kenya, East African Community, East African Agriculture and Forestry Research Organisation.

Munn, R.F. (1973) *The East African Literature Service.* UNESCO Bull. Libs. XXVIII, No. 1.

NAS (1972) *Scientific and Technical Information for Developing Countries.* National Academy of Sciences, Washington, D.C.

Njoragc, D.L. (1981) Personal communication.

OECD (1971) *Information for a changing society.* Paris.

Robertson, A.B., Achilladelis, B. and Jervis, P. (1972) *Success and failure in industrial innovation.* (Project SAPPHO). University of Sussex.

Schlie, T.W. (1973) *Some aspects of regional-national scientific relationships in East Africa.* Ph.D. disseration, North-Western University, Evanston, IL.

Zipf, G.K. (1949) *Human behaviour and the principle of least effort.* Addison Wesley, Cambridge, MA.

# Marketing Technological Information

*Kjeld Klintøe*
*Dansk Teknisk Oplysningstjeneste, Copenhagen, Denmark*

TO ENSURE THAT INFORMATION is identified, is transferred and used for furthering industrial development is a problem which has occupied numerous individuals, associations, committees, institutions and authorities for years in industrialised countries as well as in less developed countries. Marvellous equipment has been developed and sophisticated systems designed to identify specific information hidden in large and small data banks. Time and again we have heard that this is how you can find it, this is how you should do it. If only men were rational and systematic and behaved according to the theoretical or scientific model of man as an intelligent and well educated human being, the transfer problem would have been solved many years ago.

However, the majority of men and women in industrialised countries as well as in developing countries are not made that way. They are emotionally influenced practitioners operating at various levels with different tasks, involved with day-to-day problems, with improvements and innovation, with making progress and achieving practical results, which eventually will be appreciated and reported as industrial development. They are more occupied by doing this than they are interested in information technology—even in books and bibliographics.

Their genius as entrepreneurial professionals would be killed, if we forced upon them systems and equipment they do not desire.

To apply even effective information technology demands a kind of disciplining opposite to their way of optimal functioning and their sense of being rational.

They want technological information, i.e. knowledge, not bibliographical references. They need assistance and service from intermediaries who are temperamentally qualified, flexible and motivated to adapt themselves to the processors' and decision-makers' world; imaginative enough to identify sources, clever enough to apply primitive or sophisticated information technology directly relevant to the problem area of concern to the entrepreneurs, industrialists and planners.

Technological information is what makes progress 'tick'.

As defined by the FID/II Committee on Information for Industry, 'technological information' is *Knowledge* (of any kind, in any form)—technical, economic, marketing, management, social etc., which by its application will further improvements and innovation.

This makes knowledge a commodity—the intellectual raw material—which has to be marketed, and evaluated for those who want progress, but who, for various reasons, have difficulty in identifying what they need to know, finding it, evaluating what is most relevant, and applying it.

Science and knowledge are of community value only by application—only knowledge (technological information) relevant and appropriate to the need and to the level of understanding of the recipient will be applied. Therefore we must know how to identify potential users, how to estimate their needs, assist them in translating needs into demands, transform demands into wants, and further them into requests; and by processing these steps prepare the potential user for perceptivity towards application and conversion of technological information into practical results.

It is my understanding and firm belief that the concept of an Information Service for Industry providing pragmatic assistance is most important in furthering industrial development and socio-economic evolution.

The names for these services are numerous: extension services, advisory services, information services, information counselling, market/technology intelligence service etc. They are all primarily oriented towards stimulating, serving and assisting potential users on their individual conditions and terms—they do not bother them with information technology, they serve them; users do not want a bibliography, they want to know.

The main steps in organising such an Information Service for Industry are:

(1) Identify the locality to be served.

(2) Carry out an industrial demographic analysis by activity, i.e. products manufactured, services offered and level of sophistication.

(3) Carry out a programme of visits to interview the management on technical and commercial operations and plans for the future, and where their problem areas lie.

(4) Discover if they are familiar with local, functional specialists or centres of specialised knowledge, with other centres in the country, in the region or internationally.

(5) Discover if they use them, benefit them or if there are barriers to communication.

(6) Assist in formulating requests for technological information for a problem area, and the form in which the client wishes to receive relevant and appropriate information.

(7) Offer to procure and repackage the information.

(8) Scan the infrastructure of possible centres of specialised knowledge within the local, national, foreign or international community.
(9) Apply the most effective 'information technology' i.e. personal calls, telephone calls, letters, telex, telecommunication with data banks (bibliographical or factual), etc. for procuring the information.
(10) Evaluate the few pieces of material in which can be identified the requested, technological information and repackage, i.e. rewrite abstracts from documents or write guidelines on how to evaluate documents that will match the user's level of understanding.
(11) Present the user with the material and assist him in applying the identified technological information.
(12) Leave him to his experiments, but never leave him alone in the desert—follow up, follow up and follow up, until application has taken place.

That is how to market technological information as a contribution to the growth of competence which is essential before improvements, innovation and transfer of technology can take place with success—before we can speak about industrial development.

To sum up: "Your decisions are only as good as your information" and
"Your information is only as good as the integrity of your source."

# Multinationals, Intelligence and Development

*S. Dedijer*
*Lund University, Sweden*

## AILING, OBSOLESCENT CONCEPTS

LIKE THE PROVERBIAL mule-driver, the current world development crisis—to which no country in the world is immune—is clouting everyone over the head to attract our attention to the following facts of life in the 1980s:

> Only those social systems—nations, governments, enterprises, organisations—that concentrate on developing their own organised, collective capability to identify and cope with their internal and external problems will survive and grow.
>
> To develop such capability each of them must re-examine critically the concepts, beliefs, models, approaches stored in their brains for collecting, processing, evaluating, using or withholding information in dealing with their problems.
>
> Each of them must look ahead in order to identify and cope with their rising problems before they become failures, crises, disasters or catastrophies.
>
> Their citizens must be encouraged to have the psychological and social courage to re-examine critically their own concepts, beliefs and those of others for obsolescence and ineffectiveness to describe the rapidly changing world.

To start I suggest that among ailing, obsolescent, inadequate concepts for LDC's development one will find management information systems, scientific and technical information systems, comprehensive information systems, and even informatics.

First, these concepts are being rejected as "organising principles" in the developed countries. Starting in the 1960s many managers of enterprises saw in computerised management information systems the miracle tool to solve their problems. By the 1970s, as Russell Ackoff pointed out in his famous article 'Management Misinformation Systems', there arose a much more realistic appreciation of the limits of such systems. On a national scale there was an extremely strong effort between the 1970s and 1980 to build scientific and technological information systems (STI) as a "miracle tool" for innovativeness. International bodies like UNESCO, and the UN Conference on Science and Technology for Development in 1979 devoted considerable effort and attention to STI urging the LDCs to establish such systems as essential for national growth. In 1978 the American Library Association, in its report "Into the Information Age—A Perspective for Federal Action on Information", was pointing out that the concept of STI was "an ailing and possibly obsolescent concept" and was proving "incomplete as an organising principle". The Association urged that the "problem identification and solving" approach be adopted as an organising principle for the use of information technology. The same can be said of the comprehensive information system approach advocated by some United Nations organisations, as for example, the United Nations Center on Transnational Corporations, as the basic tool for the developing countries in dealing with the MNCs.

Second, it is my impression that all the above concepts are advertisements aimed at selling information industry technology to the LDCs. The informatics industry salesmen and interest groups in LDCs make us believe that to acquire overnight the necessary problem-identifying and solving capability it is necessary that enterprises in LDCs should acquire the management information systems, governments the STI systems and the whole country the informatics technology and its software.

The latest critical assessment of the problem of effectiveness of the informatics systems is the book "Decision Support Systems: issues and challenges", (G. Frick and R.H. Sprague, editors, 1980), the proceedings of a symposium on the subject at the International Institute for Applied Systems Analysis in Vienna. In the introduction the editors point out "... early attempts to 'throw technology at the problem' have revealed that such a direct approach has serious limitations. In fact, it appears that significant integration of advances in both technology and a set of related disciplines will be required, in order to apply information technology intelligently and effectively to the class of unmet problems and needs facing decision makers in organisations."

Looking ahead for such ailing and obsolescent concepts to the future we must ask:

> What is this integrative, interdisciplinary concept required to describe the organised capability of government agencies, enterprises, organisations and of whole nations to identify and solve their internal and external problems in a rapidly changing world?

## A Harmful Belief

Before defining such a concept, it is necessary to consider one ailing and obsolete belief prevalent in the LDCs about multinational corporations. In preparing this paper I was amazed how ingrained in the United Nations systems, agencies and votes was the belief that 'multinationals are devils' which must be exorcised on every occasion with proclamations, resolutions and ideological incantations. Such a belief is counterproductive: it encourages the continued exploitation of the LDCs by the MNCs.

What is an effective way for LDGs to regard multinationals?. Multinational corporations, with headquarters in one country and one or more production subsidiaries or affiliates in one or more foreign countries, are one of the most important and powerful world wide social innovations invented by man in this century. It is the latest manifestation of Karl Marx's insight about 19th century capitalism: "Science and profit motive drives capitalism: forward! forward!"

Most MNCs originate in the developed, i.e. the OECD countries, but they exist and operate in every single country in the world, with the possible exception of Albania. By 1973, according to UN information, COMECON states had signed over 600 contracts with the MNCs. A large number of them have contracts with P.R. China and Yugoslavia. According to UNIDO more than 1000 firms from the developing countries such as Argentina, Brazil, India, S. Korea, Hong Kong, Mexico, etc. are operating in several countries abroad, and thus fill the definition of MNC.

An elementary statistical and analytical survey would show that at present the MNCs play an important role in world finance, in the production of goods, processes and services, in world trade, in the creation and transfer of new pioneering world technology based on the latest scientific concepts. The MNCs will play as crucial a role in the new economic, technological and social world order as they did in the old one.

The MNCs have been operating—some times for decades—in individual LDCs. They have contributed to such countries all the inputs considered essential in development planning in socialist countries: foreign investment, learning and know-how, technology, training, improvement of import–export balances etc. Yet a survey shows that in countries with dozens of MNCs operating for decades the development plans contain no mention of the possible roles of MNCs in development. Such is the harmful effect of the "devil" belief about the MNCs: it prevents the LDC's elites from asking the fundamental question: How can we utilise most effectively with least harm to ourselves the MNCs for our development?

## An Organising Principle for the 1980s—Social Intelligence

To turn from ailing concepts and beliefs to the future oriented ones as formulated in the above questions, it is essential to perceive: the capability to identify and solve problems in psychology, or in brain, computer, management or policy sciences requires intelligence.

Walter Lippman was the first to observe the growth of "organised intelligence" in 1922 as the result of growth and distribution of knowledge in society. John Dewey saw in "Organised Intelligence" in the 1930s the tool for dealing with accelerating world change. We are now witnessing the rise of social intelligence, i.e. the growth of capability of government agencies, enterprises, organisations, association, nations and their citizens to identify and solve their internal and external problems.

In the social intelligence approach, every social system is considered as a *persona* and an individual. A human social system has its own personality with specific aspirations, goals, intentions, traits, culture, problems and a more or less developed intelligence organ performing its intelligence function. The intelligence of each social system is the more or less well co-ordinated intelligence of its component subsystems, which are as often in conflict and competition with each other in pursuing their specific goals as in co-operation in the pursuit of the common goals.

The three basic components of social intelligence at this stage of its development are:

BI: biological intelligence, i.e., the brain and its software (self, mind).

AI: artificial intelligence, i.e., all man-made hardware and software (technology, social innovations, knowledge) performing or enhancing the biological intelligence functions or parts thereof.

GI: governmental intelligence, i.e., the capability of decision-making subsystems of a social system learnt through experience and formulated intelligence doctrines helping it to adapt to internal and external changes.

My current search for a theory of social intelligence points to the following propositions:

First: the existence of equipotential functions,* that is identical functions performed by different systems and by different processes, as illustrated by the following analogous examples:

| *Function A*: Travel under water | *Function B*: Capability to identify and solve problems, i.e. intelligence |
|---|---|
| Performed by Systems: | Performed by Systems: |
| $a_1$ man diver | $b_1$ Biological Intelligence |
| $a_2$ shark | $b_2$ Artificial Intelligence |
| $a_3$ submarine | $b_3$ Governmental Intelligence |

* I believe I owe the idea of "equipotential function" to Prof. Donald Michie of the Machine Intelligence Research Unit, Edinburgh University.

At present, what I call, the Hubel-Martino Reductionist axiom is valid for the BI, AI, and GI components of the social intelligence:

The human brain, the computer and the social system are "machines" that process information and all three work with signals that are roughly speaking electrical.

The development of organised, collective, social intelligence is being generated at present by

- increased complexity and integration tendencies within the world system and its rapid changes,
- the extremely rapid growth of knowledge or information industries based on the application of science,
- the growing distribution of knowledge and distribution of power within each social system and globally giving rise to increasingly distributed intelligence systems,
- the growth by means of research and innovation of integrated knowledge of biological intelligence, artificial intelligence and governmental intelligence.

At the micro level of a social system its intelligence function will tend to grow with increasing awareness of the threats and opportunities for growth generated by its environment, by its capability to co-ordinate its internal goals and conflicts, by the awareness of its internal complexity and by the belief of its members in rationality as an intelligence component.

It may be conjectured that all these factors are stimulating the growth of social intelligence as an emerging phase in human evolution in general and of human intelligence evolution in particular.

## INTELLIGENCE—THE ESSENTIAL TOOL FOR DEALING WITH THE MNCs

USING THE SOCIAL intelligence approach rather than comprehensive information systems and others like them as the organising principle for our problem we start by asking:

What are the basic problems of the less developed countries (and as a matter of fact all countries) in dealing with the foreign multinationals?

How can these problems be expressed in terms of threat-problems and opportunity for development problems to be dealt with by the existing problem identifying and solving capability of an LDC?

To deal effectively with these problems one must ask the complementary questions:

What threat and opportunity problems do the MNCs perceive when engaging in activities in the less developed countries?

What is the MNCs capability to identify and deal with such problems and how does it compare with the similar capability of the LDCs?

These four questions are closely related in all phases of interaction between an MNC and a LDC. The following table lists some of these threat-opportunity problems both for the LDCs and MNCs in their mutual interactions:

| LDC goal: Development | | MNC goal: Profit, Expansion | |
|---|---|---|---|
| Threat to LDC[a] | Opportunity for LDC[b] | Threat to MNC | Opportunity for MNC |
| Foreign ownership | Foreign investment | Political instability | Market |
| Power plays | Learning, know-how | Nationalisation | Cheap labour |
| Flexibility (transfer pricing) | Technology acquisition | Terrorism | Tax incentives |
| Powerless trade unions | Creation of jobs | Lack of work habits | Raw materials |
| Restrictive practices | Ancillary industries development | Agressive corruptibility | |
| etc. | etc. | etc. | |

[a]The list of threats to LDCs from the MNCs is taken from a publication by Shell Company in Sweden: "Multinationell Företagsamhet", Tekniska Komersiella Ekonomiska Fakta, No. 17, 1973.

[b] The list of opportunities for LDCs by MNCs is adapted from *Nestle in the Developing Countries*, Nestle Alementana S.A., Vevey 1975.

There are several phases of interaction between a less developed country and a multinational corporation. In each of these phases the interacting actors use or should use their available capability to minimise the threats, and increase the opportunities the interaction offers. How large is this capability to minimise threats and increase opportunities available to the MNC as compared with the less developed country? The power of the MNCs has several dimensions: purely economic (financial strength, marketing capability etc.) and technological (the ability to translate the latest scientific results into new, high-use-value products, processes and services).

The most important dimension of an MNC's power, however, is its capability to adapt to new social, political, economic, technological environments. This capability consists in the accumulated expertise of its personnel who can:

– procure long range, future oriented basic and operational intelligence knowledge, that is, action oriented, timely, problem identification and solving knowledge of the threats to their survival and the opportunities for their growth.

– protect their own essential secrets and penetrate the secrets of others: competitors, governments, new social environments—all in the line of problem solving.

In Table 1, one finds the phases of interaction processes between an LDC and an MNC as described in literature, with an estimate of their respective intelligence knowledge capabilities.

In the "Estimated Intelligence" column we have marked the portion that information technology as an intelligence resource plays in the total organised capability. The rest of this capability consists of such crucial intelligence resources as secrecy and its management, strategic and tactical research, personnel trained to acquire intelligence, intelligence doctrine based on acquired experience, guiding their intelligence know-how.

In each phase of interaction both actors utilise their more or less well organised intelligence to optimise their opportunities and minimise the threats involved by using the available intelligence resources, of which informatics, or the artificial intelligence component is only a small part.

The difference, the gap between the estimated intelligence capability of the LDC and the MNC can be defined as the extent of LDC's ignorance in relation to the MNC. If taking advantage of an unorganised, ignorant opponent in the process of choosing him, in negotiation with him, in monitoring his activities and in readjusting the relationship with him is exploitation of the opponent, then the LDCs are being exploited by the MNCs. To decrease the advantages the MNCs have it is not sufficient to concentrate on informatics technology. To decrease this advantage the LDCs must develop creatively and innovatively their own intelligence capability.

## HOW DO 'THEY' DO IT?

IT IS SUGGESTED that a less developed country can speed up the development of its intelligence function about the MNCs and its national intelligence function in general, by initiating a study of how another country, Sweden for example, has developed its intelligence about the MNCs, and how the MNCs develop their own, and digesting and applying the results of such a survey.

An elementary search of the Lund University library catalogue on Swedish studies on MNCs operating in Sweden produced several dozen books and papers published between 1974 and 1980, amounting to over 2000 pages packed with facts, data, statistics, case studies, policy proposals etc. From these studies one can draw—among other—the following observations:

First, the interest groups and organisations actively involved are diverse—i.e. government ministries and agencies participating in policy-making and implementation, negotiation, economic and social planning, foreign relations and monitoring of economic activities; government and university study centres; legislative bodies; managers of MNCs' firms and their competitor firms, trade unions, association of employers, media, critics, etc. No single organisation or institution has the monopoly of concern about the MNCs or "responsibility" for them.

**Table 1**

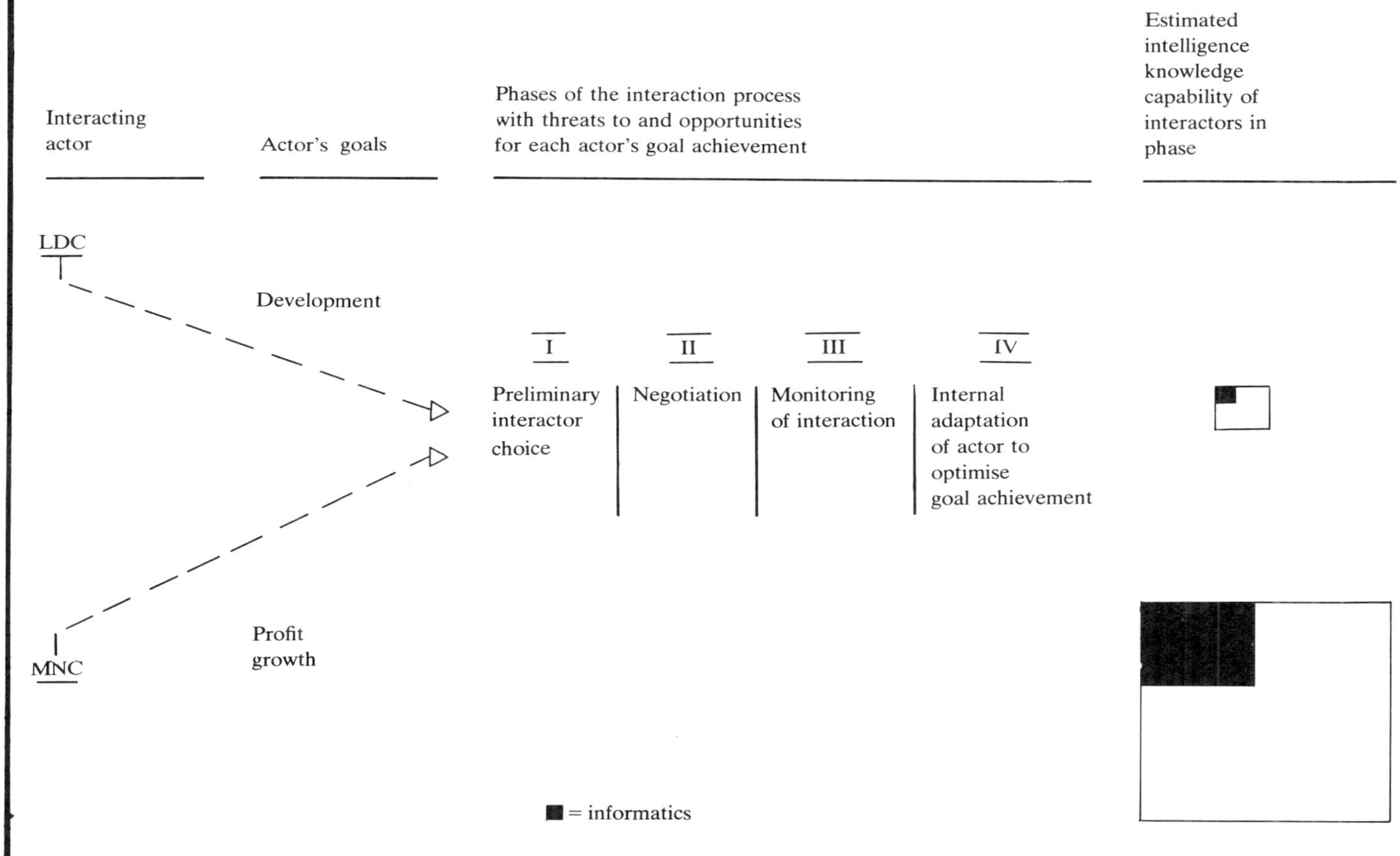
Interacting actor
Actor's goals
Phases of the interaction process with threats to and opportunities for each actor's goal achievement
Estimated intelligence knowledge capability of interactors in phase
LDC
Development
MNC
Profit growth
I
II
III
IV
Preliminary interactor choice
Negotiation
Monitoring of interaction
Internal adaptation of actor to optimise goal achievement
■ = informatics

Second, dozens of individuals participated in preparing such studies. In this they learned the skill of acquiring and analysing data about individual MNCs and of significant cases of threats or opportunities they offered to Sweden. Numerous banks, enterprises, government departments, universities in Sweden employ individuals with current knowledge about the MNCs. All these individuals of various ideological commitments can be said to constitute an 'MNC's Intelligence Community'—they know each other, and often communicate with each other seeking help and giving it when a particular case arises.

Third, most of the members of this informal 'MNC's Intelligence Community' in Sweden are dissatisfied with the state of knowledge about the foreign MNCs operating in their country. At present there are several dozen projects underway in universities, government agencies, trade unions aimed at improving Swedish knowledge of MNCs.

This knowledge network on MNCs is only one informal network of intelligence making part of the total "nervous system" whereby Sweden learns about the changes in the world outside and in itself. As Björn Tell said in a recent paper: "Sweden is one of the most open information societies in the world. The planning function in Sweden is widespread and the planners try to get information from a variety of sources. The prevalence of team organisations, research groups, working committees, participatory democracy in state boards and in the management of industrial firms, assures an opportunity for many to air different views. The interacting individuals reinforce each other's creativity and productivity upon which decisions can then be based."

Thus Sweden can be taken as a good example of a country where the production and distribution of knowledge and the consequent distribution of power is leading to a national system of distributed intelligence. This is an example of a general trend which no LDC country wanting to industrialise can avoid. The sooner a country identifies the barriers and constraints from the past to the development of such a national intelligence system consisting of distributed intelligence centres throughout the society, the quicker will it be able to deal with the MNCs effectively.

### The MNCs Own Intelligence Capability

In spite of an intense search I could not find any literature showing that researchers in LDCs and in United Nations agencies dealing with the MNCs have asked themselves the following question:

> How do the MNCs develop their own intelligence capability about their environment—including their competitors, and the LDCs with which they do or plan to do business?

I have found that many researchers and decision makers from LDCs assume that there is no answer to such a question because of the "extreme secrecy of multinationals". In reality those who are not blinded by this idea will find a whole series of open sources of materials answering the question. Here I shall only list six sources of information about MNCs intelligence capability which I found in my library:

(1) *Business Intelligence bibliographics.* Example: "Business Intelligence and Strategic Planning", L.M. Danies, Graduate School of Business Administration, Harvard University, 1979. The bibliography contains references to hundreds of articles on competitor intelligence, anticipatory intelligence and about intelligence in the MNCs.

(2) *Current articles about MNCs' intelligence.* Example: L. Kraar's article in Fortune, March 1980, "The Multinationals Get Smarter about Political Risks".

(3) *Consulting firms.* Examples: Stanford Research Institute Business Intelligence Center and the Economist Intelligence Unit offer their customers special, confidentially prepared studies about competitors and MNCs. The former even teaches how to develop such intelligence.

(4) *MNCs' publications for their employees.* Example: D.E. Noble, "Ignorance is the Limiting Factor", editorial in No. 1 vol. 1, of *Correlations*, an Engineering bulletin from Motorola Inc. 1978. The article discusses the importance of intelligence awareness by Motorola engineers and how it can be improved.

(5) *Government archives* in many countries, for example in the USA, contain enormous quantities of information about the MNCs which they (MNCs) are obliged to provide by law. In its negotiation with the bauxite and aluminium multinationals in 1971–72, Jamaica used these archives with great benefit.

(6) *Conference proceedings.* Example: on June 9–11, 1980 the OECD Development Center in Paris held a conference on "Knowledge Industry and the Process of Development" dealing with intelligence for development. Among participants were representatives from such companies as IBM, Nestle, and consulting services like the Economist Intelligence Unit, each of them contributing papers on the subject.

To this might be added another crucial source on intelligence capability of MNCs and how it is developing: interviews with MNC officials. In preparing the paper on which this note is based I had talks with five MNCs' managers on the following subjects:

– the "intelligence culture" of the corporation

– the problems of an executive's personal intelligence (what is his own personal style to identify problems, to obtain timely, reliable knowledge on them in his daily work etc?)

– personnel intelligence (what criteria to apply in choosing personnel in general, personnel operating in the LDCs in particular, and how to sensitise them and train them to fulfill the company's intelligence needs?)

– marketing intelligence about investments in the developing countries.

It was surprising how openly the MNC managers will discuss such questions, provided one does not ask about current operations, and are willing to give some ideas in exchange during the interview.

How to use all these sources to improve the intelligence related to a specific problem of the threat or the opportunity kind a decision maker from an LDC has in relation to a MNC is a creative act for which no recipe can be given, but on which he must learn to exercise "the most powerful weapon—the mind".

## HOW TO DO IT?

A NUMBER OF RULES useful in developing an effective intelligence for interactions with the MNCs can be established.

Assuming that in the strategic development plans the possible contributions of MNCs are assessed, the central task of the intelligence organisation is to devise ways to optimise these contributions and minimise the threats the MNCs represent. The operational plans for such intelligence in various phases of interaction with MNCs should be related creatively to the threat-opportunity problems.

As in the liberation war and anti-colonial struggle, LDCs should start to develop their MNC intelligence using what intelligence resources they have. In the liberation wars the enemy had all the modern weapons which the LDCs did not have but by using what they had developed extremely creatively.

To start with the dream of using the most modern informatics technology is to start on the wrong foot in developing the intelligence necessary to exploit the MNCs in the development process. The organising principle for acquiring informatics technology in relation to MNCs should be that of problem identification and gradual improvement of technology in dealing with them.

It is essential to survey the ignorance about the MNCs in terms of questions the MNC interest groups ask about them, and those they do not ask. The increase of awareness of this ignorance about the MNCs is a difficult but an essential task in improving this intelligence.

The development of intelligence incentives in all interest groups is a continuing task for the LDC decision makers, as is the gradual improvement of intergroup exchanges of intelligence about the MNCs.

One may consider the idea of selecting one young individual to be responsible for basic and operational intelligence about each MNC.

The establishment of a MNCs intelligence coordinating body without having centralised power to deal with them should avoid the practice of a Centre on Multinationals with a formal or informal monopoly of intelligence technology, research capacity, regulation power etc.

The mapping and management of secrecy in each interest group and among groups should be a task for the whole intelligence community on MNCs.

Finally, learn from the experience of others with intelligence, such, for example, as the principle England operated on in the darkest days of World War II: "Intelligence—the weapon of the weak".

## SECTION 7

# Computerisation in Different Industrial Sectors: Surveys and Case Studies of Advanced Applications; Informatics and Small and Medium-Scale Enterprises

Informatic applications in particular industrial sectors and especially in smaller enterprises are considered. The applications discussed range from complex realtime production control to basic use of personal microcomputers in assisting management decision-making, and while the main emphasis is on applications in less developed countries, projects from more advanced environments are also included.

# Application of Informatics to Production Management and Prospective Uses of Microcomputers

*S.V. Deodhar*
*Ballarpur Industries Limited, New Delhi*

## APPLICATION OF INFORMATICS TO PRODUCTION MANAGEMENT

PRODUCTION MANAGEMENT in an industrial environment includes monitoring and controlling the various parameters in a manner that would result in, say, the reduction or elimination of wastages, maximisation of machine throughputs, minimisation of overhead expenses, optimisation of energy consumption, controlling the input costs, inventory control etc. The common factor in all these areas is the need to capture relevant and reliable information and to process this within a timeframe required by the specific application. The application of informatics to production management got a shot in the arm with the advent of microcomputers. Individual microcomputer machines built around the modern microprocessor chips can now be designed so that from the view point of hardware as well as software, each machine is dedicated to meet the requirements of a specific application and hence does this with least cost and maximum efficiency. With the present level of development in this technology it is possible to design such dedicated microcomputers at an attractive price with a high performance. These mini/micro computers are also comparatively easy to understand and operate by shop floor supervisors, who would ultimately determine the utility of these machincs. In this paper, I have tried to present some aspects of the science of production management in general and the interplay of this branch of science with that of computers, in particular. What I have not dealt with except by way of a passing reference, is the basics of Informatics and the design criteria of the dedicated microcomputers referred to earlier. It will be noticed that this presentation has a particular orientation towards an integrated pulp and paper plant. This was inevitable to a certain extent as for the last ten years I have been associated with this industry in India. None-the-less most of what I have to say on this subject applies equally well to other Production Management situations found elsewhere. I have also touched upon certain specific sub-areas of this application as case studies.

## PRODUCTION CONTROL PROCESS

THE HIGHEST EFFICIENCY in production is obtained by manufacturing the required quantity of the end product to the required quality at the required time by the best and cheapest method. Production control essentially comprises basic organisation, pre-planning, checking on material, defining manufacturing processes, allocating production time/output factors, co-ordinating inspection and determining optimum means and routes for despatch and delivery. There are three basic divisions of the production control process namely: planning, tactical immediate control, and corrective information feed back.

Computers are used as tools in meeting objectives in each of these three divisions. The steady reduction in the cost and the improvement in the reliability of microelectronic equipment of all descriptions is bringing the techniques and benefits of computer based calculations into the reach of an ever increasing number of firms in every industry and although not all may agree out of prejudice or ignorance, it is possible to conceive a situation of 'zero waste'. All too often we have seen a miracle of today turning into a commonplace occurrence of tomorrow. One only gets correct answers to one's problems if one learns correctly to formulate and pose the right questions. There are no shortcuts which could enable us to escape that discipline.

## DESIGNING A PLANNING AND CONTROL SYSTEM

WHEN DETERMINING production programmes, it is necessary to consider the availability of men, machines and materials.

Thus production planning and control covers—scheduling, labour control, material control and despatch. It is equally important to consider the effects of poor or non-existent production control. These can include excessive work-in-progress, excessive stocks of materials, poor employee morale, inaccurate delivery dates and low machine utilisation. By designing a planning and control system for a company, management is provided with an organisational tool and prior knowledge of achievable performance is obtained. The task of production control then is to ensure the maximum utilisation of the equipment with management techniques such as forecasting, production smoothing, inventory control and simulation. This problem can be seen at three different levels. Firstly, there is the strategic problem of ensuring that the product is produced at the correct mill in order to maximise the contribution as a whole. The second is that of production smoothing within a mill. Finally there is the problem of ensuring that individual machines are loaded in the most efficient manner at the mill. It will be seen that the time spans involved at these three levels decrease stage by stage. At the first level the time spans involved vary anywhere between one to five years. Allocation of products between mills must obviously take account of the following areas: machine capacities; forecasts of product demand; tariff barriers (if any); transportation costs; and mill production costs, selling prices, etc.

At the second level, the time spans drop to not more than 1 year. Here one tackles issues like—how much to manufacture of standard line items and how much of special individual customer requirements. Then again, either the product can be made for stock in anticipation of orders, or one can wait for sufficient orders for standard lines to accumulate so that a reasonable quantity can be produced in one batch. This can ease the scheduling problems but it does mean that a considerable amount of money may be tied up in finished stock. Forecasting future demands, quantifying economic run lengths, adequate control on stocks are some other factors which must be considered at this stage. The effects of set-up costs and stock holding costs on the total product costs will show a behaviour pattern similar to Figure 1.

At the third level, one deals with local problems like deckling* and ensuring that when customers' orders for standard lines and stock orders are being produced in one batch, all orders are combined together in such a way as to minimise the trim loss. In fact minimisation of trim loss has been one of the most popular computer applications not only in paper mills the world over but also in other places like steel rolling mills, sheet glass factories, plywood factories etc.

Management usually requires two classes of information viz, generalisations prepared at relatively infrequent intervals to help them in the economic assessment of the mill's operations, and particular information prepared at frequent intervals to indicate what areas of the mill's operations require their immediate attention. These would either show a decline in efficiency requiring that action be taken to arrest the decline or show an improvement in efficiency indicating the results of actions previously taken.

Throughout the range of computer systems there should be one philosophy i.e. keep it as simple as possible. The stages suggested to mills contemplating the introduction of production control systems are therefore—

* A paper industry term relating to the edges of a batch of paper.

**Figure 1.**

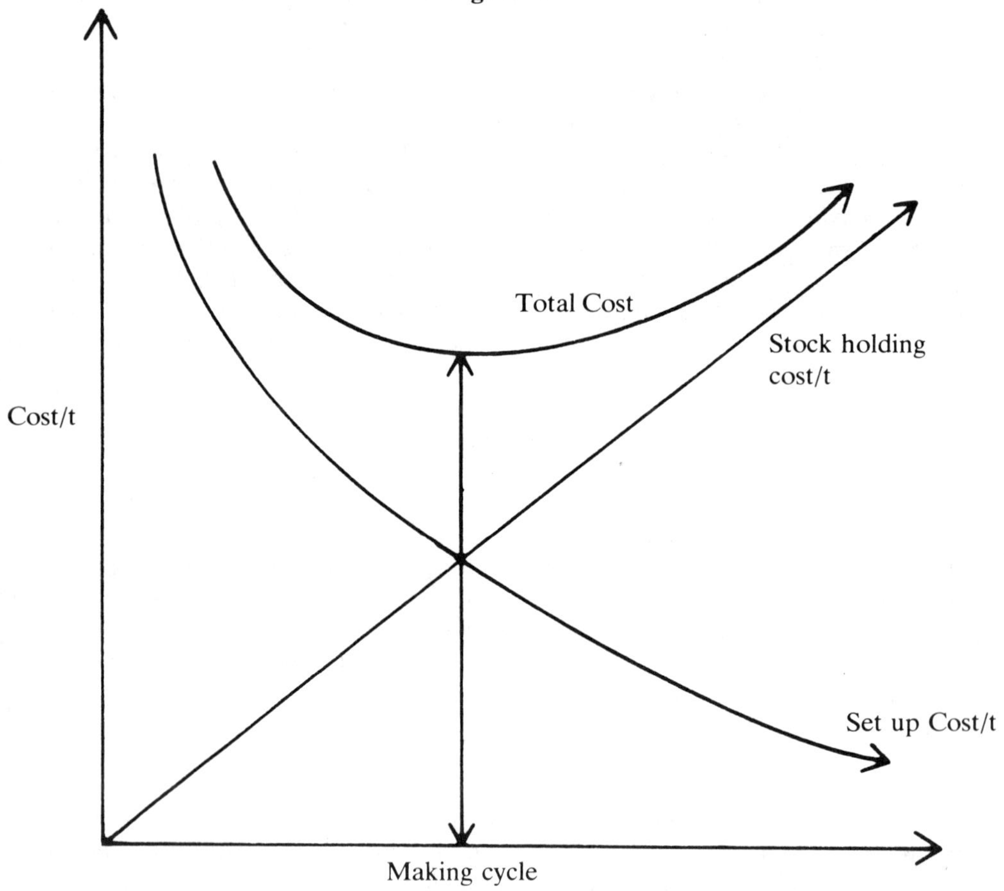


(i) install a simple manual system.
(ii) conduct acceptance trials.
(iii) improve and if possible perform calculation and update, using a computer.

It is felt that once a system can be seen to work it can then be computerised and the information trusted.

# THE SYSTEMS APPROACH

A SYSTEM can be defined as a group of resources, usually men and machines, set up to perform a given function with a definite objective. The production facilities of a firm are one such system and production control is concerned with the operation of this system. The setting of production policy must be related to sales policy as a part of a joint corporate planning exercise. In setting these policies it should be recognised that the actual business received by the firm will deviate from the forecasts. Policies should be determined so that the firm can operate in a robust manner in a fluctuating environment. For instance, if a sudden increase in demand outstrips current production capacity, the firm can either accept all orders allowing its delivery dates to lengthen; or it can discriminate against a certain class of orders so that its delivery performance to other classes of customers is not affected. Again, it is difficult to determine the best policy on such matters, but lack of any policy will almost certainly lead to frustrations and inefficiencies in the operation of the firm. The setting of objectives is the most important part of the systems approach. Too often, one finds systems which have just happened, rather than designed purposefully to do a particular job as effectively as possible. Attempts to improve these systems often get frustrated because there is no objective against which to measure the improvement. The systems approach helps to focus the attention on the importance of setting the objective.

There are basically two ways of providing a given level of delivery service when there is a measure of uncertainty and fluctuations in the demand pattern—

(1) Holding finished goods stock enabling orders to be met without waiting for production.

(2) Providing the required flexibility which reduces the production lead time.

The most difficult part of the sequence of operations described in this section is determining the stock level, excess production capacity or mix of both which would give the same level of service. The specialised computer systems packages based on simulation and optimisation techniques which have been developed by mathematics and operational researchers are of great help at this stage of the exercise. There is no obvious solution because the cost of providing service by stock or surplus capacity are of the same order. For example, taking a paper machine costing Rs.2000/tonne of annual output and material and operating cost of, say, Rs.1600/tonne; in designing a system for 20,000 tonnes per annum throughput, the cost of various alternatives would be:

(1) Increase machine capacity by 10 per cent. The capital required is 20,000 × Rs.2,000 × 10 per cent i.e. Rs.4,000,000. The interest and depreciation burden of this capital would be about Rs.800,000 per annum.

(2) Increase stock holding by 6 weeks' output. The working capital required is

$$20{,}000 \times 6/50 \times \text{Rs.}1{,}600 = 3{,}840{,}000.$$

Capital for warehouse space at Rs.200/tonne of stock held is Rs.480,000. Interest on this capital and depreciation on the warehouse would be around Rs.600,000 per annum.

If we already have extra machine capacity, we may only require labour force increase to give the extra capacity. It is important therefore that all these alternatives are carefully considered.

## CONTROL AND PLANNING DECISIONS

We are still a long way from the day when customers' computers will communicate with the manufacturer's computer which will in turn produce production schedules for each machine and despatch instructions to the storeman. Until that day comes, we will rely on computers to take decisions as per a predetermined logic and on humans to refine these decisions further by use of their higher level skills. The schematic of such a decision process is shown in Figure 2. Here actual performance is compared with required performance and based on the knowledge of system behaviour, control action is taken to bring the actual in line with the required. As an example, the required performance might be a certain capacity for a finishing operation at minimum cost. Control action could involve recruitment, overtime, incentives.

The other decisions will be in the planning systems such as scheduling a batch on a particular machine. With these decisions there is no direct measure of output against which the effectiveness of a particular operational decision can be measured. The mechanism of these decisions is shown in Figure 3. An event occurs such as the receipt of an order and action must be taken to schedule the order based on the policy and knowledge of the system operations. Such a policy might state the minimum size of batch which should be made. The only measure of the effects of the decisions will be the throughput of the system, the specific level provided and the cost over a period of time. Based on this, revisions of policy can be made. Whatever the type of decision, designing a system for efficient operation involves

(1) identifying the decisions which need to be made.
(2) identifying the information required for these decisions (included policies).
(3) deciding who should make the decisions.
(4) providing systems to collect, prepare and distribute the information.
(5) design of decision aids.

If one's critical analysis of the system has been carried out thoroughly, identification of decisions should pose no problems. Conceptually, following a single order through the firm is a useful way of ensuring that no decisions have been overlooked. Placing oneself in the position of the decision maker is a good

**Figure 2. Mechanism of control decisions**

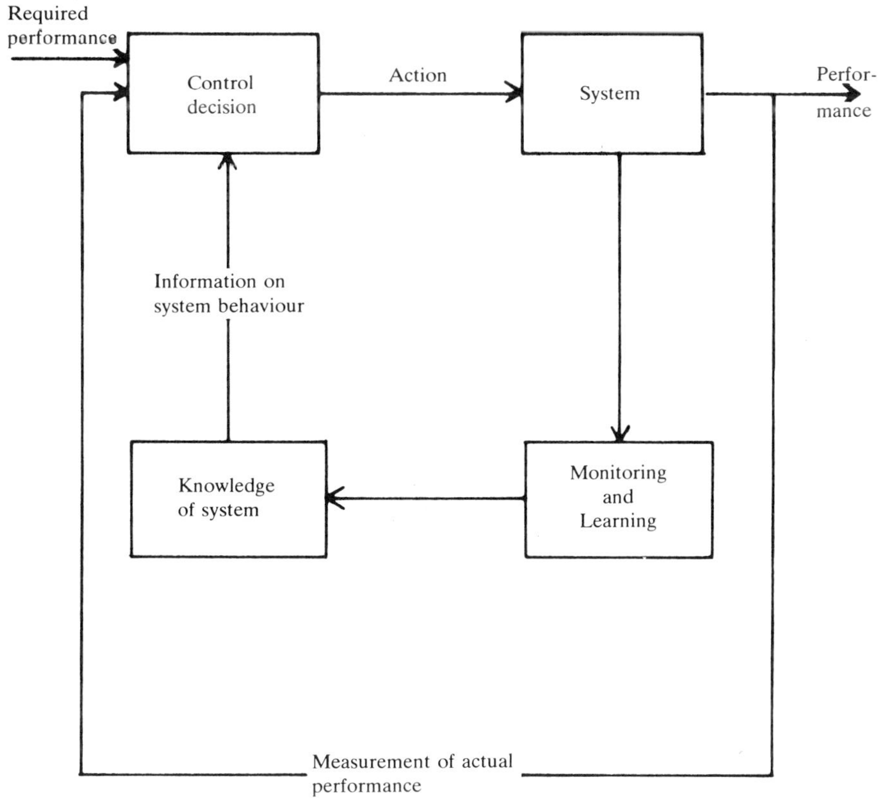


way of determining information needs. The information will include the information on the individual order, information on how the order interacts with others, information on resources (including costs of alternatives), and information on policy. Decisions on who should make the decisions should take account of the desirability of reducing information flow; and, the fact that some information which is required for decision makers can only be obtained informally by close contact, should not be overlooked. This is an area where much still needs to be done. However when the decision and the information analysis is carried out as above, the system to transmit the information can be designed with or without a computer.

To summarise:

(1) Define and analyse the system in relation to the wider system of the company. Particularly attend to:

(a) What is the system under consideration?

**Figure 3. The mechanism of planning decisions**

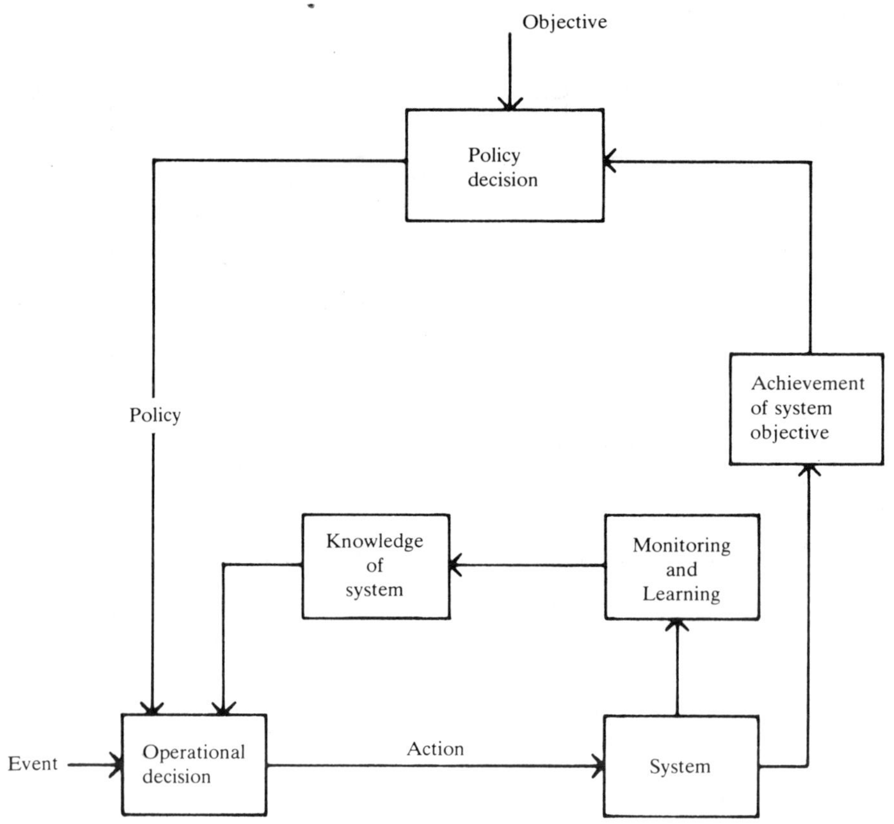


(b) What is the main function of the system and what subsidiary functions are included?

(c) How does it interact with other systems?

(2) Formulate the objectives for the system. Particularly attend to:

(a) What are the main variables which can be used to define the system objectives?

(b) How should policy be set on these objectives in relation to overall company objectives?

(c) What will be the criteria by which we decide on the best system?

(3) Design the basic system structure, covering:

(a) What is the expected environment in which the system will operate?

(b) What alternative systems are there?

(c) What are the economic implications of the alternatives?

(d) What are the control implications of the alternatives?
(e) What is the best system?

(4) Design for operation, including:
(a) What decisions will need to be made to operate the system?
(b) What information is required to make these decisions?
(c) What information systems are required to collect, prepare and present the information required?
(d) What decision aids should be used?

(5) Implement the selected system.

This approach can have a big impact in situations where it has been applied; some of the greatest benefits come from the breaking down of the interdisciplinary barriers which the approach implies, and from the unifying influence this would have on the management. Figures 4 and 5 present the above ideas pictorially.

**Figure 4. Flow-block diagram of the system**

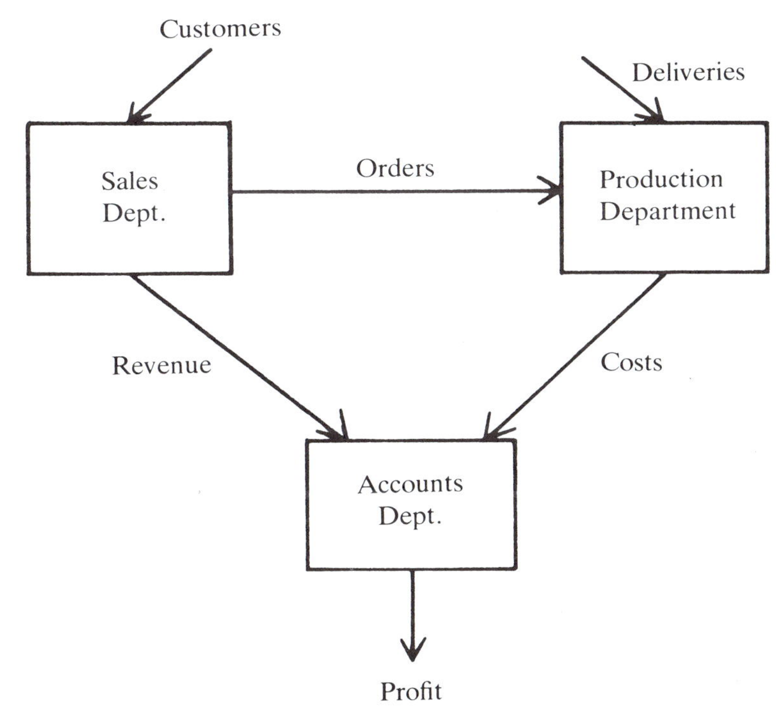

**Figure 5. Balance between the contribution from and cost of a service**

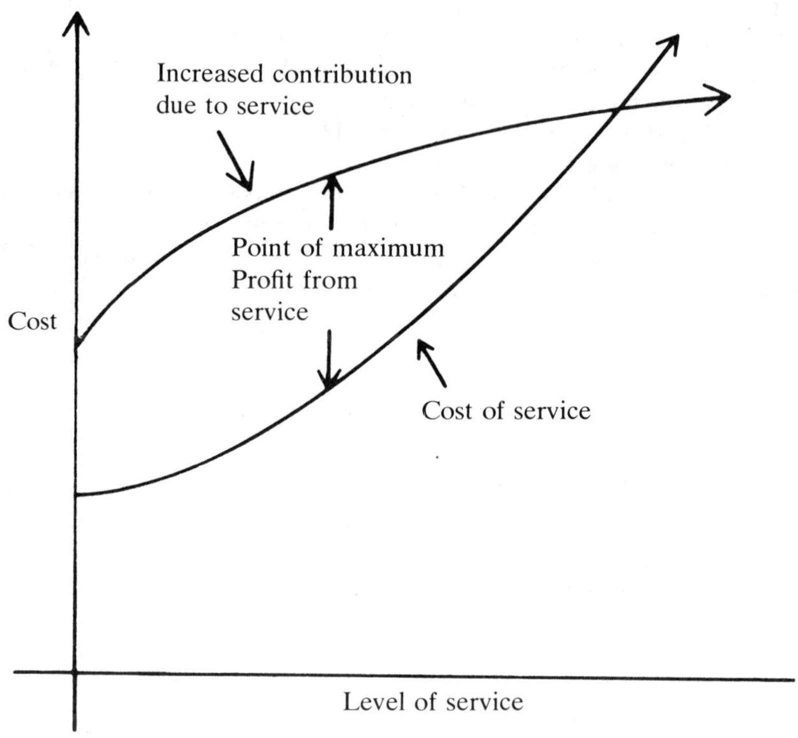


## PROBLEMS OF IMPLEMENTATION

Most manual production control systems suffer from the problem of obtaining sufficient and accurate information quickly enough for effective action, remedial or otherwise. The problem of coordinating this information is also great as many production systems are highly fragmented with different departments working to different priorities frequently being unaware of the overall production requirements. In these circumstances, therefore, the task of deciding precisely what is the correct information can be very difficult. As a cynic once put it—the future of a company is often being determined by the way in which clerks handle customers and orders with no policy to guide them, while the executives try to run the company from meaningless information.

Problems of this nature must be understood before implementing any computer system, otherwise even worse confusion can and will result. However using a computer as a central source of information for the organisation, through establishing master files of data such as production batches, current stocks, machine capacities, etc., has tremendous advantages, not least because of the necessary discipline which it exerts on all departments operating the system. Given that a computer system must provide financial benefit to the company, there are four attributes of a computer which enable this to be achieved. A computer can:

(1) process large volumes of data
(2) with great accuracy
(3) of considerable complexity
(4) at high speeds.

These are usually the areas of greatest weakness in manual production control systems where the sheer size of clerical effort required to provide and coordinate all the information to run the company effectively usually makes such a course of action uneconomic and unpracticable.

It is vitally necessary that a new computer user shall gain experience initially in a safe and inexpensive way. He should not try to implement complicated systems immediately but adopt a step-by-step approach towards comprehensive company systems. However, one should guard against a piecemeal or disjointed approach. The best policy to adopt is to define the long term data processing objectives; for instance the implementation of a comprehensive production control system which minimises stock levels and work-in-progress, reduces throughput times, optimises machine utilisation, and ensures as far as possible the meeting of delivery dates and then working towards this step-by-step, by starting off with a stock control system or an order breakdown system and adding further sections as the previous ones become fully justified. Attempting to put in a fully integrated production control system straightaway without gaining experience with simpler systems first would very probably result in highly expensive failure.

If a very long term data processing application is envisaged in production control, it is probably advisable to go for initial establishment of information for the system e.g. order breakdown, netting and batching, stock control, etc. It is very important for all levels of management to be involved in the implementation of a computer system, no matter how trivial the initial application may seem to be. Particularly in the larger applications, it is necessary for the senior management to make clear policy decisions defining the overall strategies which are to be adopted and for the middle management to be involved at a more detailed level in decisions concerning the design of the system. The principle should never be forgotten that computer systems are run by the line management of the company and not by the computer staff whether they be in-house or at a bureau. The data processing staff are there to provide a service to line management who still have responsibility for ensuring their departmental systems are operating effectively.

## FINANCIAL IMPLICATIONS

FINANCIAL IMPLICATIONS of the production control system can be assessed by first setting out the production budget. Typically the budget figures would relate to a week's performance and are derived through research into the previous 12 months' actual performance on one hand and market requirements for the future duly modified to fit within the broad management policies, on the other. Thus a build up of production is provided through:

(1) Allowable time loss, analysed to component parts.
(2) Available paper making time and loss time.
(3) Deckle, substance and speed.
(4) Breakdown and paper making time expressed as percentages.

These figures must be determined for each machine and each product thereof. The build up of production in this manner facilitates:

(a) The measurement of actual performance against standards with variances analysed by the factors of production.
(b) The presentation of changes in the production pattern.
(c) The final approval of the budget.

Certain other issues also need to be considered when it is intended to use computers to assist in this exercise. For example:

(a) Which paper machine should fill which orders?
(b) What would be the best machine to curtail in a period of over capacity?
(c) What would be the economic effects of undertaking specific rebuilding projects on specific paper machines?
(d) What would be the economic effects of making specific changes in the product-mix?
(e) What type of new machine would best serve future demand?

These and other issues can be dealt with adequately if the computerised system is developed as a multipurpose data bank, extracting the data needed to allocate demand to paper machines by the use of well established and powerful techniques like 'Linear Programming' and 'File request management system'. However, as more working experience is gained, even the non-linear characteristics of certain relationships should be identified and the computer model improved further by employing the technique of 'Simulation'. Typically, the L.P. model structure can be divided into the following main row categories—

(a) Economic measures
(b) Raw material requirements
(c) Paper machine capacities
(d) Miscellaneous constraints
(e) Demands.

The size of the matrix would vary according to the detail. Data requirements would cover—

(a) Material flows—

**Figure 6. The Production System**

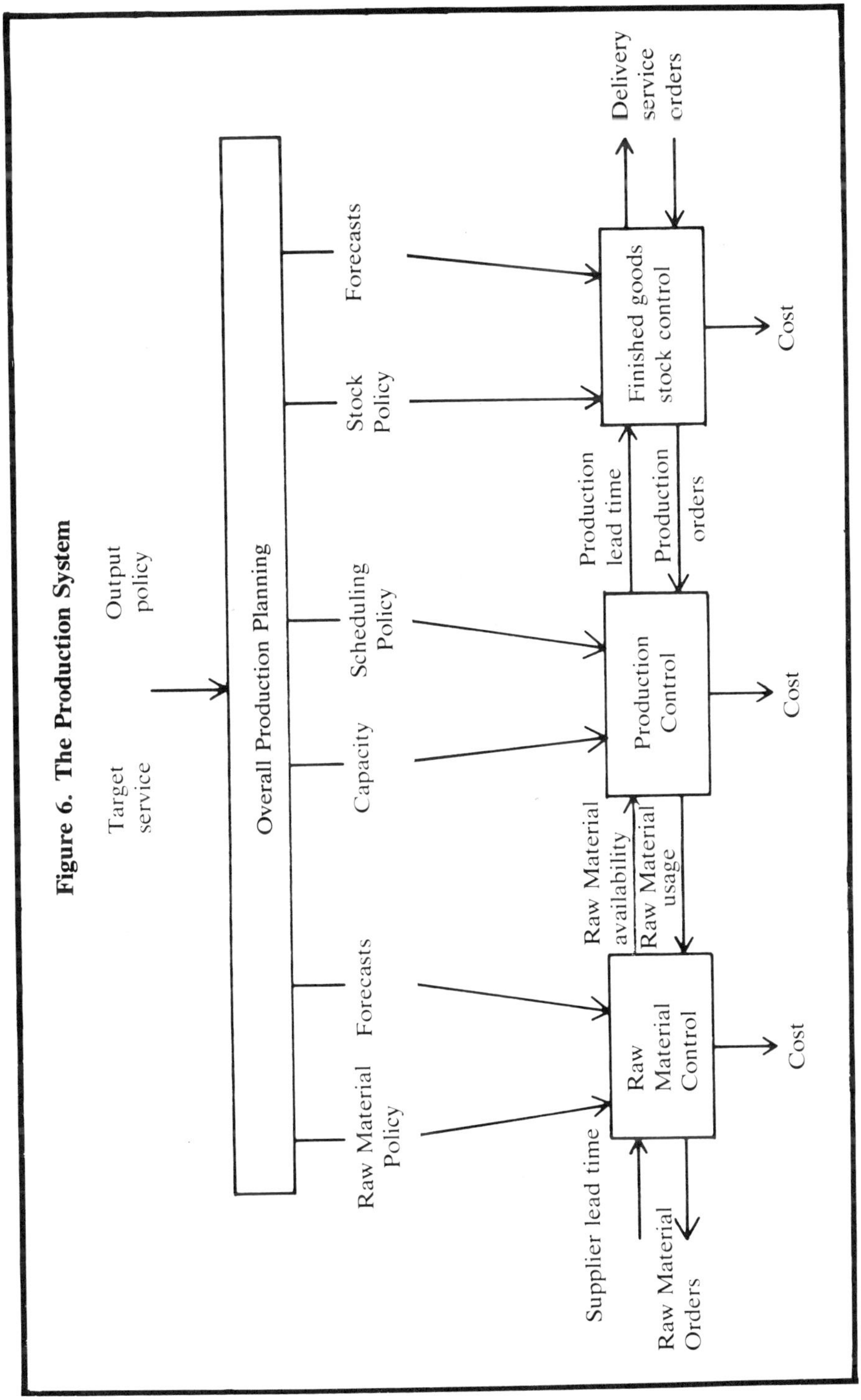

Demand attributes.
—Production descriptions (grade, basis, weight, colour, etc.)
—Tonnage requirements
—Delivery points
Paper machine attributes.
—Raw material availability
—Paper machine capacity
—Location
Demand attributes/Paper machine interaction.
—Production rate
—Raw material requirements.

(b) Cash flow—
Demand attributes.
—Price
—Discounts and rebates
Paper machine attributes.
—Raw material cost
—Chemical additives cost
—Paper machine operating cost

**Figure 7. Interaction of production and other systems**

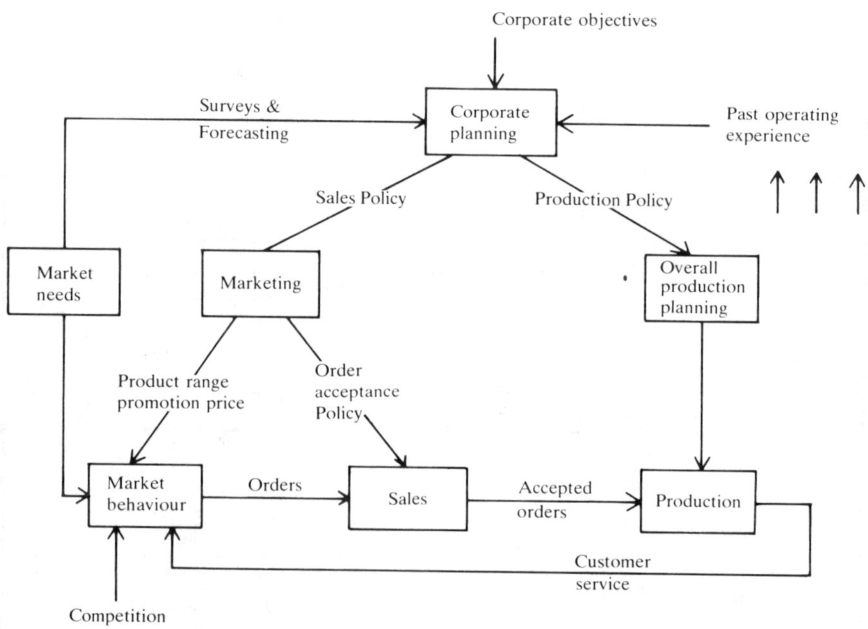

Demand attributes/Paper machine interaction.
—Freight cost
—Some paper machine costs
—Miscellaneous costs like handling, finishing, etc.

In fact the concept of common data bank can be so employed that the bank would be useful to other computerised systems dealing in the areas of—

(a) Sales and standard cost analysis
(b) Profitability analysis
(c) Marketing statistics
(d) Tax accounting
(e) Credit management.

Figures 6 and 7 show a typical production system and its interaction with other systems.

## TOTAL INFORMATION AND CONTROL SYSTEM

COMPUTERISED SYSTEMS can also be applied to other areas of operations, notably:

(a) Stock preparation
(b) Stock blending
(c) Control of refining
(d) Basis weight control
(e) Moisture control
(f) Headbox and other control loops
(g) Grade change control
(h) Bleach plant control
(i) Coating control
(j) Finance and accounts
(k) Inventory control.

It can generally be said that computerised systems have the potential for offsetting increasing labour and material costs through improved manufacturing efficiency. An ideal solution to the total problem may involve computerisation in selected areas leaving the rest to merely real-time data logging or controlling only a few of the process variables. Each company and each mill represents a unique functional activity with individual economic and technological needs. This should influence the selection of methodology and areas for computerisation. For example, grade change programmes are generally not justified on machines that make infrequent grade changes such as for newsprint and linerboard, yet good returns have been obtained on small fine paper machines—partly because they make frequent grade changes. Thus it is towards relieving the economic shortcomings of the mill that computerisation should be aimed. A superior understanding of the system should result from prior research and project implementation and is usually credited with contributing significantly to improvements. Another important benefit resulting from process computerisation is the reduction in process variables as evident in the lower standard deviations, which, in turn, improves the yield.

Because of the relative characteristics of sales, variable costs, contribution, fixed costs, and pre-tax profit, a 5 per cent increase in production translates into a much higher, say 12 per cent, increase in pre-tax profit. When higher production is obtained without a proportional increase in raw material consumption, the effect on profit will be even more marked. Thus a 5 per cent increase in production, as a result of improvements in cooking and bleaching, accompanied by a 2 per cent decrease in variable costs, may give rise to about a 25 per cent increase in pre-tax profits. The leverage realised by a production increase is determined by the relative magnitude of the contribution and fixed costs. Thus the desirability of computerising a mill cannot always be assessed solely in terms of accounting principles. In fact, the final decision to go ahead with a project has, like many business decisions, often involved an act of faith in the future.

Figure 8 shows how a process control system provides total information and control.

**Figure 8. Process Computer System provides total information and control**

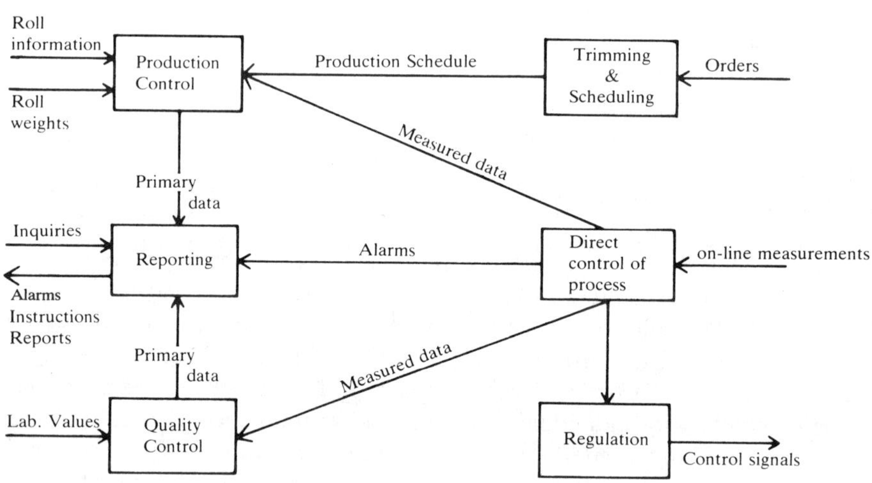

# Evolution of Computer Integrated Manufacturing Systems

*T.R. Crossley*
*University of Salford, United Kingdom*

## INTRODUCTION

THIS PAPER is concerned with the evolution of those manufacturing industries which are involved in the production of discrete items of high-technology mechanical devices in small quantities. By definition, manufacturing industry involves organised activities. The organisation extends to all phases of the activity from the identification of a saleable product and a suitable market, through the actual manufacturing processes, to the sale of the product and the provision of after-sales services, and it involves many extremely complex relationships.

## MANUFACTURING SYSTEMS

THE MANAGEMENT of an engineering concern strives to control the interplay between personnel, materials and machines with a view to optimising the activity as a whole. Traditionally, this control has been determined by a largely empirical approach. In the early 1960s, however, the idea of scientific engineering was conceived and the techniques of systems engineering were applied to manufacturing for the first time. The set of procedures employed by management, the information networks and the equipment available within a company may be regarded as parts of a system responsible for turning raw inputs into material product outputs. This so-called manufacturing system is concerned with all elements of design, planning, control, machining, assembly and testing processes. It consists of a system in its own right made up of individual sub-systems connected by paths for communication.

Changes in economic policy and the rising costs of materials and labour together with the social trends producing shortages of blue collar workers have forced changes in the way many companies approach the problems of manufacture. The purpose of a manufacturing system is to convert inputs into outputs. Its performance in achieving this may be evaluated in terms of two criteria; the production efficiency measured in terms of labour and machine productivity, rate of stock turnover, delivery performance, etc., and flexibility, which is a measure of the ability of the system to respond to changing demands and resources. Many companies, in particular long-established concerns now operating in areas of rapid technological advance, are finding that their manufacturing systems, evolved from intimate man-machine relationships of the 19th century, are no longer efficient in the face of increased competition, or flexible enough to keep pace with changing markets and technology. Machine tools and personnel have been added to increase capacity, ageing equipment has been replaced by more productive plant and new skills have been developed but very rarely have changes been made to the actual manufacturing system laid down when the manufacturing unit was established.

## ACTIVITIES AND COMMUNICATIONS

THE ACTIVITIES WITHIN an engineering manufacturing system may be divided into five broad areas:

(1) Product identification, specification and design
(2) Production scheduling and forward planning
(3) Production planning and control
(4) Actual manufacturing processes
(5) Inspection and quality control.

These areas are interconnected by routes for the flow of products and information. If the system is to function effectively it must have facilities and methods not only for controlling the physical flows but also for the timely generation, collection and communication of information, and many of the problems in manufacturing are caused by poor communications between areas of activity.

The traditional conveyor of both physical items and information is man. It is now held that the degree to which it is possible to optimise the performance of a manufacturing system is directly related to the degree to which communication within the system can operate without human intervention. Planning for optimisation in manufacturing is, therefore, synonymous with automating the manufacturing system, from the design concept to the finished part. The area of manufacturing which has attracted most attention to date is the most labour intensive area, the actual production process, and it is in this area that the most significant advances in automation have been made.

## TYPES OF PRODUCTION

BEFORE CONSIDERING the extent to which the automation of production processes has developed, it is necessary to identify the different types of production process employed in manufacturing engineering. There are three main types of production, although all three tend to be closely associated and may overlap in many circumstances.

### Job production

Job production describes the method by which single articles are manufactured. All engineering concerns, whatever their nature, are involved at some time or other in job production be it the manufacture of small components required for maintenance of plant, the production of prototypes or tools, small jobbing contracts for other concerns or large-scale job-type production such as shipbuilding.

Usually a wide range of general-purpose, versatile machinery and equipment is available together with a staff of highly-skilled personnel and a permanent store of standard materials and components to permit the manufacture of as great a variety of work as possible at short notice. The fluctuating demands on a job production system make it necessary for the system to be highly flexible and change rapidly to suit each particular job. This is usually made possible because individuals, or small teams, are given responsibility for parts, or the whole of the job from beginning to completion and, therefore, the communication problems caused by transfer of authority, information and goods from one section of the system to another can be avoided.

### Batch production

Batch production may be defined as the manufacture of a product in small or large batches or lots, by a series of operations, each operation being carried out on the whole batch before any subsequent operation is started. Batch production is by far the most common method of working in manufacturing industry and it is estimated that approximately seventy-five per cent of all parts produced by the metalworking industries are produced in batches of less than fifty.

Batch manufacture has several major disadvantages caused by the delays and movements between operations. These communications problems include:

Large amounts of work in progress develop which involve large capital investments.

Large production storage areas and generous transport facilities are needed and a very effective planning and control system are needed to meet production deadlines.

Comparatively long production periods are needed due to the time that each batch has to wait before proceeding from one operation to the next.

Batch production probably presents the greatest problems in manufacturing due to the combination of poor efficiency and communications with the need to maintain a high degree of flexibility for a continuously altering plan of work output to be applied.

### Flowline production

Flowline production is the manufacture of a product by a continuous series of operations, each article going onto a succeeding operation as soon as available. Flowline methods are usually only applied when components are required in very large numbers over long periods of time. The manufacturing system tends to be very rigid, and depends heavily on large financial investments on capital equipment which is designed and arranged, with knowledge of the type of component to be produced, to operate at optimum efficiency.

## MECHANISATION AND AUTOMATION

THE WORD AUTOMATION, which is generally used when referring to increasing the efficiency of a manufacturing system, is used to describe the automatic handling of workpieces. The earliest examples of mechanisation in engineering were the use of multi-spindles and power feeds in machine tools. Later developments included sequence-controlled machines which, once set up, could produce large numbers of identical components faster, and at higher rates, than manual machines.

By automating work handling, machine tool and information feedback systems, it has been possible to attain very high levels of efficiency in manufacturing. In general, the method of obtaining this efficiency has resulted in very expensive systems, rigidly designed for the production of specific items and only of use in the mass production sectors of industry.

## NUMERICAL CONTROL

THE EARLY AUTOMATIC MANUFACTURING systems were based on special-purpose machines and work handling equipment designed and constructed to carry out a single job with little or no variation allowed, with the specific purpose of obtaining high output of accurately made products. Such systems are of little use in batch manufacture.

One of the most significant attempts to applying the technique of automation into batch manufacture was the introduction in the 1950s of numerically-controlled machine tools.

The advantages of the NC machine over conventional equipment in batch manufacture include:

The ability to produce components of consistent geometry and quality at high rates for long periods so reducing scrap and rework.

The use of long control programs and automatic tool changers make it possible to combine many conventional operations into one NC operation.

Rather than relying on jigs and fixtures for geometrical information, as do conventional machines, the NC machine obtains all the required information from the control program.

As NC machines became more sophisticated it was necessary to employ aids to their programming. This led to one of the first applications of computers in manufacturing.

## COMPUTER-AIDED MANUFACTURE

THE DEVELOPMENTS in computer technology during the early 1960s led to a rapid expansion in the number of applications of computers. About ten years ago, the use of computers had advanced so far that the term Computer-Aided Manufacture (CAM) was coined to describe the application of computers in manufacturing systems. More recently, there have been many applications of computers in batch manufacturing. These may be divided into the off-line processing of data pertinent to produce design and manufacturing planning, and the on-line control of production processes and information.

### Off-line applications

Off-line applications are those in which the computer is remote from the manufacturing system and operates independently of it. It is convenient when considering CAM systems to consider planning and control as separate sub-systems. The computer-based production planning systems that have been developed typically include program modules and procedures for:

Requirements planning
Capacity planning
Scheduling.

Typical computer-aided production control systems involve modules and procedures for:

Purchase control including replenishment of raw materials, stock control and purchase order control.

Production and assembly control including materials control, load control, inventory control, tool control, job control and dispatching.

The off-line type of applications are usually the result of applying the computer to the manual procedure. The speed and accuracy of the computer permits the optimisation and simulation of plans which would be impossible in the time available using manual means.

**On-line applications**

On-line systems may be divided into two sub-systems: those for monitoring and information systems and those for control of manufacturing processes. The purpose of monitoring and information systems is to register and report production data. This may be data concerning active, idle and breakdown times of different machines, inventory transactions or job status and is collected automatically via direct connections to the computer. These systems provide management with up to date information regarding the status of the manufacturing resources and so increase the flexibility of the manufacturing system by easing the decision-making processes.

There have been many developments in computer-based control systems which include:

The sequence-control of a production line, which may involve the knowledge of production data including the number of pieces produced, cycle time, and idle time.

Computer Numerical Control (CNC) of machines which is numerical control in which the hard-wired conventional control is replaced by a minicomputer programmed to perform the control functions.

Adaptive Control (AC) of machine tools in which the computer is used to measure, for example, cutting forces and speeds and to control the axis motion and spindle speed accordingly so as to maintain the optimum metal removal rate.

Direct Numerical Control (DNC) of machine tools which is the connection of several NC machines to a central digital computer for part-programme distribution and storage.

The characteristic feature of on-line systems is that a dedicated computer is used in real-time mode; that is, the computer is available at any time on demand to perform its function.

## INDUSTRIAL ROBOTS

DEVELOPMENT of general-purpose industrial robots was commenced in the late 1960s, and rapid progress was achieved which enabled them to be put into practical use in the first half of the 1970s. It is expected by the Japanese Industrial Robot Association, JIRA, (which was established in 1971) that robots will be put into wider use in the future and that their full-scale dissemination throughout the developed industrially-based countries will take place in the 1980s.

Industrial robots are commonly defined as manipulators which have a high degree of freedom and which perform versatile movement functions. The Table shows the classification of industrial robots and their respective definitions as approved by the Terminology Standardisation committee of JIRA in 1974.

**Table 1. Classification of Industrial Robots**

| Name | Definition |
|---|---|
| (1) Manual manipulator | Manipulator which is directly operated by man. |
| (2) Sequence robot | Manipulator, the working step of which operates sequentially in compliance with preset procedures, conditions and positions. |
| (2)-1 fixed sequence | Sequence robot as defined above, the preset information of which cannot be changed easily. |
| (2)-2 variable sequence | Sequence robot as defined above, the preset information of which can be changed easily. |
| (3) Playback robot | A robot which is taught first a certain working procedure through operating it, so that the robot itself memorises the procedure, then it can continuously repeat its operation. |
| (4) N.C. robot | Manipulator which can execute the commanded operation in compliance with numerically-controlled information such as positions, sequences or conditions. |
| (5) Intelligent robot | A robot that performs various functions itself through sensing and recognising capabilities. |

**Advantages of Robots**

(i) Industrial robots enable improved productivity, particularly as an effective means of automating small batch production which could not be achieved by existing special-purpose automatic machinery and equipment. This advantage can only be obtained from highly-flexible working functions characteristic of industrial robots. That industrial robots have such functions is clearly indicated by the fact that they are able to easily meet changes in specific work required with the passing of time—for instance, their operation programmes can be easily modified to cope with model change-over (time flexibility), and the spatial modification of their working and movement path can be easily made (space elasticity).

(ii) In the case of mass production, where product redesigning requires an enormous amount of time and money for remodelling, industrial robots can save much time and money.

(iii) Industrial robots can be diverted to other applications or be transferred to different plants.

(iv) Industrial robots are capable of 24-hour operation and thereby greatly enhance the efficiency of expensive plant and equipment.

(v) Industrial robots enable substantial changes to be made in production volumes.

(vi) Unlike human beings, industrial robots are free from fatigue of simple duties performed over long hours and reduce the number of defective quality products caused by such fatigue.

(vii) Industrial robots can increase the service life of tools such as welding devices and can economise on the use of materials (for example, paints by precisely repeating the given motions).

(viii) Industrial robots help prevent industrial accidents and occupational diseases often caused by working in dangerous environments and under unfavourable conditions. This is one of the greatest socio-economic advantages to be gained.

(ix) Industrial robots help reduce economic losses caused by workers leaving their jobs due to working under unfavourable conditions.

## COMPUTER INTEGRATED MANUFACTURE

COMPUTERS are being used in ever increasing numbers in manufacturing systems but in a somewhat disjointed fashion. New applications have solved particular problems but the overall contribution of the computer has often been less than that forecast. It has also become apparent that it is not enough to superimpose computer technology and techniques onto traditional manufacturing systems.

The most promising concept for solving the problems of efficiency and flexibility is the Computer Integrated Manufacturing system. Such a system would be based on work stations, interfaced by automatic handling systems, which have been designed from the floor up to efficiently interface with the digital computer. All aspects of the manufacturing activity including detailed design, specification, manufacturing engineering, materials management, production of parts, assembly, test, warehousing, sales and service, would be controlled by individual modules of computer software, all of which would be linked together in a hierarchical system. CIM is a total technology which will involve tremendous amounts of software and will be evolved over a fairly long period of time.

# Automation in the Automotive Industry

***P. Flapper***
*DAF Trucks B.V., Holland*

## INTRODUCTION

DAF Trucks' main activity is the manufacture and selling of trucks of its own development and design. In the struggle for survival it is essential that products of high quality are delivered, and that service given after sales is adequate to ensure a reliable functioning of the products. Annual production totals 17,000 vehicles of various kinds, such as trucks, trailers, buses, and container transport. Also main components i.e. cabins, axles, chassis, and engines are delivered both in Holland and abroad. It stands to reason that the use of automated information is a vital element to ensure an efficient and alert management at all levels of the company.

The company has a corporate systems and automation department which operates central computers and supports a number of decentral minicomputers. Central computers are linked to these decentralised units and to the computers of subsidiaries abroad. A special timesharing computer serves a great number of users in the marketing and sales organization as well as in the engineering departments. Moreover for engineering and manufacturing purposes a number of smaller desktop computers are in use to perform technical calculations of various kinds.

Automatisation follows the main functional processes in the company and covers the following areas: research and development, manufacturing, marketing and sales, parts, and finance and economics. Apart from corporate systems of a general nature like financial planning, wages and salaries, and a pension system, which are not related to the automotive industry in particular, the following main groups of information systems are dealt with in this paper: systems supporting research and development; systems supporting manufacturing and assembly; and systems supporting sales activities, truck performance and after sales services.

These three topics help illustrate the use of automated information processing in the truck industry and cover the present state of the technology used in some applications. Systems support for the control of a number of processes, in our opinion, is a must for survival in the coming decade.

## SYSTEMS SUPPORTING RESEARCH AND DEVELOPMENT

COMPUTERS AND COMPUTER PROGRAMS are used to perform very complex calculations in specialized branches of industry such as aviation. However more and more computers are becoming a daily tool to engineers, draftsmen and research people. This development has been stimulated by ever decreasing prices of hardware and the availability of adequate software. It is expected that this trend will continue. Typical of the requirements of a system operating in a technical environment is that it must be interactive. Research and development imply a great number of alternatives from which one has to choose. Consequently the system should be able to react quickly in a dialogue with the user. Another typical requirement of technical applications is the need for graphical presentations of calculation results, including the actual production of technical drawings of parts. DAF is presently active in two of these applications: technical calculations in timesharing and automated research by means of realtime systems.

Technical calculations consist of calculations related to a vehicle, by programs specifically written for that purpose, e.g. performance calculations, simulation of the braking system, simulation of ride and vibration behaviour, and calculations using standard software based on the finite element method (FEM) to tackle complex structures on strength, rigidity, and vibrations of chassis or engines. Realtime systems handling trial runs have to deal with an enormous variety of values to be measured by various measuring-equipment and an even greater variety in results when processing these measurements. To complete this task the computer system should be compatible with a number of instruments and be easy to program for one specific application. DAF uses such systems for measuring and analysing the combustion and compression in engines and the noise produced by a vehicle. As stated above the trend to integrate automation in the daily process will continue. It will neither be limited to calculation and analysis nor be confined to control of machines or processes only, but extended to technical design and drawing, using database information necessary to the engineer like norms, materials, and past experience of designing.

## SYSTEMS SUPPORTING MANUFACTURING

### Main Data Base Structure

In 1976 DAF decided to implement a manufacturing parts list, in addition to an existing engineering parts list. At the same time an end-product structure was introduced, based on options per product type. A basic end product type together with a set of selected options can be expanded into an engineering components specification or into assembly instructions. Both expansions result in a total set of assembly parts, constituting the end-product. In the same way an assembly part has both an engineering and a manufacturing structure. Applications of the manufacturing parts list are:

Maintaining assembly instructions per work centre and per assembly line and assigning particular instructions to a truck order.

Maintaining the parts production sequence, having opportunities to define extra levels for operation and for material management purposes. Both parts lists are modification controlled.

### End-Product Order Control

*Order entry*: orders are entered into the system by the Sales department in 4-weekly batches. The number of types and options in a batch are checked and deviations beyond certain limits are printed and discussed by Production and Sales planning departments. In this meeting also sub-orders are discussed. Delivery dates from Production to Sales are established and are input to the system.

*Lineset*: linesetting is performed in weekly batches. Using the assembly lines offsetting schedule, delivery dates for all assembly line components are derived from the final delivery dates. At the same time a material availability check is performed taking into account the available economical stock of predefined critical assembly parts.

*Line instructions and material supply*: for each assembly line the derived weekly component batches are prepared. For each line order, an instruction specification is printed, showing the type, relevant option(s) for that line, sub-assembly information and additional information e.g. chassis and engine numbers, etc.

The sub-assembly instruction is compiled from the assembly-instruction data base. At the same time the required assembly parts are determined and totalled in weekly batches, while some high volume parts are kept in daily batches. Material picking lists are printed and stock on-hand is automatically decreased to reserve parts for assembly. Shortages are reported. The reserved assembly parts are stored in a work-in-process-file so that parts can be booked out for work completed later.

*Line order completions*: when a line order has been completed, a report is input to the system. The work-in-process-file is read and the relevant assembly parts are sent to the material control and costing systems for booking off, invoicing and efficiency calculation purposes. Final assembly line completions are also sent to after sales service systems to prepare service documents.

*Costing*: in the costing system all assembly parts are calculated quarterly to obtain new cost prices, although a fixed cost price is used throughout the year for invoicing purposes. Based on this fixed cost price, the assembly parts concerned, and value added per assembly line (dependent on product type) the order cost price is calculated for each assembly line separately and totalled for an invoice to the Sales department. Order efficiency calculations are performed per assembly line, by hand. Very soon this will be automated and extended to work units per assembly line.

*After sales information*: because the after sales organisation has a separate parts list, based on spare parts to be sold, a transformation of production information into spare parts terminology is provided for automatically. A list of indexes to exploded views is produced, by which the service people find the relevant spare parts. At the same time, a list of specific order information is produced to support the warranty department, such as chassis, engine, fuel pump and key numbers. Copies of these lists are sent to the internal transport department, quality control department, etc.

## Material Control

*Material requirements planning (MRP)*: a master schedule of assembly parts is updated periodically and serves as input to this system. The initiation of the master schedule is described below. At present the MRP system is still a gross requirement planning process; net requirements and time scheduling are established manually. Within a couple of months a net requirements planning system will be implemented, which also performs shop-order and purchase-order suggestions, which are, altered if so wished, input to the order system. Net requirements are based on the master schedule, existing stock and orders already confirmed by suppliers, and a calculated lead time. The start will be on a fixed frequency fixed quantity basis as used in the factory nowadays. At a later stage other methods such as economical order quantity or part period balancing will be supported by the system.

*Purchase order control*: purchase orders are "blanket" orders to cover supplies of a full year at least. Delivery schedules against these orders are generated and updated periodically. A schedule contains both firm quantities and indications of future requirements. Receipts are checked against these schedules and transactions are fed to the stock control and to the accounting systems. Soon, quality control features e.g., sample test results and analysis reports, will be added to the system. Information regarding late or early deliveries is generated weekly, to enable quick corrections. A sub-contracting file is kept to register sub-contracting activities in the manufacturing process. The orders are handled like purchase orders.

*Shop Floor Control*: manufacturing orders are controlled throughout the various operations concerned. Related to manufacturing orders are the material requirements, received from the manufacturing parts list. Material issues are manually input and compared with the stored requirements. Shop floor documents are generated from the operations file, linked to the manufacturing parts list. Work in progress control, order status-reporting and efficiency calculations are performed per order, per operation and per workstation.

*Stock Location Control*: the system provides the opportunity to register stock-movements in every required detail. As already stated, the interface with assembly control is fully automated.

*On-line inquiries*: a constantly growing number of inquiries on a continually increasing number of terminals (VDU) is supporting clerical and production personnel offering information asked for from the data base.

*Material planning*: the gross requirements of assembly parts are built up from three different sources:

> The first period of four weeks consists of the material required by orders already under assembly order control. This material is already reserved and booked from stock on hand.
>
> The second and third period of four weeks consists of orders already planned (delivery date) but not yet in the assembly process.
>
> The fourth up to the 13th period are generated from sales forecasts which means estimated quantities per type and per option within type. Because some assembly parts are selected by a combination of options, quantities of those combinations are calculated by the system taking into account the recent turnover in those combinations.

The real orders will in time of course deviate from this prognosis and therefore a safety stock has to be calculated. This last subject is largely based on experience and adequate formulas have not yet been introduced in the system. It is however under serious study.

## SYSTEMS SUPPORTING SALES ACTIVITIES, TRUCK PERFORMANCE AND AFTER SALES SERVICE

### Topec

To back-up its sales organisation, DAF Trucks has developed a number of programmes under the heading of Topec, which stands for Truck Optimum Performance and Economy Calculation. Our sales organization has access to these programmes from anywhere in the world. Connection can be obtained by using a portable terminal via the telephone network. This direct link provides rapid availability of information. The programmes are interactive and can be executed in the Dutch, French, German or English languages. One of these programmes is T.P.P.—the Truck Performance Program. It has been developed to meet the increasing demand for information on vehicle performance to be expected, e.g. tractive effort, gradeability, maximum speed, acceleration time, and an increasingly important factor, fuel consumption. Processing the input, the program makes use of a library, in which essential data about components of the product line are contained, e.g. for engines, number of cylinders, bore and stroke, torque curve and the iso-diagram. Similar data is stored for gearboxes, axles and tyres. From the type of component given in the input, the system is able to obtain these essential details from the library and produces economic calculations when asked for. Upon completion of the input the vehicle specification is returned, so that it can be verified. For each gearbox ratio the following data is calculated: vehicle speed in km/h, tractive effort available at the wheels in KN, surplus engine power in KW, maximum gradeability in motion, acceleration time when shifting up and accelerating from stationary. If the results are not satisfactory, changes can be made in the vehicle specification and calculations repeated. If asked for, a fuel consumption diagram is produced. Another Topec-feature enables the client to make his choice (on subsequent levels) from the various alternatives shown on the screen of a colour terminal resulting in the picture and description of the vehicle that corresponds with his wishes.

**Visar, vehicle consumption indicator and shifting advice** (Registered trade mark)

This is an aid to save on fuel consumption and to adopt a style of driving that guarantees the lowest possible fuel consumption. By means of a microprocessor, relevant sensor data like vehicle speed, revolutions per minute of the engine, and engine load are calculated to ascertain the actual fuel consumption. This consumption is shown on a fuel consumption indicator with high and low scales. As in Topec, Visar uses the iso-diagram. This iso-diagram can be considered as the fingerprint of the engine, giving the lowest possible fuel consumption. Per type of engine an iso-diagram has been stored in the microprocessor. Whenever necessary, shifting advice is given to indicate that consumption will improve by shifting up or down.

**I.T.S. International Truck Service**

I.T.S. is a service to drivers that encounter difficulties abroad. In the case of a breakdown abroad drivers may have difficulties with language, money, specific parts etc. Daf I.T.S. is available 24 hours a day, 7 days a week, to give the necessary assistance. By just phoning I.T.S. Holland, all steps are taken to ensure a quick repair and to bring the vehicle on the road again in the shortest possible time. These services include guarantee of payment, supply of special parts, appointment of authorised repair dealers etc. Data on breakdown-reports received and dealt with, are registered, with all relevant information such as geographic location, type of defect, previous occurrences of this fault, repair dealer, type, chassis number, etc. Relevant data recorded is used to report on frequency by day, and by year, spread over locations, kind of defects, etc. Results are used by the warranty department, quality control department, and the parts and costing department.

# Technological Options and Banking

*P.V. Cremin*
*Bank of Ireland, Dublin*

## INTRODUCTION

MONEY, IN ALL ITS FORMS, is a core concept of modern society. Fundamentally, money is information which defines the relative material position of someone viz-a-viz someone else. Basically, it is information stored on the medium of paper and metal. To harness the revolution of informatics, all we need to do is exchange the paper medium of money for the electronics medium.

A conventional bank looks like that portrayed in Figure 1. It consists of a head office with physically remote branches radiating from it and, radiating from each branch, are individual customers. Because of the necessity for integrity and honesty in dealing with physical money, the calibre of staff in the head office and the branches must be high. Therefore, staff costs in banks are disproportionately heavy. The nature of the business means that all the branches must be located in the shopping or business centres of the cities and towns. They are, therefore, invariably expensive premises. The very simple model of a bank, as in Figure 2, suggests itself. The total rectangle is revenue and the two constituents are costs and profits. It is not to scale. The major portion of costs is made up of staff and premises. The model helps to illustrate the impact of staff and premises on the overall profit of a bank.

Because of the vicious escalation of staff and premises costs, all banks in the Western world find themselves confronting a dilemma. (See Figure 3). In Europe and certainly in Ireland, the conventional bank products of current accounts and deposit accounts are becoming less and less economical to the extent that a large number of accounts are loss-making. This situation results from the high cost of processing transactions. The revenue and costs curves in Figure 3 are converging. If one attempts to generate more revenue by selling more accounts, the costs

**Figure 1.**

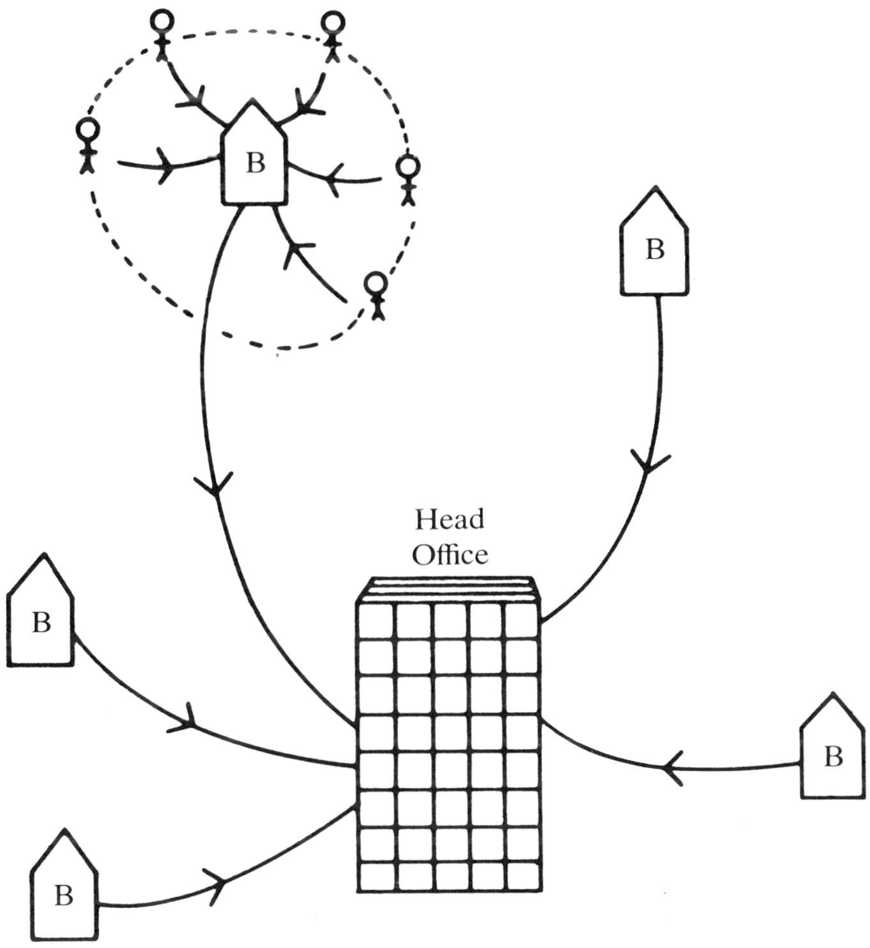


curve goes up. Since its slope is basically greater than the revenue curve, one only exacerbates the problem.

## AVAILABLE TECHNOLOGY

I HAVE RECENTLY had occasion to look at the financial or banking products of nine manufacturers. One rather striking feature occurs repeatedly. In spite of the

**Figure 2.**

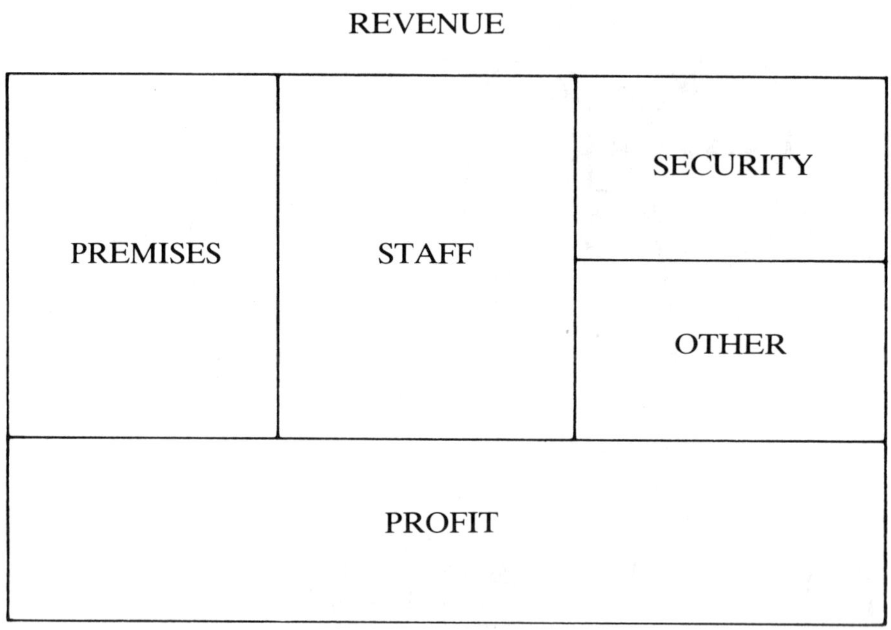


major advances at the basic technology and component levels, there are no new services or products emerging in the financial field. Technological advance is being directed into costs, reliability and speed where the improvements have been very dramatic.

It is still the traditional defensive role of data processing, i.e., to automate existing processes and to convert labour-intensive activities into capital-intensive ones, that prevails. Technological advance is not being used as a basis for new products or services. Yet the dilemma of all banks, the converging revenue and costs curves, suggests that a radically new approach is required either through new products or by a major restructuring of the type of bank portrayed in Figure 1.

The process through which technological advances get through to the man in the street is worth examining and is shown in Figure 4. The requirements of the end user are mapped onto the banks through a process of representation, assumption, market surveys, etc. The process can have a one-way or two-way flow. Through a similar process, the banks' requirements are mapped onto the manufacturers through a marketing mechanism. There is, probably, two-way flow involved. In turn, the manufacturers impose their requirements on the component makers who, in their turn, employ the advances made at the fundamental

**Figure 3.**

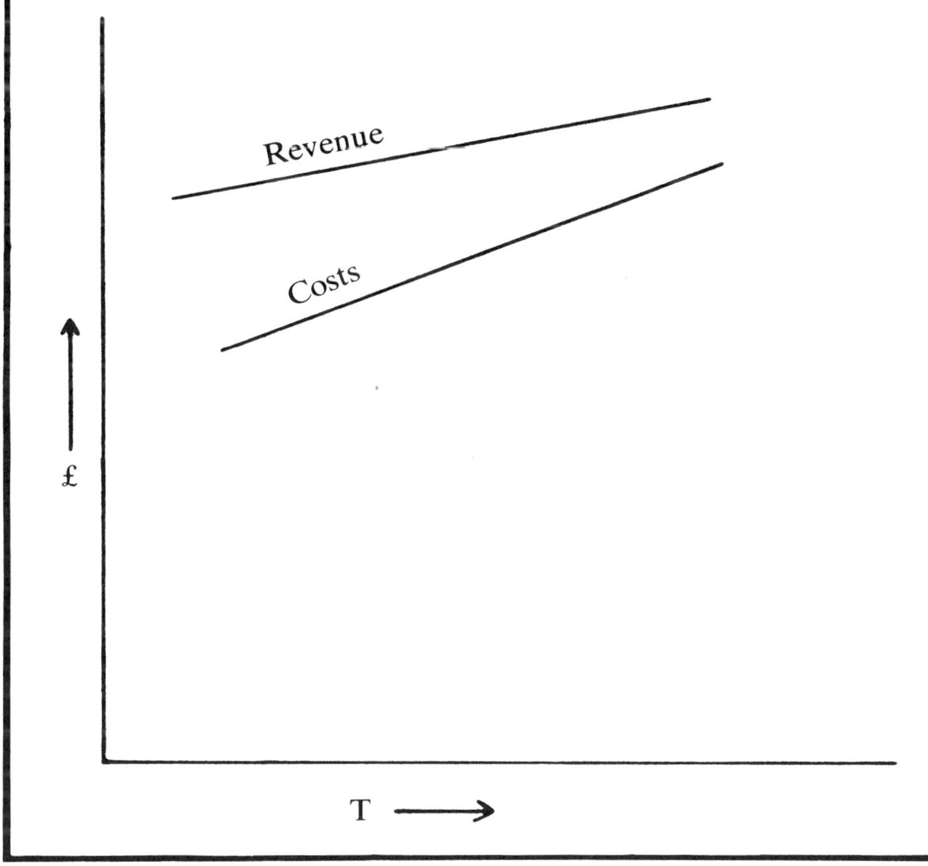


technological level. It is a reasonably long, vertical, communications process. Many of the upper levels involved have a marketing or commercial orientation. The rapidity and complexity of the microelectronics revolution have precluded those upper levels from an understanding and an appreciation of the power of the technology. Hence, I suggest that this is the reason for the slowness in the production of new services and products.

For the majority of manufacturers, the users and banks, the top two levels of my model in Figure 4 are specifically in North America or in the more advanced countries of Western Europe. I have already said that appropriate solutions were slow in coming forward or did not come forward at all because of product life cycle, diffusion of technology, etc. But the solutions, relevant or irrelevant, have

**Figure 4.** Technological Diffusion

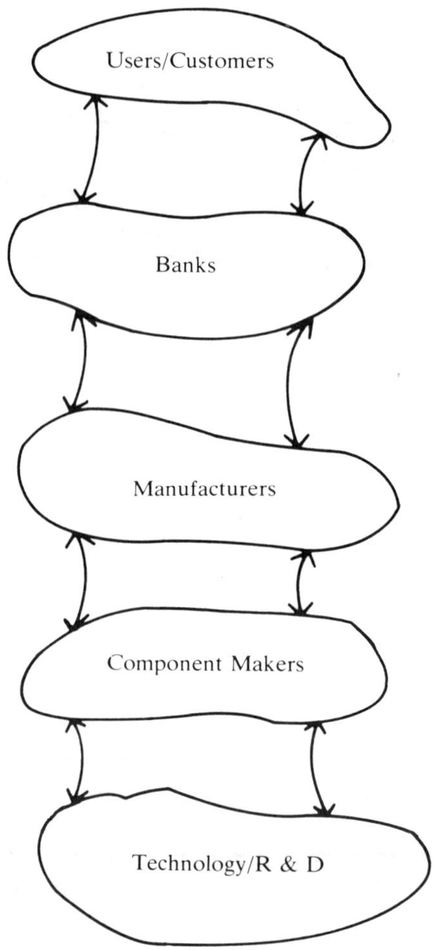


been designed for a North American or European market. The problems of other countries and other banks, will be defined or, indeed, re-defined to match those existing solutions.

My model in Figure 4 could have another level added to it as follows. It represents the LDCs, and shows how they do not fit in the process of diffusion of technology up to business and commercial applications. I will put some technological data through this model at the end of the paper which, I hope, will prove its veracity.

## WHICH TECHNOLOGY?

THE WORD TECHNOLOGY is loosely applied to particular sectors. Hence, one talks of online technology, offline technology, distributed intelligence, plastic card technology (or more recently plastic card culture as it now epitomises a way of life), viewdata technology, etc. Figure 5 is a personal and arbitrary categorisation of different aspects of technology and their applications to banking. It attempts to plot the relevance of each type of technology in that it shows the point in time from which it is generally available to the point at which it is no longer relevant for economic reasons, obsolesence, etc. For instance, one could centralise and automate a bank's accounting system successfully and without risk from the mid 60s. But from the mid-70s onward, online technology, and terminal systems allowed one to put intelligence in branches and make them less dependent on central computers. Because of the investment involved, the centralised system would still be relevant until the early 80s.

A bank will have to develop a very wide 'look angle' in order to position itself correctly so that it can avail of developments. The look angle can be demonstrated as follows. The narrower one's look angle is, the less options or flexibility one has.

Because of the awareness of rapid economic and technological change and the availability of new methodologies like technological forecasting, we can take a fairly wide look angle, with some confidence, if we choose to do so. Because of the investment involved, the relevant duration of each technology and the

**Figure 5.**

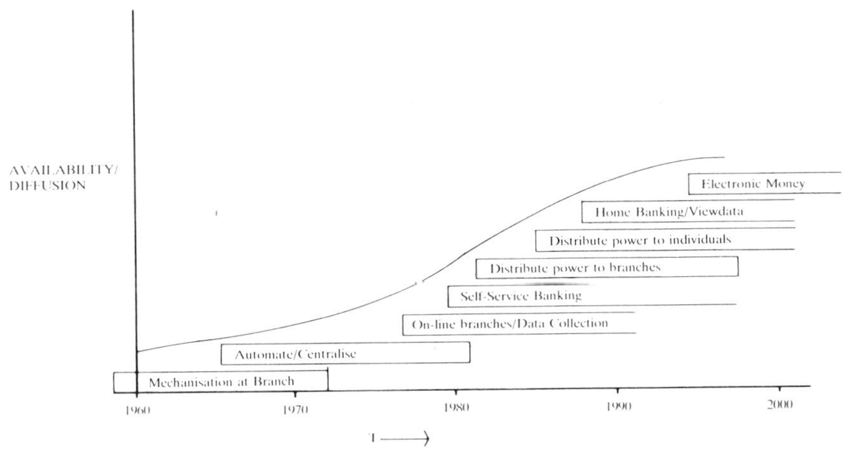

dramatic impact of each step up in the diagram of Figure 5, one can ill-afford not to use a very wide look-angle.

For example, if one looks at the technological options available in 1981 with a narrow look angle, as demonstrated, one would see and consider the following options:

On-line data collection and report distribution to branches;
Some straightforward self-service banking, e.g. automatic cash dispensers;
Distributed intelligence and information availability at branches.

If any one or a combination of these options are implemented, it will have the effect on the revenue/costs curves convergence as illustrated in Figure 6. It achieves a stay of execution rather than a commutation.

**Figure 6.**

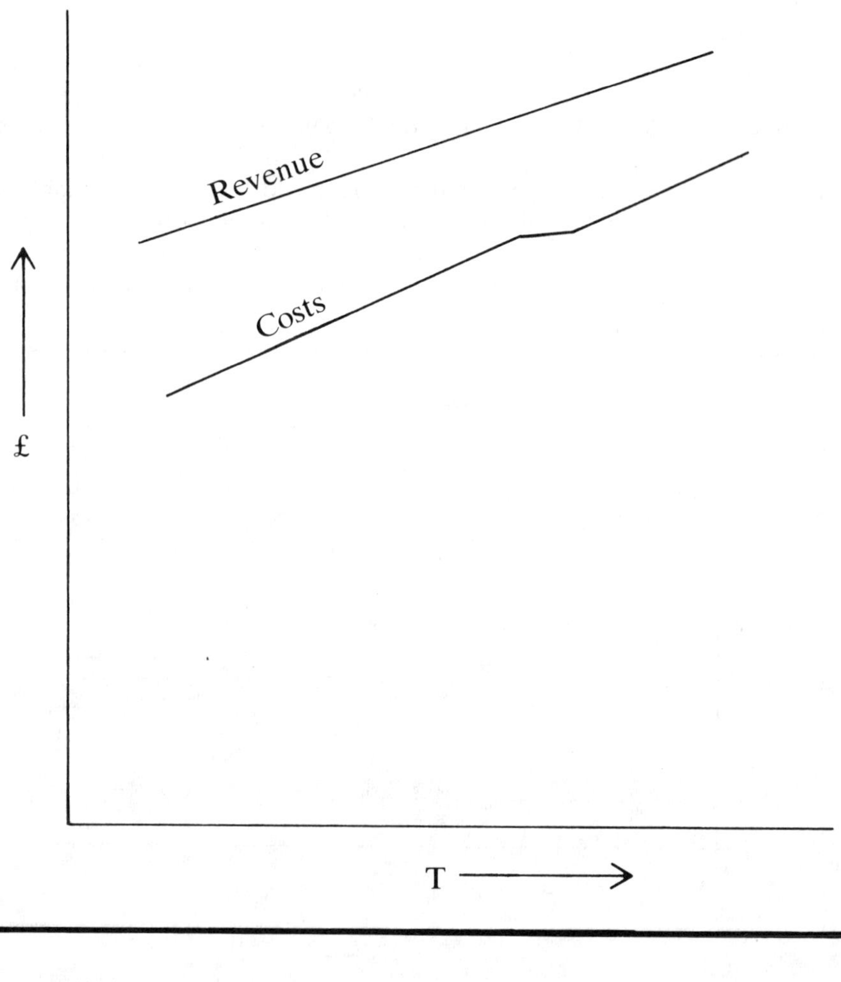

It achieves no more than this because it does not address the basic structural problem of uneconomic products and high staff and premises costs. The revenue model of the bank would look like that of Figure 7. The premises and staff costs would be contained or even marginally reduced as a proportion of the whole rectangle because increased automation at branch level would allow an expansion in business without an increase in staff or premises costs. Hardware costs will increase substantially. Generally, the revenue profile of the bank will not have changed much. The centre of gravity of the bank is still very much enclosed by the staff and premises costs.

On the other hand, if one takes a wide look angle, as demonstrated, one would consider the likely technological options of the 90's some of which are included in Figure 5.

If one looks at the potential of home and office banking that can be achieved through the use of domestic television sets and telephones plus viewdata technology, it is difficult not to conclude that it will become all-pervasive. Using this type of system, it would be possible for a bank customer to dial a telephone number from his office or home and look at his statement on his own T.V. screen connected to a bank-owned private viewdata system. Another example could be communication at transaction level. Instead of using the present paper-based cheque payment system, a customer could instruct his bank, from his T.V. set, to

**Figure 7.**

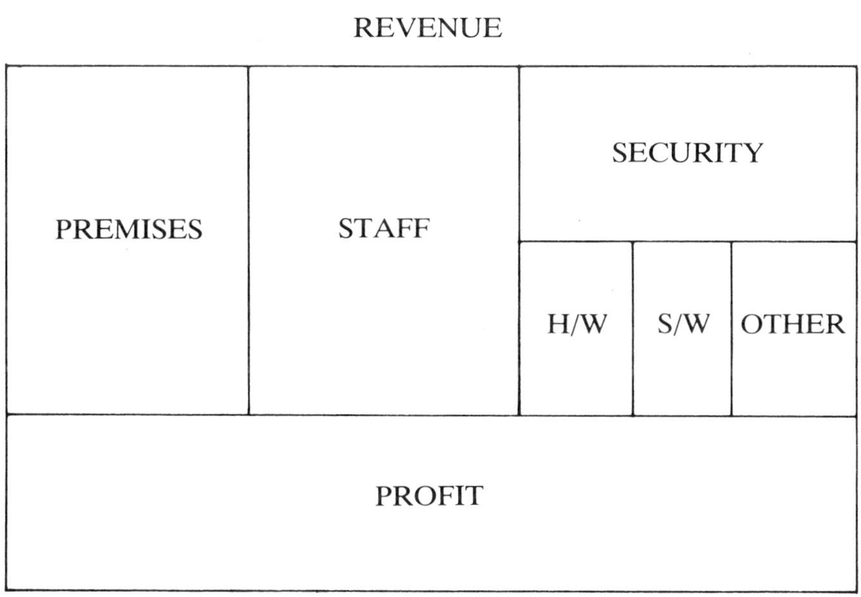

transfer funds from his account to the credit of a third party account. It would be possible for a customer to make a loan application through an interactive program. Or order money which would be despatched by registered mail. Or, indeed, he could set up his own standing orders and amend them as he saw fit.

We have not introduced electronic money. An earlier technology included the ability to distribute memory and processing power to individual members of the population. For instance, the chip card is already in existence in several forms. One can have value input to it and use it as an electronic wallet. The French banks are just about to use it as a cheque substitute in point-of-sale applications. One of the major advantages of this distributed power is that it enables offline electronic funds transfer, this attribute is particularly important to countries that do not have a highly developed communications infrastructure.

If one could visualise, for the moment, an untrammelled jump from the bank portrayed in Figure 1 to the bank portrayed in Figure 8, some very dramatic changes occur in the financial structure of a bank. The bank in Figure 1 was a labour-intensive service industry with disproportionately high staff and premises costs. In the bank in Figure 8, all the customers have terminals, i.e. T.V. sets, which they have paid for themselves. They access a large central bank computer system using their terminal and the telephone network. The customer pays for the telephone call as well. The bank of Figure 8 is a self-service, capital-intensive industry with very low premises and staff costs. Our earlier financial model would now look like Figure 9. The bank of Figure 8 could now be described as an information industry and could survey other than financial information. These real possibilities are offered to show that it will not be sufficient to say that one will be responsive to change. One must anticipate change and position one's self correctly to be effective.

I will stretch our look angle one last bit. It now includes electronic money. Some foolhardy souls in the '60's forecast the cashless society. I think they were carried away by the general euphoria and mystique surrounding computers. It was not possible to map out the steps to a cashless society from the '60's. With the advent of microelectronics, one may cautiously predict a near-cashless society. If a microcomputer can be imbedded at low cost in, for instance, a card (and it can), then every member of the population can be given one. Using a relatively cheap transaction unit, value can be passed from one card to another. If one wanted, one could insinuate a bank into that transaction process. But it could be a private transaction between two people and enabled by them both. Two of the main reasons for the evolution of banks was to safeguard and to distribute physical money. Electronic money transfer between personalised microcomputers obviates those two reasons. By pursuing this train of thought, very interesting scenarios can be generated.

I said that I would put some data through the technology diffusion model of Figure 4. The chip card has come from the component level of my model or maybe a little bit above it. It will have many powerful applications, particularly in the electronic funds area. It is cheap and it will allow secure, offline working. It should be ideally suited to LDCs. However, many major computer manufacturers

**Figure 8. Bringing the Bank to the Customer**

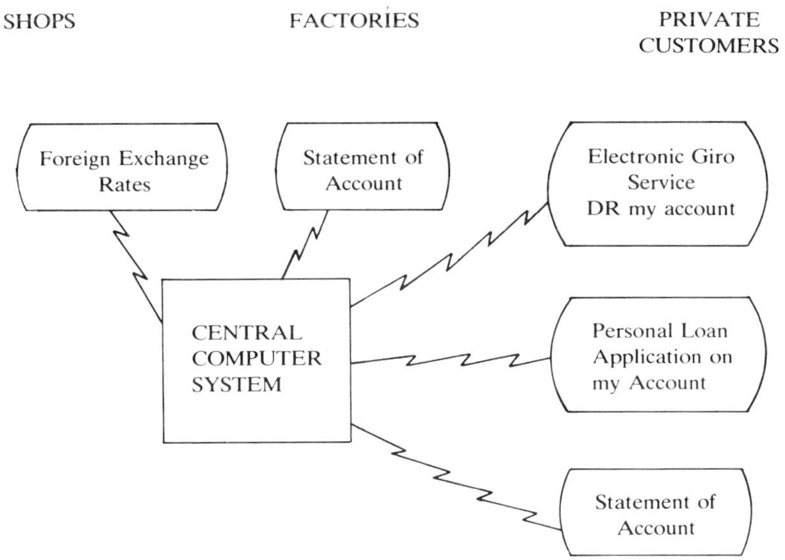


**Figure 9.**

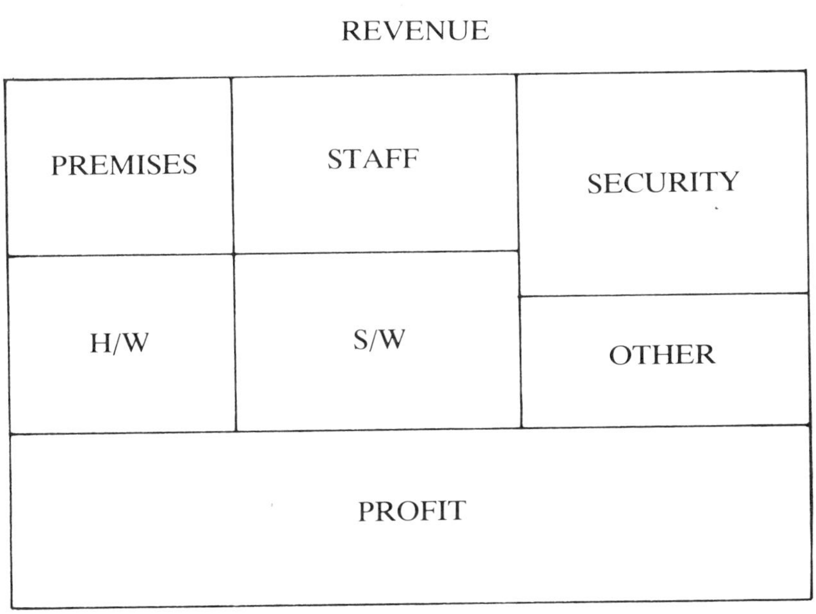

have made a major investment in online systems. A large part of their product line is activated by conventional magnetic stripe cards. Equally, there are many banks with massive investments in automated teller machines and plastic card technology. This investment and commitment has led to pressure for the chip card to conform to existing plastic card standards. This will constrain its development. For instance, if the chip card can be thicker than standard, it can be made more robust and can incorporate conventional edge contacts rather than surface contacts. This, in turn, would result in cheaper and more robust transaction units. There is also pressure to include the conventional magnetic stripe on the back of the card to enable it to be read in current card activated devices. The result of these pressures could be an emasculated plastic card substitute rather than the personal, cheap, portable and powerful payments instrument that it is.

# Appropriate Informatics in Industry: A Third World Viewpoint

*Utpal K. Banerjee*
*Computronics, India*

## INTRODUCTION

THE 1980s HAVE MARKED the beginning of the Third Development Decade during the Twenty-Seventh Session of the UNDP Governing Council in Geneva in June 1980. It was estimated that the cereal deficit among developing countries would be 88 million tons by the year 2000, or two and a half times the deficit of the mid 1970s. In 1980, more than 600 million men, women and children were experiencing, every day, a deficit in minimum caloric intake, while 800 million were condemned to use impure water and 500 million were denied basic shelter. It was further estimated that over the next ten years, the labour force will increase by an average of 45 million new workers each year, placing enormous new pressures on the capacity of developing countries' training systems. All this is apart from the problems of over population, demands for education, poor condition of health and almost no protection of the environment (Rateau, 1981).

Looking closely, one can discern wide diversity among the developing countries themselves. In the OECD study on "Facing the Future" (OECD, 1979) the following five main elements were highlighted:

(1) diversity with regard to the volume of population ranging from hundreds of millions to hundreds of thousands or tens of thousands;

(2) dispersion of average national revenue per capita: 34 very poor countries, with 1.2 milliard individuals, had an average revenue ranging from 70 to 250 dollars in 1976. At the other end of the scale, 27 countries have an average revenue of between 670 and 2700 dollars for a total population of 440 millions;

(3) wide disparity in the rate of economic growth during the past 15 or 25 years, ranging from $-1$ per cent to rates higher than 4 per cent;
(4) appreciable differences in the internal distribution of revenues;
(5) large variety in the availability of natural resources. The case of the oil producing countries illustrates the importance that the evolution of the world economy can give to this factor.

Thus, it can be seen that a great deal remains to be done regarding development and the reduction of the inequalities among nations.

## SITUATION IN INDIA

OVERPOPULATION, leading to a legally enforced overstaffing, has become a way of life in industry. Taking Kenya as an example, Clayson (1980) described the industries as overstaffed by American or European standards. In America, productivity studies consider the substitution of capital for labour or even the elimination of labour by better scheduling. In Kenya, these options are not always possible. In fact, in many cases the national goals are directly opposite to those in American industry: the substitution of labour for capital is encouraged as well as an increase in staff through a reduction of scheduling. This is neither a topsy-turvy world nor an irrational one; it is a world to be taken seriously and on its own terms.

As regards the use of computer-based information systems, there are two computer service bureaux in Nairobi which help to run a few computerised inventory control systems in Kenya. While stock levels only are calculated, there is no prediction of future sales or re-order supplies. There are, however, several isolated Operations Research type (OR) techniques being used in Kenya, e.g., online inventory control for spare parts in the automotive industry; least-cost animal feed-blending; production scheduling in the textile and sugar refining industries; risk analysis for new tourist hotel complexes; and online government social security and police records retrieval. But each of these techniques was imported directly from outside Kenya. Without adequate training personnel, Kenyan firms have been unable to adapt these imported techniques to other purposes or to use them for teaching new skills. In several cases, internal antagonism to imposed organisation changes concomitant with OR techniques killed their implementation. In Africa as in America and Europe, imposed OR will be branded as yet another form of "imperialism" if it is not supported by all those workers directly and indirectly involved.

The above industrial situation is typical of many Third World Countries where there are few ongoing computer-based information systems or institutionalised OR activities within the industrial environment. In order to understand the role of 'appropriate informatics' in such an environment, the following hierarchical levels have to be visualised in industry:

(1) *Operational control level*: the organisation is identified with supervisors and the like who execute decisions and use resources in conformity with rules in a real-time perspective. Their interactions are limited to work-centres based on readily available information which is tailor-made to problems. The decisions are programmed and highly repetitive in an environment of certainty, e.g., acceptance or rejection of turned-out jobs on lathes, ordering for material based on re-order levels and re-order quantities, etc.;

(2) *Managerial control level*: divisional or departmental profit-centres are operated on set rules, resources allocated and performances measured on short-term time perspective. Information is adequately integrated on a historical basis. Decisions are not programmed, although repetitive in an environment of risk, e.g., product-mix, layout of plant, etc.;

(3) *Strategic control level*: The corporation or divisional top management sets objectives to determine resources for the system on a long-term time perspective. Considerations are for the whole organisation, based on predictive and not so accurate information. Decisions are not programmed and are on a one-time basis, in an environment of uncertainty, e.g., location of a new plant, introduction of a new product, etc.

## CONCEPT OF "APPROPRIATE INFORMATICS"

In India, as in Third World countries, quite a few industrial computers can be used at the operational control level, with data-sets comprising transactions, working and master files. At the managerial control level, some additional external data is needed and computer capability can be used for producing monitored reports (prepared at specific time-intervals) and triggered reports (calling for executive intervention). Less common at this level are the demand reports (of an investigative and inquiry nature) and planning reports (involving the environmental data). Strategic control can be achieved at the topmost level of the hierarchy, although seldom helped by computer-based data-banks, since the latter need to maintain a large amount of environmental data besides the organisational data bases (Srinivasan, 1977).

The basic ideas in "Appropriate Informatics" are then as follows:

(1) Automated control of industrial process can be tried out only at the operational level where a high degree of instrumentation is possible and where response time is very short. Here digital computers can be used in a real-time environment and man-machine interfaces can be built up to gather data on operational problems and to transmit control commands either by operators or automatically through computer devices in a repetitive manner;

(2) Since decisions are not programmed (although repetitive) at managerial

control level, automatic computer control should be discarded in favour of operator guidance. If permitted response-time is medium, data-processing can be on-line, with off-line communication of control instructions for action. If, however, a long response-time is permitted, data-processing can be in a batch mode, with still off-line communication of control instructions;

(3) At the strategic control level, the off-line demand and planning action described above are yet to become sophisticated in the developing countries. To that extent, off-line use of computers are justified for specific problem-solving.

Some illustrations of appropriate informatics have been attempted in the succeeding paragraphs at operational control, managerial control and strategic control levels, particularly as they have been either planned or practised in India. The *leit motif* running through all these applications are that the role of human labour and skill have to be kept in view wherever possible, and the computer has to be brought into the picture more for providing operator guidance than for automated control. The viewpoint applies to all developing countries.

An integration between operational control and managerial control can be a stepping stone to strategic decision-making at the corporate level. Such an approach can be illustrated typically for an integrated steel plant as shown in Figure 1.

## ILLUSTRATIONS OF OPERATIONAL CONTROL

IT WILL BE EVIDENT that if a high degree of instrumentation is possible, automatic control of industrial processes can be introduced. A few examples are furnished of these extreme cybernetic processes in the succeeding paragraphs.

*Industrial process control*: Data is collected on-line for the industrial process through automated sensors, and also from direct access files by a dedicated computer or micro-processor. Appropriate control programs and sets of control variables needed are built into the memory of the computer. Optimisation algorithms provide the bases for adjustment of the control variables to bring the industrial process up to a desired optimal performance.

For instance, an IBM-1800 computer is used for controlling every LD operation in the "heat" of steel making shop number 1 at the Bokaro Steel Plant. The process control function is carried out around the clock by the thermochemical model which uses the criteria of temperature of the steel at the completion of the blowing of oxygen and the endpoint carbon, for successful end-results. The model runs through the following stages:

(1) Initial charge calculation based on quality and quantity of steel for every "heat";

**Figure 1.**

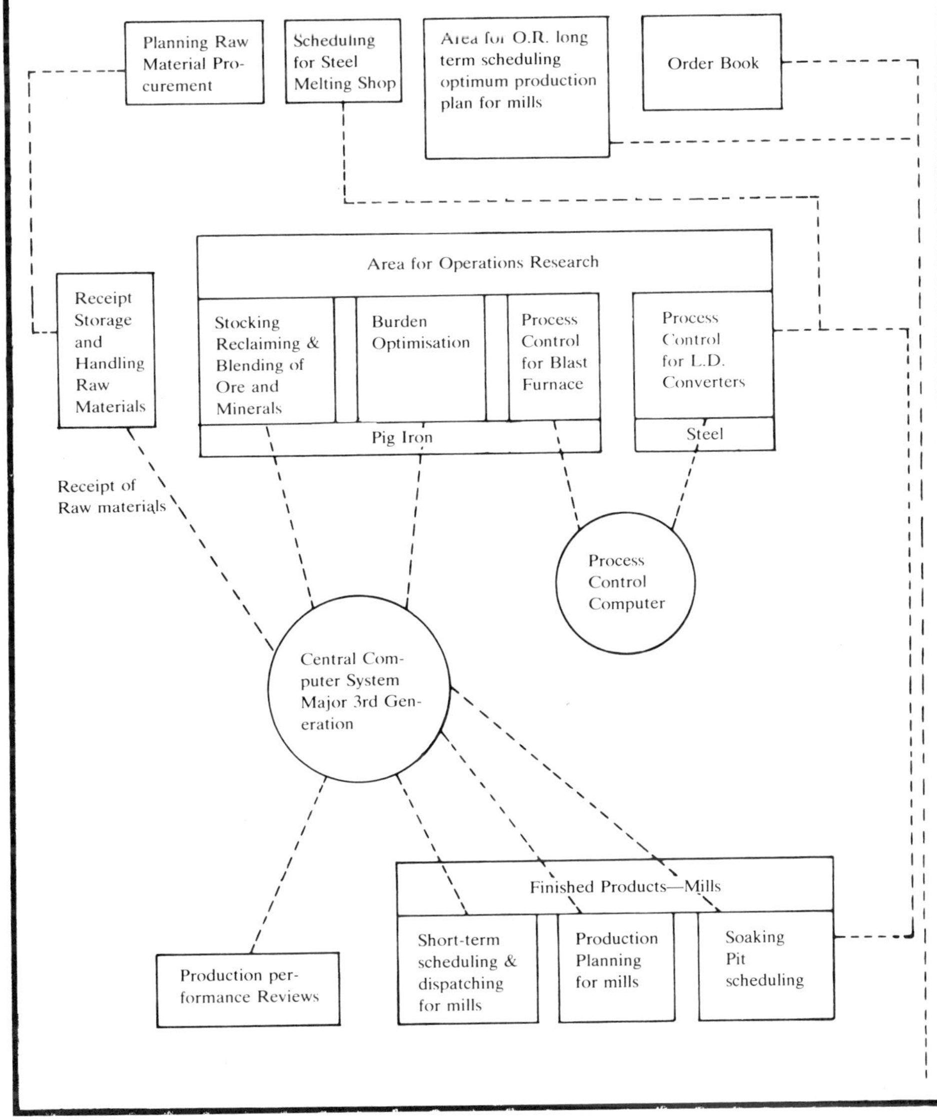


(2) Calculation of blow requirements (including bath additions);

(3) Corrective action in case the blow did not result in desired specifications of steel;

(4) Calculation of ladle additives, e.g., ferro-manganese or ferro-silicon;

(5) Self-correction based on the results of the present 'heat' to make adjustments to the marginal drift in the values of coefficients.

The last feature actually ensures that any deviation in the values of the coefficients is taken care of by an adaptive feed-back loop.

The important parameter in the above kind of application is the 'response-time' of the process control, since the computer has to control its environment by receiving data, processing it and returning the results sufficiently quickly to affect the environment at that time. A response-time of less than a few minutes is often needed, which justifies the real-time computer use. The algorithms for controlling the process are usually well-defined, thus making it possible to calculate the 'worst case' response time in order to assess the risk-margin with some precision. If the computer is not dedicated to the process, it must be capable of handling multiple interrupts, as the transaction needing a fast response-time has to break into programmes doing less immediate work. Files or data-bases become significant only where tables or data sets have to be maintained in the computer's memory, usually in the case of optimal control calculations.

The computer-based control helps in blowing oxygen on molten pig-iron in order to oxidise carbon, silicon, sulphur and phosphorus at a temperature of 1500°–1600°C inside the vessel, while adding fluxes through hoppers. The response-time for the entire process is 20–25 minutes. The cost of the system including all peripherals is Rs.54 lacs ($590,000) at the 1.7 million tonne level. The estimated saving in terms of improved performance and production is Rs.1 crore ($1.1M) per year, through reduction in tap to tap time, reduction of reblows, reduction on consumption of ferro-alloys and reduction in off-heats. The other intangible benefits are better co-ordination, improved lining life and integrated information control and data logging, resulting in better understanding of the process and for future development.

*Thyristorised Shear Control*: This is yet another application of on-line process control in the steel industry. The system consists of the following:

(1) Automated sensors for measurement of the length of. a semi-rolled red-hot billet while in motion;

(2) Thyristor control system for regulating the speed of a 1.2 MW flying shear motor; and

(3) Computer programs and necessary interface for cutting billet lengths to minimise end-crops.

The shearing operations in a 1 million tonne billet mill at the Bhilai Steel Plant are controlled using a TDC-312 computer. The on-line control system measures the length of the semi-rolled products and estimates the length of the final rolled products. It then determines the optimum billet length to produce minimum end-crop. Finally, it generates and supplies the desired signal value to the thyristor control unit in order to regulate the flying shear speed for obtaining the desired billet length.

The complete system, developed at a total cost of Rs.10.5 lacs ($115,000) (without the computer cost), is expected to save Rs.1 crore ($1.1 M) every year through decrease in end-crop lengths.

*Machine Tool Control*: Here the control program is implemented through off-line data processing by a digital computer, to achieve the following benefits:

(1) Machining of components of great complexity with tolerance conditions and repeatability;
(2) Flexibility and economy in the production of simple and complex parts.

The above benefits relate to increased efficiency and better service, and are hardly linked with saving of manpower.

The functions of the numerically controlled lathe machine are as follows:

(1) The product to be machined is dimensionally defined in an engineering drawing, along with material type, surface-finish and tolerance;
(2) The complete description of all motions and machine-functions required to fabricate the part is manually specified in the programme and the off-line data-processing by computer creates the control paper tape with detailed instructions;
(3) The machine is loaded with the job and the proper cutter. The control paper tape is now read on-line and the feed-back control system executes the command by guiding the cutter in the predetermined path.

It is to be noted that the entire job can be undertaken by an on-line, real-time computer which can generate the necessary cutting instructions after executing the part programme instructions and the results can be communicated on-line to the numerically controlled machine so that the material is moved accordingly.

Benefits in the above control process are; assured quality of performance on a repetitive basis, elimination of operator errors and feasibility of all operations being done at a lesser skill level once the basic programmes are properly written.

A few more examples of computer-based on-line process-control systems in India (data through sensors, on-line data-processing, on-line feed-back control, short response time) are as follows:

India's largest computer-based control system for monitoring the fast breeder test-reactor at the atomic power plant at Kalpakkam near Madras;

Computer-based control system with telemetry-interface operational at the Sriharikota Rocket Launching Facility for data acquisition and analysis;

Computer-based data-logging and control system under installation at Gujarat State Fertiliser Company at Baroda.

(N.B.: The fabrication and supply of the TDC computer systems for the above three process-control applications are by ECIL, Hyderabad.)

## ILLUSTRATIONS OF MANAGERIAL CONTROL

SINCE THE DECISIONS are not programmed here (although repetitive), automatic computer control cannot be attempted. Data is collected from source documents manually and conveyed to the computer through terminals. Two alternatives are possible for operator guidance:

(a) Data-processing is on-line, with off-line feedback of control instructions for action, with a medium response time;
(b) Data-processing is off-line, with off-line communication of control response for action, with a long response time.

A few examples are furnished of both types of guidance system in the succeeding paragraphs which are planned in an on-line mode on Burrough's 6800 series computers at TISCO and Bokaro Steel in India.

**On-line management control**

(i) *On-line production planning and control system* can be of use in any large industry. It has a high pay-off potential in an integrated steel plant where the utility is particularly at the slabbing mill, hot rolling and cold rolling mills, and the merchant mills. Major control objectives are visualised as follows at Bokaro Steel (Technical Services Division, 1978);

Effective order processing ensuring best customer satisfaction;
Improvement of yield;
Reduction in mix-up of finished materials and in-process inventory;
Reduction in work-in-process inventory;
Optimum planning of quality and sizes in rolling programmes and campaigns to improve life of rolls, ingot moulds, etc;
Reduction in fuel consumption.

Effective benefits, going by the experience in the West and Japan, are expected to be the following:

1.3 per cent improvement in yield;
10,000 tonnes reduction in mix-up losses;
20 per cent reduction in work-in-progress inventory; and
Reduction in fuel consumption in furnaces, along with improved ingot mould and roll-life.

The annual financial benefits are expected to be between Rs.2.1 crores ($2.3 M) and 4.5 crores ($4.9 M).

(ii) *On-line inventory control system* can provide a large industry with corrective, follow-up and review facilities. Control of inventory-levels and triggering of automatic purchase-orders are feasible in order to cut out excessive lead-times and to generate "alerts" for undue delay in supplies or deficient stocking or excessive stocking. In TISCO, more than 100,000 items are handled at Jamshed-put, 20,000 at mines and 20,000 at collieries. This gives rise to nearly 40,000 transactions per month. At Bokaro Steel Plant, over 120,000 items of inventory are maintained with transactions nearing 15,000 per month. It is proposed at both these plants to locate terminals in the stores to collect details of transactions and generate alerts for exceptions. This will facilitate minimum corrective action and provide a facility for controlling inventory level and cutting out excessive lead-time.

A benefit of around Rs.1 crore is considered feasible at each Plant by reducing lead-time and by bringing down stocks to a 9 months consumption level from the current average of 12 months.

(iii) *On-line raw material control system* is useful where such raw material as iron ore and coal have varying compositions over time and where sophisticated equipment such as stackers and reclaimers are available for oreblending purposes. Blending is a must for consistency in sinter and coke composition, otherwise the blast furnaces behave erratically due to wide variations in the burden composition. Because of large amounts of materials and their quality fluctuations, proper blending at the storage silos requires quick processing of large data volumes using mathematical models and simulation runs.

The system at Bokaro Steel is planned to have terminals at the blending yard, storage silos and inward exchange yard. While maintaining master information on all raw material stacks and silos, the computer is to get feedback information on wagon arrivals, material analysis and details of wagon and rake contents. Control instructions, based on statistical models, are to predict the blend and composition of material going to sinter plants and blast furnaces in order to improve the latter's operations. Even a 1 per cent gain in productivity of the blast furnace should result in 20,000 tonnes of additional production per year.

## Off-line Management Control in a Batch-Mode

(i) *Computer-based manpower planning system* planned for 3,200 executives and 30,500 non-executives at TISCO. The master information relates to dates of birth and joining, grades and designations, marital status, family particulars, blood group, accommodation and bonus eligibility, qualifications and training. All changes in their status are to be furnished as feedback information, in order to cover control areas like recruitment, growth planning, training, refresher courses, welfare facilities, trade conversion training, and so on, regarding individual employees and their specified needs.

(ii) *Computer-based structural design planning system* is meant to assist in the design phase of substantial construction work at Bokaro Steel, involving an additional 1.3 lacs tonnes of structural work and 2.2 lacs tonnes of equipment erection for the 4.0 MT stage. Master information relates to design specifications, needs of the plant and indigenous manufacture and supply of sub-assemblies and components. Based on feedback information on design progress, changes in requirements and performance evaluation, control results are required by solving structural engineering equations.

(iii) *Computer-based maintenance planning system* aims at not merely recording the status of major equipment and planning for their components and spare parts (to be made available in time for preventive maintenance), but also indicating control schedules for the desired shut-downs and subsequent maintenance activities, consumption of spares and progress milestones.

## ILLUSTRATION OF STRATEGIC CONTROL

THE OFF-LINE DEMAND and planning reports, envisaged above, are yet to be stabilised in India. To that extent, an integration between operational control and managerial control is a stepping stone to strategic control at corporate level.

Rastogi (1979) attempted a cybernetic analysis using a systems dynamics approach, to some ill-defined problems and policies at TELCO and Ashok Leyland. While this could be termed strategic control in industry, his analysis suffers from a lack of quantitative rigour and a facile assumption of communication of all feedback information, without specifying where and how precisely it is linked with control instructions. Srinivasan (1977), on the other hand, illustrated a few information indicators for strategic planning in industry, such as, "economic trends and forecasts", "political trends and developments", "social trends and forecasts", "competitive information", "technological trends and forecasts", market data and internal data for planning. These indicators are outputs from a data-bank, which has not so far been created in any Indian industry but, with progress in communications and computerisation, is well within the limits of realisation. Once automated, even for a long-term time-horizon, these indicators can considerably help strategic control which is currently dependent on manual information.

## CONCLUSION

TYPICALLY, operational control processes in industry are most susceptible to improvement with recursive learning situations. In fact, the statistical and O.R. models can be progressively made both sophisticated and comprehensive, as logical fallacies and missing data are identified with passage of time. At the managerial control level, the learning situation is almost wholly human, although it can cope with advantage with better monitoring and planning reports from the data-processing system. 'Appropriate Informatics' will pave the way to such judicious applications which can bring the best results to industry in the Third World.

**References**

Clayson, J.E. (1980) *How Relevant is Operational Research to Development? The Case of Kenyan Industry*. J. of the OR Society **31** (4) 293–299.

OECD (1979) *Facing the Future*. Organization for Economic Co-operation and Development, Paris.

Rastogi, P.N. (1979) *An Introduction to Social and Management Cybernetics*. East-West Press, Delhi.

Rateau, O. (1981) *Informatics: Evolution and Use in the* 1980*s, with Particular Reference to Developing Countries*. In *Computers in Developing Nations*, North Holland, Amsterdam.

Srinivasan, A.V. (1977) *Management Information Systems*. Mimeograph, Admin. Staff College of India, Hyderabad.

# Thoughts on Informatics and Small- and Medium-scale Enterprises in the Third World

*Maurice Goldsmith*
*Science Policy Foundation, United Kingdom*

## CONCEPT OF AID

Now, at the beginning of the 80s, as we move into the third development decade, the concept and nature of aid is being redefined. No longer do we continue to cultivate the fantasy of aid as essentially a good deed. That is best described as 'tied charity', which few countries seeking self reliance see as anything but a neo-colonial device. That is out, completely unacceptable. It is not aid, but support, that is needed. Support for what?

To ensure that the poor do not become poorer, but are provided with a road leading away from their poverty; that the hungry are provided with food, not purely as a charitable handout but on the basis of an innovative social process which teaches them how to grow food themselves; that the wanderers, dispossessed and homeless in their hundreds of thousands, are provided with secure personal roofs; that the illiterate, numbered in millions, are provided with means of learning to become an unchallengeable and vital national resource.

In this task, industrialisation is a key, an important element in the process of modernisation and diversification. It contributes to increased employment, and to the establishment of a modern, outward-looking group of managers and skilled technicians and workers. But—a note of caution—it can contribute, also, to serious economic and social imbalance.

The Third World (TW) countries are a highly differentiated group. Thus, according to income levels:

(i) Least Developed, e.g., Bangladesh, Tanzania, Yemen, Guinea, Upper Volta;
(ii) Other Low Income (less than $400 per capita) e.g., Egypt, Indonesia, Pakistan, India, Vietnam;
(iii) Low Middle Income ($400–1000) e.g., Korea, Morocco, Philippines, Tunisia, Bolivia, Swaziland, Jordan;
(iv) Upper Middle-Income ($1000–2000) e.g., Brazil, Yugoslavia, Mexico, Argentina, Turkey;
(v) High Income (over $2000) e.g., Israel, New Caledonia;
(vi) OPEC Members.

## INDUSTRIAL DEVELOPMENT

THE INDUSTRIAL DEVELOPMENT record of the TW countries *as a group* during the past 20 years has been impressive, whether we use the criterion of manufacturing value added, or the share of the national industrial sector in the GDP, or the share in world industrial production (Table 1).

**Table 1.**

| | Percentage Average annual growth 1960–1976 | | Percentage Share of Industry in GNP | |
|---|---|---|---|---|
| | Industry | Agriculture | 1960 | 1976 |
| Low-income developing countries | 6.0 | 2.1 | 17 | 24 |
| Middle-income developing countries | 7.6 | 3.1 | 32 | 37 |
| Industrialised countries | 4.9 | 1.3 | 40 | 38 |

Industry has been growing much faster than agriculture: about three times as fast in the low-income countries, and more than twice as fast in the middle-income countries. But this growth has been uneven among different countries and country groups. For instance, the rise of the share of TW countries in world industrial production has been from 14 to 19 per cent between 1963 and 1977. By contrast, the poorest countries are falling relatively further and further behind.

## SMALL AND MEDIUM SCALE ENTERPRISES

Small and medium enterprises (SMSEs) play an important part in the productive structure in the First World (FW). Usually they make up 90 per cent of the total number of companies, and provide from 30 to 60 per cent of total employment. They are prominent in: metalworking, capital goods, textiles and clothing, food, furniture, ceramic products, non-metallic minerals, and publishing.

In the main they are concerned with a domestic market, although there are signs of growing international involvement. For instance, they account for from 15 to 30 per cent of the subsidiaries of parent companies of European origin (primarily West Germany, France, the Netherlands, Sweden, Switzerland and the UK) operating in Brazil, Mexico, Peru, and Venezuela, and for a similar percentage of the licensing agreements made by European firms in those countries.

What part do the SMSEs play in TW countries? Clearly, a role more important than that in FW countries.

In a country such as India the small scale industrial sector contributes from 40 to 50 per cent of total industrial production.

The agricultural sector in most TW countries contributes more than 50 per cent of the GNP.

Thus, special attention has to be paid not only to the small business entrepreneurs, but also to the farmers. Together, they provide employment for over 60 per cent of the less sophisticated, lowly skilled, and mainly illiterate.

## INFORMATICS IN THE THIRD WORLD

At the moment, informatics in TW countries serves the research and academic groups, big business, and large-scale industry, who already understand the value of information in their work.

If informatics is to support national development it must concern itself with the small and medium scale.

Informatics is relevant immediately as an information system for the newly industrialising countries (NICs—Spain, Portugal, Greece, Yugoslavia, Brazil, Mexico, Hong Kong, Korea, Taiwan, Singapore.) It is a highly heterogeneous group with respect to geography, per capita income, and development policies. But, it is also characterised by rapid growth in the level and share of industrial employment, expansion of export market shares in manufactures, and real per capita income levels approaching those of some of the advanced FW countries.

What advantages do SMSEs in TW countries offer? These are some suggestions:

(i) Their technology in terms of labour intensity and small scale of operation is most suitable to domestic conditions;
(ii) Their management style is national and traditional, therefore acceptable;
(iii) They are disposed to engage in joint minority ventures with foreign enterprises;
(iv) This involvement may secure more diverse sources of technology.

Informatics as a key area of knowledge in the service of the small and medium scale enterprises in TW countries is of significance only if applied to the attainment of national objectives.

This means that a greater awareness of the implications of informatics must be developed at government level. And government must have a plan for national development, that is science and technology policies, with informatics as a key tool.

We know very little about the appropriateness of the transfer of knowledge in the area we have been considering. More factfinding and much research is needed, such as:

(i) Gathering of information to provide a general picture of the phenomenon of informatics transfer;
(ii) Indentification of industries in which special attention to the needs of SMSEs might be of national advantage;
(iii) Study of the organisational forms which SMSEs might require to have introduced to take advantage of informatics;
(iv) Study of the impact of informatics transfer on the SMSEs;
(v) Study of the public policy implications for government.

Informatics might provide that factor in the world economy which will enable those countries only now entering the industrial development scene to secure in harmony with the FW countries growth rates leading to economic independence and the end of aid.

# Making Computer-based Problem-solving Techniques appropriate to Smaller Enterprises in Developing Countries

***J.E. Clayson***
*Management Consultant, Paris*

## OPERATIONAL RESEARCH

BECAUSE IN THE WEST much money has been spent towards designing complex computer models that seek optimal solutions to complex industrial problems, there is a tendency to assume that this approach to informatics is equally appropriate for developing countries. A closer look, however, at the history of operational research (OR)—the discipline that encourages formal mathematical modelling in government and industry—reveals that these large, complex models have a very limited applicability since their great costs can be borne by only the richest organisations.

OR was created in England during World War II to analyse logistical aspects of modern warfare. Since then its applications have multiplied and nearly every large government agency and enterprise has had some kind of OR department.

For those not familiar with the discipline of operational research, let me give a standard definition that is used by the British Operational Research Society: "Operational Research is the application of the methods of science to complex problems arising in the direction and management of large systems of men, machines, materials, and money in industry, business, government, and defense. The distinctive approach is to develop a scientific model of the system incorporating measurements of factors such as chance and risk, with which to predict and compare the outcomes of alternative decisions, strategies or controls. The purpose is to help management to determine its policy and actions scientifically".

Until recently, OR people have been an intellectual and organisational elite. They held advanced degrees in hard sciences and have reported to top officers in the organisations where they work. This has given them authority to cut across departmental boundaries and to build models reflecting the complexities of a total situation. They have also been paid large salaries. This kind of OR is enormously expensive.

If a mathematical model is built to encompass many different organisational goals—e.g. profit, rate of return, growth, job creation, environmental impact, political influence, etc.—along with a description of how each of these goals is affected by the variables that describe the interrelationship and dynamics of the individual elements of the organisation, the number of equations the model contains soon becomes enormous. In order to manipulate these equations towards finding the best solution, heavy demands must be made on computers and computer staff. The collection of specialised data needed to support these models and the time necessary to verify their appropriateness all contribute to their cost even before they are applied. Obviously, unless the savings generated by their application are substantial, the entire exercise is not financially justified. And since, characteristically, large models of this type generally produce savings of less than 10%, it is only in the largest industries that these savings are enough to cover their costs. In addition, operational research that is only practised by an elite group may fail to improve the problem-solving skills of the managers, upon whom the ultimate success of many firms rests.

## SIMPLE MODELS

FORTUNATELY, there is another, simpler approach to operational research. This alternate approach is based on the notion that the search for 'optimality' is not realistic in most contexts: it is either too expensive, takes too long, or requires data that are too difficult to gather. This alternate approach to OR concentrates on building much simpler models that can be applied by the line managers to solve the bulk of their everyday problems. The resulting solutions may not be optimal, but they are quickly produced and workable. Such models are usually based on the manager's own rules of thumb—his 'heuristics'—that he has developed through experience for finding his own answers to problems. In this way, the very "querky" aspects of a real production situation can be accommodated with a manner and ease that the more formal mathematical modelling techniques do not permit.

There are other attractive features to this approach as well. Because it links the manager directly to terminals, micro-computers, or programmable calculators, it allows the manager more freedom to find his own answers. This manager-centred OR not only produces faster results but it seems to provide an incentive for the manager to experiment and improve upon his own heuristics.

Having said all this, it is interesting to note that the idea that simple models may be more efficient than complex ones is spreading. Many large American firms are closing their centralised OR departments and dispersing OR activity and OR personnel throughout the operating divisions. The same trend is seen in universities and professional associations which are beginning to integrate OR techniques into their training programmes in the traditional business disciplines of accounting, finance, marketing, and production—rather than treating them as worthy of special disciplinary status.

I think that this rather subtle shift of viewpoint in Western industry has important—and perhaps even radical—implications for developing nations: simple computer models based on heuristics—that is, common-sense approaches to finding solutions—are not only more cost effective, but they provide the opportunity for many more people in a given country to improve their own skills. Skills which are self taught and self learned tend to have a more lasting impact than those which are imposed; and the possibility for continuous, incremental improvement becomes institutionalised. This is one key to industrial development.

I think the lesson for countries with limited resources is obvious: an investment in simple and heuristic models of informatics which are applicable throughout the industrial sector on a grass-roots management level may be much better value for money than investment in fancy models which will, at the very best, be relevant to the activities of only a limited number of people in only the largest organisations.

## HISTORY OF TECHNOLOGY

At this point in our discussion it might be helpful to have a look backwards in time. The history of technology has shown that new technologies arise whenever new tools are created which are easy to use, rely on existing skills, and are applicable to a wide variety of situations. Let me illustrate the point I am trying to make here by citing two well-documented cases from the history of technology: the brace and bit, invented in the 1420s by an unknown Flemish carpenter or shipwright, and the steam engine, invented by Thomas Newcomen in 1712. Both of these machines had spread throughout Europe and were being applied to a wide spectrum of industries within a matter of years. And this without benefit of modern communications.

The invention of the brace and bit was revolutionary insofar as its usefulness was recognised simultaneously in sectors ranging from ship-building, to millwrighting and the construction trades. In the same fashion the atmospheric steam engine was diffused from the mining sector, where it was first used to pump water, to such diverse industries as textiles, brewing, and municipal sewage disposal. In both cases, and following our rules of thumb for the rapid diffusion of technology, these tools were successful because they required few new skills to operate, while at the same time mobilising those skills towards new outlets and applications. The tool itself created its own demand for an environment in which change and development were self-generating.

## MICROCOMPUTERS

Today's equivalent of these revolutionary, historical tools is the ubiquitous pocket calculator. Like its antecedents, the pocket calculator is relatively cheap, simple, fast, and fun to use. In addition, the pocket calculator—like other successful tools—encourages the new user to experiment and adapt it to other situations. This constant experimentation creates its own demand for incremental improvements in the technology itself. Thus the simple hand calculator lays the groundwork for the spread of a more advanced calculator with a built-in memory and additional mathematical functions, which leads in turn to programmable calculators, and most recently to micro-computers.

These micro-computers have the same potential for mobilising skills useful to industrial development as the historical examples given. But, their potential will not be exploited until such time as their utility becomes self-evident. What is needed to bring this about is the creation of programmes sufficiently easy to use and versatile, that their diffusion is assured. Most of the software on the market today is specific to such problems as payrolls, stock control, and invoicing procedures. These are specialised routines requiring specialised skills for their application, and which do nothing, consequently, to improve general management skills.

## UNIVERSAL PROGRAMME: TABULA

During the past year, as part of a course I am teaching in applied operational research, I have been experimenting with the design of programmes which will go beyond single, specific applications, to something much more universal.

How have I gone about this? First, I had to identify a management task so common and universal that any tool developed to assist in performing it would be assured a wide diffusion.

It seemed to me that the manager's—any manager's—most basic task is to *order data.* He must do this, to a greater or lesser degree, each and every time he wants to evaluate current procedures, compare alternative ways of doing things, predict future demands, or to document various financial aspects of the organisation. The manager's problem, as I see it, is to take the raw data and to find how to assemble and present it in a meaningful way.

One of the most effective presentations is done in tabular format, where the different elements are laid out side by side and can be compared at a glance. In this case, the picture is worth—literally—a thousand words because it combines the temporal and spatial dimensions of the problem in one display. Double-entry bookkeeping is the classic illustration of this construct.

Many managers already have rules on how they like to solve problems through the tabularisation of data. This is a form of heuristic model building and, as I have already argued, it is at this level that the computer's assistance is most effective. In other words, this is the critical stage in which to introduce a tool capable of mobilising and amplifying existing skills.

Having selected as my management task the tabularisation of data, I set about designing a programme which would help. It is called TABULA.

TABULA is a computer routing that can be applied to all the problems inherent in the tabularisation of data. It uses a micro-computer with screen and printer. Its input requests, operation, and output requests are all in a "friendly" conversational mode. The programming of TABULA is relatively simple and it can be applied to all the business disciplines from accounting and finance to production and marketing.

TABULA accepts data in the form of numbers, words, and simple arithmatic; it manipulates and orders this data according to the user's heuristics, and then converts the results into a visual display on the computer screen. Once the shape of the picture has been fixed for example in rows and columns—the user has the freedom to experiment with different inputs, and he can see instantly what the impact will be. Or, he can experiment with his own heuristics by playing with the way the data are manipulated and displayed.

Besides the conventional row-and-column format, TABULA can transform the same data into other graphic modes such as curves, histograms or maps, thus providing the user with a choice of modelling tools.

In my OR class we have successfully tried TABULA on the following kinds of problems: standard accounting exercises and reports, inventory control, production planning, sales forecasting, warehouse location, and product distribution. We have even been able to convert French accounting reports into their American equivalents. Having defined the conventions of each national system and fed these heuristics into TABULA, the routine was able to produce for us the desired balance sheet on demand, while also allowing us to investigate a whole series of what-if questions just by changing the data input.

The real strength of TABULA, or any other routines of this general type which can be developed, would seem to be their obvious and immediate utility for managers. TABULA can be mastered without learning a special language or having any particular knowledge of computers. And once TABULA is installed it should become the vehicle for the continual introductions of ever more sophisticated methods for the manipulation and display of data.

An informatics which can promote self generating change and improvement, at minimal cost, is obviously appropriate to the needs of developing countries, and I think we must exploit its potential.

# A Proposal for a Technical Information System for Transfer of Technology to Small Industries in Tanzania

***D.K. Rulagora***
*Small Industries Development Organisation, Tanzania*

## INTRODUCTION

THE UNITED REPUBLIC OF TANZANIA lies in the East African region on the Indian Ocean. It covers an area of 939,361 sq. km and has a population of 18 million people. More than 14 million are peasants who live in the rural areas. The economic growth is 4.5 per cent, whereas the population growth is 3 per cent annually (Bureau of Statistics, 1979).

In the current Five Year Development Plan (1976–1981), there is more emphasis on industrial development. The industrial sector will spend over 27 per cent of the country's total capital investment. Small industries are encouraged in Tanzania for the following reasons:

(a) to provide inputs for agricultural development;
(b) to produce goods and services to meet the basic needs of rural population;
(c) to enable peasants to be productive throughout the year as agriculture is seasonal;
(d) to provide employment opportunities to school leavers whose influx in the cities has increased the urban population by 50 per cent in the last five years;

(e) small industries use less capital and they can be managed by local entrepreneurs;
(f) they can utilise local resources and use less energy;
(g) to eliminate the social and economic imbalance between rural and urban areas;
(h) to upgrade the skills of traditional technology;
(i) to provide interlinkages between large industries and other sectors (Ministry of Finance and Planning, 1976).

On the other hand, the "Knowledge Capital" has become a dear commodity to small industries to the extent of costing 65 per cent of the cost of a plant. This has been confirmed by a UNIDO technical adviser who said ". . . . the absence of knowledge becomes one of the prime hindrances in SIDO's performance" (Ghosh, 1980). In this context technological information is the life blood of industry for the following reasons:

to reduce the cost of software,
to simplify scientific and technological innovations,
to tap the necessary technology due to information explosion,
to enable decision makers to choose appropriate technology,
to offer services required for technical training and management of industries (UNIDO, 1973).

## FUNCTIONS OF SIDO

THE SMALL INDUSTRIES DEVELOPMENT ORGANISATION (SIDO) was established by Act of Parliament in 1973. The objectives are to plan, promote and develop private and public small industries in the country. The functions include planning and research, technical and management training, marketing and supplies services (Tanzania Government, 1973). The organisation chart is shown in Appendix A.

The official definition of the small industrial sector does not specify the limits for both investment and employment. For the purpose of this paper, the small industries covered employ 3–50 people and have capital investment of $2,000–£500,000 (TS 40,000–TS 10m).

In 1975 the small industries employed 21,528 people, and the figure rose to 50,000 in 1978. While the former represented 18 per cent, the latter was 20 per cent of the total workforce in the industrial sector. The number of industrial units increased from 1,742 to 5,000 in 1975 and 1978 respectively. The Economic Survey has shown that small industries contributed 1 per cent to GDP in 1975. In 1978, the contribution increased to 2 per cent of GDP. In the current plan (1978–1983) more than TS 1,700m (£100m) will be invested in 1,067 factory-type and 2,000 non factory-type small industries. They will generate 3 per cent of GDP and employment opportunities will rise to 80,000 (Tanzania Government, 1978).

## TECHNOLOGY TRANSFER

TECHNOLOGY is said to be the most important factor responsible for economic growth. According to some scholars, the 'upward spiral' is brought about by higher technology. This leads to higher incomes, savings and capital, improved technological developments and higher productivity (Stewart, 1972; Mansfield, 1968). Technology has two components, namely 'embodied' (hardware) and 'disembodied' (software), the latter is sometimes known as "Knowledge Capital" (Dalgic, 1979).

Modern technology is sophisticated and it needs systematic planning and rational choice. Technology is not neutral, henceforth it has to be appropriate. Its appropriateness has been defined as "harmonious with social, economic and ecological conditions and it has to foster the countries' self-reliance" (Sachys, 1979; Braun and Collingridge, 1977; Schumacher, 1973).

Technology transfer is a "purposeful movement of established technology from place to place, company to company, or use to use" (Schon, 1976; Rosenberg, 1967; Spencer, 1971; Solo and Rogers, 1972). The objective of technology transfer is technological innovation, i.e. improved and efficient ways of performing things that are more economical. Such innovation would lead to invention, i.e. a process of bringing new technology into being or new technology created in a process.

Professor Allen and colleagues have identified two 'coded systems' in the transfer of technology. One is the 'acquisition code' whereby the technological information gained is applied to specific situations (Allen, 1977).

Technological innovation requires Technological Information Systems to plan and communicate it to users. This saves time, money and ensures that appropriate technology is transferred from the industrialised to the developing countries.

## PROBLEMS OF TECHNOLOGY TRANSFER

ACCORDING TO RESEARCHERS "acquiring information is costly and the demand for technology is always a demand for information . . . ." (Alenger, 1978).

Since 1975 the Organisation (SIDO) had embarked on countrywide industrialisation programmes involving factory type small industries. Several bilateral and multilateral agreements and individual contracts have been signed to supply machinery, equipment, raw materials and technical know-how. Some of the contracts are the World Bank loan for Urban Industrial Clusters, the 'Sister Industry' Programme (Swedish) and the Indo-Tanzanian Projects, to name a few.

The problems of technology transfer are summarised by Swedish researchers who have said, "at first glance it may seem that the developing country only purchases machinery (hardware components), but more important it purchases the software components, i.e. all the technological and managerial know-how required to obtain a well functioning industry".

In the case of the Swedish projects, the cost of software ranges from 30 per cent to 65 per cent. The cost of creating a job in a small industrial unit has risen by 48 per cent, TS 27,000–TS 40,000, in the last three years. The cost of technological information has increased due to the following reasons:

(a) the disclosure of technology, although it is not a new technology, costs up to 25 per cent of the price of the plant;
(b) the organisation spends £23,000 for each small firm to train its personnel abroad;
(c) the maintenance of a Swedish engineer exceeds £2,500 (TS 50,000) per month. The amount is 12 times the salary of a top executive in government and parastatals;
(d) translation of products formulae and other operating and maintenance manuals;
(e) consultancy fees (foreign consulting firms). In Appendix B the cost of software in one of the Swedish Projects is analysed (Rulagora, 1979).

For the other 'Turn Key' projects, the following bottlenecks have been experienced:

(a) short lading and breakage of components due to choice of wrong suppliers,
(b) lack of layouts and operating processes;
(c) lack of market information during project planning;
(d) manufacturers are unaware of distribution channels;
(e) lack of advertisement and sales promotion;
(f) lack of innovation makes some industries less competitive;
(g) incorrect accounting and sales forecasting;
(h) shortage of raw materials and spare parts;
(i) shortage of utilities such as power and water;
(j) transport and communication problems (Kilewo, 1978).

## A PROPOSAL FOR TECHNOLOGICAL INFORMATION SYSTEM (TIS) FOR SMALL INDUSTRIES IN TANZANIA

THE TERMINAL REPORT of the UNIDO Technical Adviser has the following recommendation: "as more small industries join ranks, it would be necessary to store the data on establishment, operation and up-date those on a regular basis. A mini computer is to be used for storing, sorting and retrieval of data . . . ." (Ghosh, 1980).

The functional units of the proposed Technological Information System are shown in Appendix C. Some of the seven units are Information, Documentation and Publication, Field Liaison and Question and Answer Services. Others are International Relations, Public Relations, Systems Analysis and R & D and Administration.

The TIS will have the following functions:

(a) to reduce the cost of software through disclosure of technological information;
(b) to provide technological innovation for economic development;
(c) to provide a Data Bank for planning and choice of appropriate technology;
(d) to provide construction layouts;
(e) to publish and distribute plant installation and maintenance manuals;
(f) to give Technical Training Techniques for in-plant training which save time and money;
(g) to undertake translations and publication of production processes;
(h) to provide market forecasting and sales promotion (Klintoe, 1979; UNISIST, 1977).

The organisational set-up will be based on matrix organisation and control, with the following four features. The science and technology policy will be controlled by the Executive Committee which will represent government ministries and organisations involved in Technology Transfer. The Management Committee will plan and review action plans. The matrix control and coordination will be in the form of vertical and horizontal relationship instead of work alignment. There will be feedback and evaluation for renewal of inputs. The systems approach in the organisation of TIS is shown in Appendix D.

The budget proposal for the project is shown in Appendix E. The local cost will be paid for by the Tanzanian Government. The United Nations Industrial Development Organisation has been requested to cover the costs of experts who will appraise and establish the project. Other facilities such as telex services and computer terminals are within SIDO and the Tanzania Scientific Research Council respectively.

## INTERNATIONAL CO-OPERATION

THE BOTTLENECKS POINTED out by the United Nations Conference on Science and Technology Development in 1979, which hinder the transfer of technological information, are experienced by the industrial sector in Tanzania. The following are more pressing:

(a) lack of communication facilities;
(b) costly software (up to 65 per cent of cost of plant);

(c) lack of a data bank for science and technology;
(d) inadequate number of technological information scientists;
(e) lack of information technology personnel;
(f) science and technology activities are not fully involved in technological innovation;
(g) lack of information systems;
(h) low investment in R & D (0.25 per cent of N.I.) (U.N., 1979).

The Small Industries Development Organisation, Tanzania, will need bilateral and multilateral assistance to carry out the following activities:

(a) planning and setting up technical information system, e.g. Industrial Information Section of UNIDO:
(b) training scientific and technological information staff, e.g. INDIS and FID;
(c) providing information services in the form of bulletins, catalogues, abstracts, journals, bibliographies, films, etc.;
(d) programming text books, audio-visual aids, e.g. Documentation Centre in Ireland;
(e) technological information exchange through conference, seminars, study tours (international technology gatekeepers);
(f) publishing for small firms, e.g. Small Firms Information Centre in U.K.;
(g) technological innovations, e.g. TORDOK (Turkey);
(h) Small Industries Consultancy and Advisory Services, e.g. SENDOC (India), CENDES (Ecuador), Small Businesses Administration (US);
(i) Appropriate technology R & D, e.g. VITA (US), ITDG (UK), GATE (FR Germany) and ITAR (UN);
(j) Documentation of patents, e.g. Patents Institute (Computer Policy Advisory Committee, 1978; Pauline, 1977; Allen, Piepomeier and Cooney, 1971).

## CONCLUSION

The importance of technological information in industry is summed up in the UNIDO report which states that: "SIDO has virtually no library at the moment, and the above-mentioned information resides in the staff member offices. Although SIDO is linked with several institutions throughout the world to collect information, they could use additional information on processes, trade catalogues . . . ." (Lafond, 1980). Technology transfer will be enhanced by the technological information system that will collect, process, store, retrieve and disseminate data.

## Appendix A: Organisation Chart for SIDO

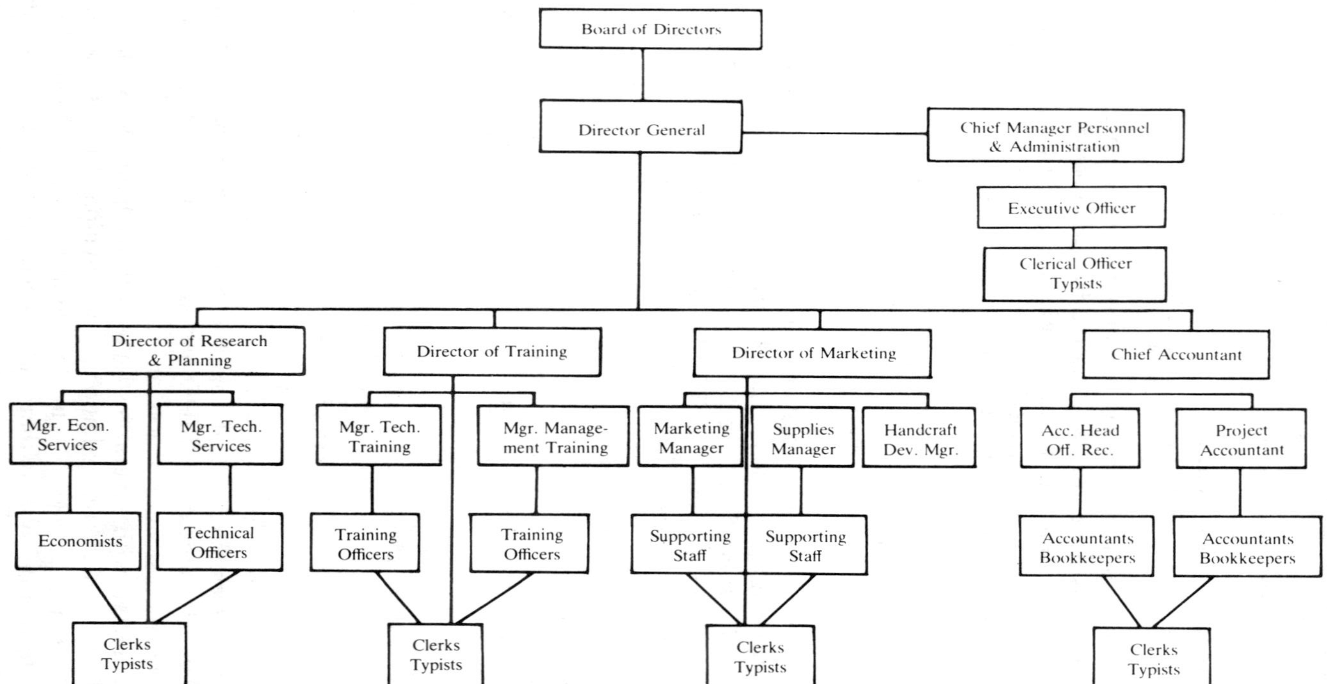


Note: (1) 20 Regional Extension Offices with small industries promotion officer (SIDO Economists and Technicians/Engineers for Research and Planning).
(2) 10 Training cum Production Centres and Small Industries Management Institute Training.
(3) 10 Industrial Estates (Research and Planning).

## Appendix B: Cost Analysis of Software for Electric Motors

| Serial No. | Item | Cost in Swedish Kroner |
|---|---|---|
| (1) | Technology disclosed | 563,000 |
| (2) | Training for six entrepreneurs in Sweden | 200,200 |
| (3) | Fee and allowances for Swedish engineers | 273,000 |
| (4) | Fare and other costs for Swedish engineers | 103,250 |
| (5) | Fare for six trainees | 84,000 |
| (6) | Local allowances for trainees | 136,800 |
| | TOTAL | 1,360,250* (£163,560) |

* This is 65 per cent of the total cost of the plant(£252,000).
Source: SIDO/Bevi Company Contract, 1978.

## Appendix C: Organisation Chart of TIS (proposed)

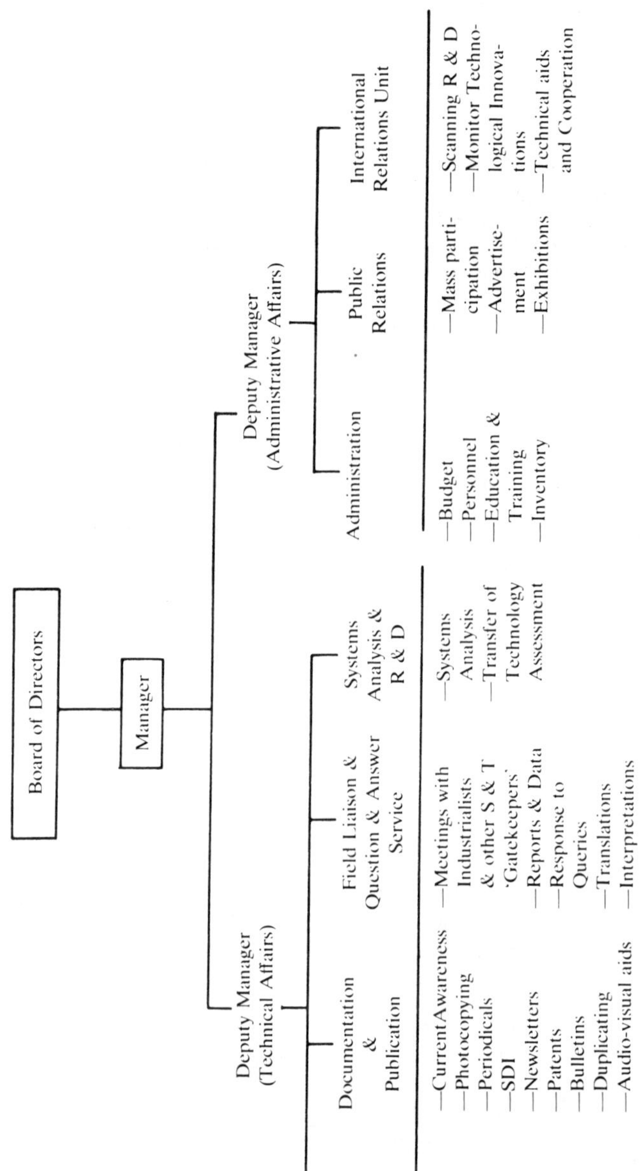

## Appendix D: Systems, Boundaries and Information Flow

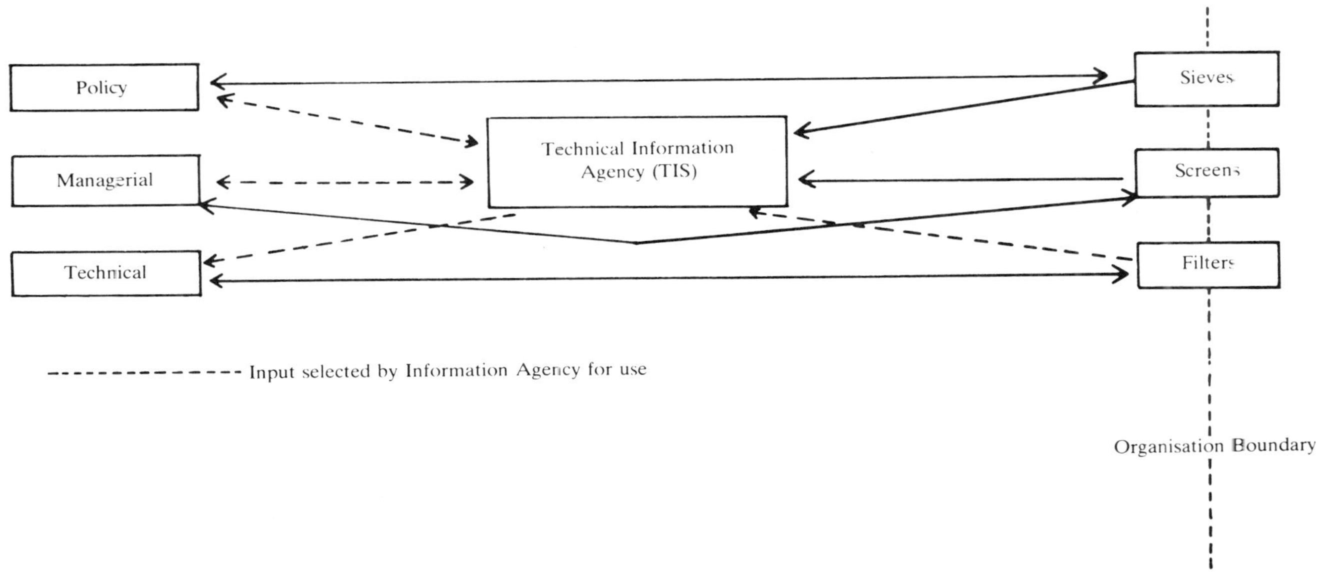


Source: Baker, Frank (1973) *Organisational Systems, General Systems—Approach to Complex Organisations.* P. 24. Richard Irwin, New York.

## Appendix E: Budget Proposal for TIS

| | | | Tz. Shillings |
|---|---|---|---|
| 1. | Construction of building, 320 $m^2$ | | 320,000 |
| 2. | Staff:* | | |
| | 1 UNIDO Adviser | 250,000 p.a. | |
| | 1 Manager | 48,000 p.a. | |
| | 2 Deputies @ 45,000 p.a. | 90,000 | |
| | 7 Unit Supervisers @ 36,000 p.a. | 252,000 | |
| | 8 Middle and Lower Calibre @ 20,000 p.a. (average) | 160,000 | |
| | 6 Supporting Staff | 50,000 | 850,000 |
| 3. | Equipment: Photocopying Machine Dublicating Machine Audio-Visual Aids etc. | | 450,000 |
| 4. | Furniture | | 10,000 |
| 5. | Telephone and Telex Charges | | 40,000 |
| 6. | Fees for Books, Journals, Catalogues | | 150,000 |
| 7. | Vehicles (cost and maintenance) | | 100,000 |
| 8. | Stationery and Postal Services | | 125,000 |
| 9. | Lighting and Water | | 10,000 |
| 10. | Travelling and Meetings (local and overseas) | | 220,000 |
| 11. | Contingencies (2.5 per cent) | | 55,000 |
| | TOTAL | | 2,330,000* |
| | | | (£137,000) |

* Based on local costs and salary scale structure in 1978 (£1 = 17 Tz. shillings).

## References

Alenger, S. *et al.* (1978) *Transfer of Technology of Small Industries Through Sister Industry Programme in Tanzania.* Chalmers University, Gothenburg.

Allen, T.J. (1977) *Management Flow of Technological Information Within R & D Organisation.* M.I.T., Massachusetts, p. 4.

Allen, T.J.; Piepomeier; Cooney, S. (1971) *International Technological Gatekeepers.* Technology Review, **73**, 37–44.

Braun, E. and Collingridge, D. (1977) *Technology and Survival.* Butterworths, London, pp. 61–70.

Bureau of Statistics (1979) *Provisional Report on Tanzania Population Census.* Government Printer, Dar es Salaam. pp. 1–5.

Computer Policy Advisory Committee of Ireland (1978) *Details of National Informatics Institutions*, pp. 6–23.

Dalgic, T. (1979) *The Role of Technical Information System in Transfer of Technology—The Turkish Case.* Seminar Paper at Trinity College Dublin, pp. 2–10.

Ghosh, A.B. (1980) *Terminal Report TF/URT/78/006*, UNIDO, p. 111.

Kilewo, A.A. (1978) *Financial Aspects in the Development of Small-Scale Industries in Tanzania.* M.A. Thesis, University of Dar es Salaam, pp. 81–83.

Klintoe, K. (1979) *Information For and Within Industry, the Activities of FID/II-c.* International Forum Information Documentation, **4**, 1, pp. 26–28.

Lafond (1980) *Report on Preparatory Mission to Tanzania DF/URT/78/019.* UNIDO, p. 7.

Mansfield, E. (1968) *The Economics of Technological Change.* W. Norton & Co., New York, p. 4.

Ministry of Finance and Planning (1976) *Tanzania's Third Five Year Plan for Social and Economic Development 1976–1981*, pp. 39–46.

Pauline, A. (1977) *Handbook of Information Systems and Services.* UNESCO, Paris, p. 4.

Rosenberg, N. (1967) *Perspectives on Technology.* Pergamon Press, New York, p. 141.

Rulagora, D. (1979) *Analysis of 16 Contracts between SIDO and Swedish Firms.*

Sachys, I. (1979) *Towards Another Development in Science and Technology.* Daghamaskjold Foundation Journal 1/1979, Uppsala, p. 28.

Schon, D. (1976) *Technology and Change.* University Press, New York, p. 20.

Schumacher, E.F. (1973) *Small is Beautiful.* Bland and Briggs, London, pp. 1–3.

Solo, R. and Rogers, E. (1972) *Introducing Technological Change for Economic Growth and Development.* Michigan State University Press, p. 4.

Spencer, D.L. (1971) *Technology Gap in Perspective.* Spartan Books, New York, pp. 5–35.

Stewart, F. (1972) *Technology and Underdevelopment.* MacMillan Publishing Co., London, p. 120.

Tanzania Government (1978) *Economic Survey (1976–1978)*. Government Printer, Dar es Salaam.

Tanzania Government (1973) *Small Industries Act 28 of 1973.* Government Printer, Dar es Salaam, pp. 1–10.

U.N. Conference on Science and Technology Development (1979) *Articles 94–95.* Vienna.

UNIDO (1973) *Information Units in Small Plants.* United Nations Publication Sale No. E.73.B5, 1973, New York, pp. 9–11.

UNISIST (1977) *Guide for Teachers.* UNESCO.

# Utilisation of Informatics in the Decision-making Process of Small Industrial Enterprises: The Case in Turkey

*Çevik Uraz*
*Aitia, Turkey*

## BASIC STEPS IN THE DECISION-MAKING PROCESS

IN SCIENTIFIC MANAGEMENT, the decision making process basically has four main steps:

(1) Problem diagnosis
(2) Developing alternative courses of action
(3) Gathering necessary data
(4) Evaluating alternative courses of action.

This fundamental decision-making process pervades all functional areas of business. Decisions are usually made only after problems have been recognised. Recognising problems is the most important step in the decision making process. Once a problem has been recognised, management, through its experience, insights and creative imagination, contemplates and reviews different means of solving the problem. When the key problem has been recognised and a number of alternative solutions have been posed, data gathering assists management in selecting one, desirably the optimum, of the alternatives. For a lot of managers in the developing countries, data gathering may appear as a routine task with most of the data required already known. It takes quite a time for this type of manager to understand the reality that new types of problems or novel solutions to old problems often call for data that are beyond their experience. Presently in Turkey, not only big industries but even medium and small ones, because of the complexity of flourishing market conditions, are forced to use more and more data based and processed information for their survival.

The final phase of the decision-making process is evaluating alternative courses of action. To facilitate decision-making, each alternative course of action has to be checked against certain criteria. Even if the most universal of these criteria are profits, there are other criteria—such as timing, risk, limitations of the company resources and the general framework of the company's operations—which have to be taken into consideration. Each decision (or decided course of action) leads to certain ends over a given period of time. Decision-making is always predictive in nature. Generally we make decisions on the premise that this new *future* course of action is the most effective. As a rule, this future time period, with which we relate courses of action is the *long run*. But in the case of Turkey, decisions of many enterprises (government or private) are made on a daily basis or short-time, depending upon decision makers' insights and their work under rapidly changing conditions and poor planning.

## INSTITUTIONAL STRUCTURE AND MANAGERIAL LEVEL OF THE TURKISH SMALL INDUSTRY—MACRO APPROACH

ALMOST ALL of the small-scale industrial firms in Turkey are one man (owner) managed. As the size of the company gets bigger more professional work is needed and used. But in small and medium sized enterprises the employment of trained personnel is a handicap because of its high cost. This trained personnel term covers also, in large scale, personnel who can use new informatics technology. On the other hand, highly trained personnel prefer to get a job in the big enterprises which can offer more career opportunities, job security and higher wages. Moreover, managers of small firms (private or government) increase their skill in small enterprises and then transfer to big enterprises. According to recent research in Turkey the need for computers within the firm and necessary manpower for managing them have been forecast as follows:

**Table 1.**

| Years | | 1971 | 72 | 74 | 75 | 1980 | 1985 | 1990 |
|---|---|---|---|---|---|---|---|---|
| Number of computers | min | | | 90 | 100 | 600 | 1,500 | 2,000 |
| | max | 78 | 82 | 120 | 150 | 900 | 2,000 | 4,000 |
| Manpower | min | 394 | 418 | 750 | 1,200 | 5,150 | 8,150 | 21,000 |
| | max | | | 1,000 | 1,500 | 7,100 | 11,200 | 28,000 |

Personnel working on different areas of computers in Turkey number 1,622, according to the Turkish State Planning Organisation, as of 1979. This shows that Turkey has a high rate of trained personnel, more than expected, in this particular field. In this case, Turkey's real problem is not a personnel problem but it is to introduce informatics technology within the decision-making process of all of the firms and thus find jobs for their personnel.

**Institutional Structure of the Turkish Small Industry**: The size of the small industry is defined by the Turkish Ministry of Industry and Technology as an enterprise employing up to 10 workers or using up to 10 horsepower of all kinds of motors for its manufacturing process. Almost all of them are one-man managed within a maximum of two lines of organisational structure. Enterprises using more than 10 workers or 10 horsepower and above are accepted as medium or big enterprises according to their sizes (Turkish Yearbook of Statistics, 1969).

**Level of Education**: The level of education and training of small enterprise managers in Turkey is generally inadequate. However, undertrained managers of the developing countries paradoxically have developed an unusual high risk taking ability which is the main variable in their decision-making process.

**Level of Technology:** Another problem of the Turkish small industry is related with the level of technology used. In general, most of the products are produced with old and outdated machines by highly skilled staff of limited numbers.

On the other hand, the use of informatic systems in the industry is negatively correlated with respect to the size of enterprise. Usually small size systems are broadly used by medium or big industries as a result of poor choice by the decision-makers. Today, for example, the large enterprises which need heavy data processing are not only using small scale systems but are also underutilising them. So, this paradoxical situation shows that, even in the big industries in Turkey, the use of adequate technology and modern methodology is biased. For this reason, the use of modern technology in almost all areas is still very costly and very difficult in small enterprises. Development of the use of information systems in Turkey is in the following order:

(1) Institutions which own their data processing system.

(2) Partnership in the use of another institution's data processing system.

(3) Ownership of data processing sy • ɜm by more than one enterprise.

As we stated above irrational variables strongly influence decision-making in small firms. Because of the lack of information channels a lot of small enterprises do not even know how to use or to reach the information resources of the country. Those using informatics work in closed systems which are not available to the others.

## DATA SOURCES IN THE COUNTRY

**Secondary Information Sources:** Secondary information sources in Turkey are generally government owned agencies, such as, State Statistical Institute, Turkish Technical and Scientific Research Institution, State Planning Organisation and National Productivity Center. These agencies in general provide general output on economic and social conditions of the country, national accounting, educational and cultural statistics, production statistics, statistics on manpower, jobs, workers, payments and forecasts (usually showing government preferences) etc.

Universities are also providing some limited services for private and government enterprises. The main task of the Universities is to provide technical manpower for the informatics field.

**Private Institutions:** Private institutions in Turkey are providing major services for business enterprises but they are still in the take-off stage regarding the services. Some small enterprises are using big firms' data processing services. It can be said that there is a growing interest in using these kinds of services due to the facilities they provide. But relations between small enterprises and service providing enterprises is detached due to lack of information and a general disbelief on the necessity for these kind of services. Private institutions using computer based services are situated in large metropolitan areas such as Istanbul, Ankara and Izmir.

**Primary Information Gathering:** Professional research institutions in Turkey have been developing quickly since 1973. Unfortunately data on their workload are not available. But the growing interest in their services is a proof of their development.

As we pointed out before, the small enterprises are dependent on big-business data resources, but in the recent decade the use of small computers by small enterprises is growing as the Turkish economy becomes more market oriented. Data needs of small enterprises cover areas such as general accounting, customers accounts, cost accounting, production planning, budget analysis, stock control, statistical research, general feasibility etc.

Small enterprises' computer languages vary from Fortran IV to Algol. Cobol and RPG are highly used languages for private business enterprises.

## INFORMATION STRUCTURE OF THE FIRMS

**Traditional Way of Data Processing:** One of the big problems in using modern data processing is the existing traditional information system. The Introduction of modern technology in the outdated systems of small enterprises is causing trouble in the problem-solving mechanism of these firms. Furthermore, there is no planned and adequate channel to transport the data obtained to the other parts of the enterprise.

**Use of Modern Technology in Data Processing:** However the use of informatics especially in private business is increasing more and more as shown in Tables 2 and 3 (Kaya Kilan, 1976 and Aktas, 1980).

**Table 2. Number of Computers**

| Year | Public | Private | Total |
|---|---|---|---|
| 1971 | 35 | 26 | 61 |
| 1975 | 52 | 59 | 111 |
| 1978 | 80 | 147 | 227 |

**Table 3. Size of Computer Systems**

| Year | Small | Medium | Large | Total |
|---|---|---|---|---|
| 1971 | 56 | 5 | — | 61 |
| 1975 | 80 | 28 | 3 | 111 |
| 1978 | 122 | 96 | 9 | 227 |

## EFFECTIVENESS OF THE DECISIONS MADE—BOTTLE-NECKS AND BIASES

THERE ARE SEVERAL REASONS for the ineffectiveness of the decisions made which could be summarised as follows:

(a) Supply shortages and high demand structure of the market is the main reason of the ineffectiveness of the decisions.
(b) Non-competitive market structure particularly in advanced technological products.
(c) Protection of the newly developing (creeping) industries, from international competition, by high custom barriers and lack of standardised products contribute highly inefficient decisions.

Bottle necks in the use of informatics can be listed as follow:

(a) Limited internal capital accumulation under high inflation and interest rates, limited savings and limited currency inflow.

(b) Limited use of skilled manpower and professional managers with adequate international experience.

(c) Late information gathering, slow decision-making under quickly shifting market conditions, late application of decisions.

Biases of information is another cause in ineffective decision-making. Small enterprises are mostly one-man owned and managed firms in Turkey. These managers who hold chairmanships and make final decisions are a real cause of biases. Lack of continuous routine reports from the middle to the top management and difficulties in reliable data gathering is another bias in the decision process.

## CONCLUSION

THE USE OF INFORMATICS in developing small scale enterprises in Turkey is developing in a problematic way. High cost of personnel and high cost of hardware due to continuous devaluations of the Turkish Lira is hindering the use of informatics in the decision-making process of the small firms. On the other hand, effective use of informatics even in big businesses of the developing countries is not very well realised yet. Services offered by the data processing companies are often misunderstood or misused or unused by the small ones. That is the main reason for irrational decision-making processes in developing countries. This kind of decision-making process is also the cause of the vulnerability of the small enterprises against quickly shifting market conditions. It is our hope that the high benefit providing role of informatics will be understood within the rising industrial elite of Turkey.

### References

Aktas, Z. (1980) *Computer Informatics and National Development.* Bilgisayar, No. 17, Ankara, pp. 36–39.

Kaya Kilan, N. (1976) *The Use of Computers in Turkey.* Bilişim No. 10, Ankara, pp. 6–17.

# A Data Base System of the Expertise at Riyad University

*A. Sharaf Eldin, A. Nouh and M.S. Medani*
*Riyad University, Kingdom of Saudi Arabia*

## INTRODUCTION

THE DATA ON STAFF MEMBERS within Riyad University are stored and processed in the same way other employee data are handled. This should not be the case. While the same general information for both parties should be maintained, there is some important additional information for a staff member that should be held. Examples are: the fields of research, technical and academic experience, teaching experience, research publications, . . . etc. Since there is a scarcity of qualified manpower in the Kingdom of Saudi Arabia compared to the very ambitious developing plans, a big portion of the staff members are hired from outside the Kingdom. Each of them has his own particularities, background, experience, . . . etc. There is a problem in identifying a particular individual within that community. Another fact is the fast movement of the staff members. Problems arose when governmental agencies requested technical support from the university. There was no way of knowing who was most suitable to provide the requested support. One solution to this was to advertise and let the people apply. This way was not effective and most of the people did not wish to participate in that way. Another problem was planning the development of the staff members as was the allocation of each member to the courses offered. All these problems made it feasible to handle the staff data through a computerised data base system. In the following article we summarise the steps taken in establishing this data base and its implementation.

*Data Base Design*

In this stage the data required by all the applications that share the data base were determined. The security for each data element and levels of authorised access to the data base were established. Finally the logical structure of the data base was set.

*Data Base Definition*

Once the design was complete, it was defined in schema using the data base description language. Then the schema processor was applied to check its validity and to create the root data set.

*Creation of the Data Base Data Sets*

A utility program, DBUTIL, was used to allocate the necessary disk space for the data sets that comprise the data base. It also initialised these data sets.

The data base was then ready for use by application programmes to store, update and retrieve the data.

## SOURCE OF DATA

A QUESTIONNAIRE was designed and passed to all of the staff members. This questionnaire covered all the information about each individual. These data are:

(i) *Personal data*: Such as name, department, faculty, nationality, academic position, languages, major and minor fields.

(ii) *Qualifications*: Such as details of all the academic degrees obtained.

(iii) *Employment*: This covers all the jobs and positions held since graduation.

(iv) *Teaching Experience*: This contains all courses taught and their duration. This information was gathered to give an idea of the teaching experience of each member and the courses he preferred to teach.

(v) *Technical Experience*: A major benefit of this data base is the ability to locate expertise for a specific job when the need arises. This data covers the technical projects each one participated in, consultancy activities, . . . etc.

(vi) *Association Memberships*: All the professional associations of which the individual is a member are stated. The type of membership is also indicated.

(vii) *Fields of Interest*: A detailed classification was prepared for the different branches of engineering and coded. It is worth noting that the normal decimal system was found inconvenient in coding these fields. That is because decimal classification is lengthy and does not reflect the exact specialisations. We invented another classification which is simpler and covers the specialisations more accurately.

(viii) *Publications and Conferences*: All the publications are covered and all the attended conferences listed. The type of publication is classified into four different categories, viz., refereed journals, non refereed journals, refereed conferences, and non refereed conferences.

## DATA BASE STRUCTURE

THE STRUCTURE of the data base is illustrated in Figure 1. The root data set is not indicated since its use is primarily for structural purposes and contains no actual data. Tables 1 to 3 give the full specification of these data sets together with the path identifying information:

**Figure 1. The Data Base Structure**

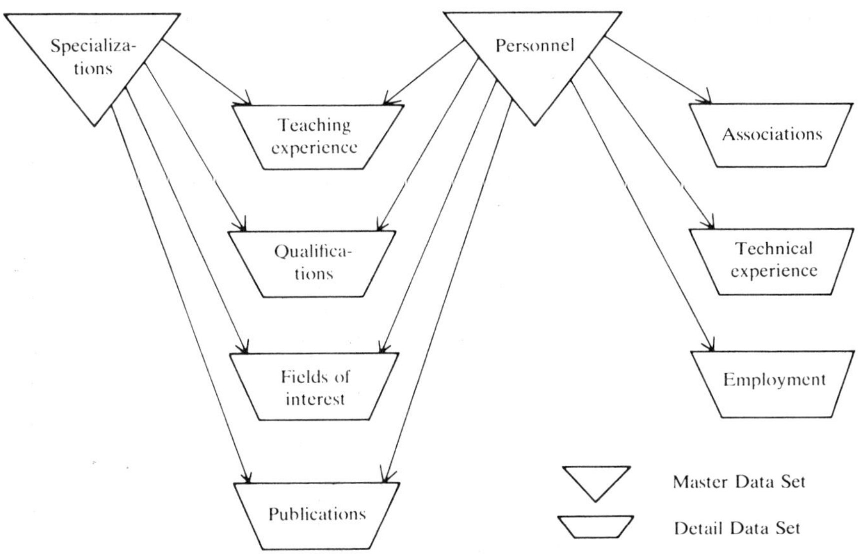


## DATA SECURITY

SINCE THE DATA stored in this data base is sensitive and contains particulars of each individual staff member, it is a must to have a powerful security system. There are

**Table 1. Data Sets**

| Sets: | Type | Item Count | Capacity | Entry Length | Blocking Factor |
|---|---|---|---|---|---|
| Personal | M | 14 | 150 | 77 | 4 |
| Univer | A | 1 | 400 | 15 | 20 |
| Special | M | 2 | 250 | 16 | 10 |
| Qualif | D | 7 | 800 | 50 | 8 |
| Training | D | 5 | 602 | 19 | 14 |
| Employ | D | 5 | 804 | 37 | 12 |
| Teaching | D | 4 | 817 | 18 | 19 |
| Experience | D | 3 | 800 | 27 | 16 |
| Assoc | D | 3 | 620 | 12 | 31 |
| Fields | D | 3 | 828 | 3 | 46 |
| Public | D | 5 | 2001 | 33 | 12 |
| Conferences | D | 6 | 2001 | 35 | 11 |

**Table 2. Path Identifying Information—Master Sets**

| Master Set Name | Associated Detail Set Name | Search Item Name | Sort Item Name |
|---|---|---|---|
| Personal | Qualif | ID-NO | |
| | Training | ID-NO | |
| | Employ | ID-NO | |
| | Teaching | ID-NO | |
| | Experience | ID-NO | |
| | Assoc | ID-NO | |
| | Fields | ID-NO | |
| | Public | ID-NO | |
| | Conferences | ID-NO | |
| Univer | Qualif | Univ | |
| Special | Qualif | Field | |
| | Training | Field | |
| | Teaching | Field | |
| | Fields | Field | |
| | Public | Field | |
| | Conferences | Field | |

**Table 3. Path Identifying Information—Detail Sets**

| Detail Set Name | Search Item Name | Sort Item Name | Associated Master Set Name |
|---|---|---|---|
| Qualif | !ID-NO | | Personal |
| | Univ | | Univer |
| | Field | | |
| Training | !ID-NO | | Personal |
| | Field | | Special |
| Employ | !ID-NO | | Personal |
| Teaching | !ID-NO | | Personal |
| | Field | | Special |
| Experience | !ID-NO | | Personal |
| Assoc | !ID-NO | | Personal |
| Fields | !ID-NO | | Personal |
| | Field | | Special |
| Public | !ID-NO | | Personal |
| | Field | | Special |
| Conferences | !ID-NO | | Personal |
| | Field | | Special |

several levels of security implemented in this data base. These could be summarised as follows:

(i) *Log-on Security*: The log-on procedure is protected via multiple passwords (Three). Illegitimate access is reported to the master console operator. These passwords are for account, user and group.

(ii) *User Classes and Passwords*: Each class of user is assigned a password which identifies his capabilities, e.g. reading only, reading and writing, etc. Each data element is given a list which identifies all the users who can read or read and write on it. Moreover, each data set is assigned read and write class lists which determine who can grant access to this data set.

(iii) *Capability Vector*: Whenever a user is entering the system, a capability vector is initiated for him according to the password given. This vector indicates which application programs the user can run.

## ACCESSING METHODS

THERE ARE FOUR DIFFERENT WAYS to access a data entry in the data base. These are:

(i) *Directed Access*: In this case, the entry is accessed using its record number. This method is used when the calling program has already determined the record number of the entry to be read. For example, this can arise when a program surveys a data set for entries satisfying some particular criteria using some other access methods and saves the record numbers of the qualified entries, then re-accesses these entries again using the saved record number.

(ii) *Serial Access*: In this method the next sequential entry is accessed. Both forward and backward serial access is permissible. Usually this method of accessing is used if most of the data entries are qualified. It is found that retrieving data entry serially and copying them to some other work file to sort and produce a report is much faster than trying to access them in an ordered sequence directly from the data base provided that more than about 75 percent of the entries are required.

(iii) *Calculated Access*: This method can be used only in association with master data sets by specifying a particular search item value. This method uses an address generation technique according to the following formula:

$$X = 1 + \text{Mod}\,(S - 1, C)$$

where

$X$ is the primary address;
$S$ is the search item value;
$C$ is the capacity of the master data set; and

Mod is the remainder function defined as:

$$\text{Mod}\,(a_1, a_2) = \text{The largest positive integer of}$$

$$a_1 - \left(\text{int}\left|\frac{a_1}{a_2}\right|\right)a_2$$

where

$$\text{int}\left|\frac{a_1}{a_2}\right| \text{ is the integral part of } \left|\frac{a_1}{a_2}\right| .$$

To illustrate, suppose that the 'PERSONAL' data set is to be accessed to retrieve the member whose number is 231, then the primary address is:

$$\begin{aligned} X &= 1 + \text{Mod}\,(231—1, 150) \\ &= 1 + \text{Mod}\,(230, 150) \\ &= 1 + 80 = 81 . \end{aligned}$$

This access method is very effective in *ad hoc* queries where infrequent access of a particular member or a few sets of members occurs.

(iv) *Chained Access*: This method is used basically in retrieving all entries in a detailed data set connected to a specific header in a master one. An example is to retrieve all the courses taught by a specific member. Both forward and

**Figure 2. Data Access Methods**

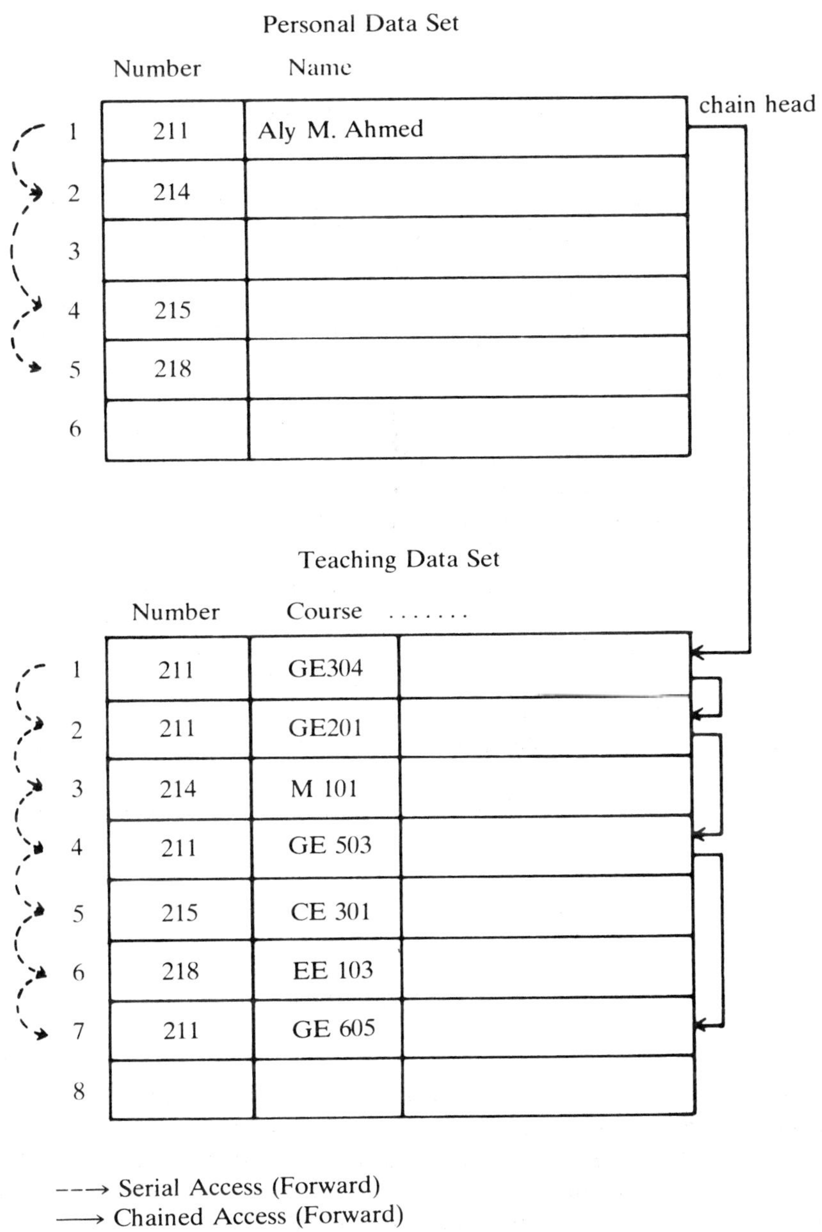

backward access of the chain is allowed. Figure 2 illustrates the access methods described above.

## SYSTEM CAPABILITIES

THE OUTPUT of the staff data base consists of different types of report and answers to *ad hoc* queries. In this section a description of these reports is given and explained.

(i) *Summary Reports*: These give summary information about different data. The information can be obtained on department level, faculty level and university level. Among these reports one can find a table showing the distribution of staff members according to their nationalities, qualifications, . . . etc.

(ii) *Detailed Reports*: These reports give detailed information concerning different data sets, e.g., a detailed report of all the staff members in a particular department giving full details of each individual.

(iii) *Statistical Reports*: To give some statistical measures about different parameters, e.g., a table of the mean value of the student/teacher ratio in each course. In this case the student registration files must be accessed since the staff data base does not contain details of the students.

(iv) *Comparative Reports*: To compare the information obtained from the current year with those obtained before. Mainly, the growth rate of different data is calculated.

(v) *Ad hoc queries*: These queries are used infrequently to search for a particular individual or set of individuals that match a set of criteria, e.g., to find all the computer specialists who can speak German and specialized in operating systems theory.

## CONCLUSION

THE STAFF DATA BASE is felt to be of great importance to the university. Although the system is implemented in a small scale in the College of Engineering it is hoped to extend it to all the colleges within the university. The fast movement and instability of manpower in the university make this data base an essential tool for decision-making in the planning and evaluation of the educational process.

This data base should be viewed as a part of an integrated data base system that covers all possible management and research requirements in the field of education in the Riyad University.

SECTION 8

# Conference Workshop Reports

There were four parallel workshop sessions at the Conference. These dealt with technology transfer (covering Sections 2 and 6), national policies and infrastructures (covering Sections 3 and 4), the implications of microelectronics for productivity and employment (covering Section 5) and computerisation in different industrial sectors (covering Section 7). The Chairmen were Sean P. Bedford, R. Narasimhan, Sang Joon Hahn and Utpal K. Banerjee respectively.

# TECHNOLOGY TRANSFER

(1) The domination of the industrialised countries in the technology field constitutes a serious obstacle to the real transfer of technology to developing countries.

(2) If the delivery of information for industrial development is to achieve the goals expected of it a number of conditions must be met. The pre-requisites for the delivery of information for industrial development must be identified; potential information users and their information needs should be identified and categorised; and the nature of the information required for selecting relevant technology identified.

(3) An awareness must be created among policy makers, advisers, industrialists, scientists and engineers, and agencies involved in industrial development of the need for information in a variety of forms. Governments, especially, should be made aware of the need to integrate industrial information policies in their industrial development policies.

(4) Users of information need impartial intermediaries to assist them to identify and evaluate appropriate sources of information, and also to assist in repackaging information to the level of understanding of users.

(5) Developed countries should co-operate with developing countries in providing access to advanced informatic technologies on favourable terms and conditions and share their expertise to promote and strengthen local technological capabilities in this field.

(6) In developing countries industrial extension and information services should be developed to interface with industrial enterprises of any size and to assist them in gaining access to data banks and other sources of technological information and expertise.

(7) Appropriate collection and dissemination structures should be established at the national level for data of relevance to the industrial development process, generated both within the country and among developing countries, and a level of informatisation appropriate to its development strategy should be defined.

(8) Developing countries should co-operate at the regional level in data collection and industrial information policies and consider the development of common instrumentalities to implement these programmes.

(9) They should co-operate at the regional level in the area of the acquisition of

informatic hardware and software and draw on the services offered by the international organisations.

(10) Priority should be given to orienting information services to the technology selection process and to utilizing the services of international, regional and non-governmental organisations for better access to information on technological alternatives. A concentration on informatics technology would assist in this regard.

(11) International organisations, such as UNU, UNDP, UNESCO, UNIDO, and IBI, play a very important role in facilitating information transfers. Such organisations should emphasise and expand specific areas of their present activities in order to assist developing countries in advancing towards full self-development in informatics technology as quickly as possible.

(12) These international organisations should in particular expand their activities in the areas of

(a) education and training in informatics;

(b) transfers of information relating to informatics.

(13) They should contribute to creating governmental awareness of the necessity for appropriate industrial and technological information policies and infrastructures. They should include informatics among industrial sectors served by their industrial and technological data banks and services, and develop networking arrangements with national structures in the Third World.

(14) They should assist in establishing national information systems supplemented in due course by information transfers from other developing countries and developed countries, in that order.

(15) It would be desirable for the UN University to organise research at the conceptual level into the broad issue of the computer age and its implications for developing countries.

(16) International organisations should mobilise international resources for the provision of co-ordinated technical assistance for the implementation of industrial and technological information policies and infrastructures by developing countries.

(17) They should also assist developing countries in the acquisition of informatics technology.

(18) International non-governmental organisations engaged in the promotion of appropriate informatics for the Third World should strengthen their regional structures and support national initiatives in sensitisation of authorities, educators and users, in the use of informatics.

## NATIONAL POLICIES AND INFRASTRUCTURES

(1) Policies to encourage transnational companies to establish manufacturing plants in order to provide employment and generate exports should have regard

to the need to ensure transfer of expertise to local employees so that an indigenous informatics industry can develop. The alternative policy of encouraging self-reliance by delinking users from foreign suppliers may be appropriate in certain countries.

(2) Appropriate informatics applications must be considered, e.g. in a developing country with high unemployment routine data processing which often eliminates jobs might be inappropriate and development-oriented applications such as water management, food production and agricultural engineering might be desirable. The problems arising in regard to producing special purpose systems using locally produced software, and interfacing with a minimum of imported hardware, must be addressed.

(3) Institutions and laws to control and regulate national informatics development and which are appropriate for developing countries must be considered.

(4) The social impact of the introduction of advanced informatics technology in developing countries must be addressed.

(5) A more pressing need exists for the transfer of expertise to developing countries than for the transfer of technology. Technology alone prolongs dependency on industrialised countries and retards the use and improvement of that technology. Measures are required to ensure that the transfer of expertise takes place.

(6) The extent to which institutes of higher education, including universities, can function as national centres of expertise must be determined.

(7) Technical Co-operation between Developing Countries (TCDC) should be used to meet many informatic needs. Developing countries and UN agencies should consider TCDC for tasks such as: installation, commissioning, writing software and training.

(8) Case studies of informatic developments would assist other countries in choosing strategies for their own informatics industry.

(9) A complete integrated indigenous informatics industry is unlikely to arise in most developing countries. A more promising approach is to develop application systems for specialised areas, particularly those related to development itself, e.g. in agriculture and health care.

(10) Major priorities in developing an indigenous informatics industry should include: provision of an appropriate infrastructure, and measures to ensure that the industry forms part of the national informatics plan.

(11) In meeting the special needs of developing countries it is important to distinguish between formal training and general education. National planning centres for informatics education should be established.

(12) An applied research unit associated with a university department and making use of postgraduate students for practical hardware/software projects could form the nucleus of a national centre of expertise. This link between practice and theory will help to integrate the university with the real life problems of the developing country.

(13) Universities in developing countries must be provided with comprehensive library and reference facilities to support informatics education. Short specialised

courses, given by suitably qualified personnel, should also be supported.
(14) Geographical location near dominant industrial countries makes special planning essential for some developing countries, e.g. Mexico, Singapore.

## THE IMPLICATIONS OF MICROELECTRONICS FOR PRODUCTIVITY AND EMPLOYMENT

(1) The environment into which microelectronic technology is introduced is peculiar to each developing country due to differences in the supply of labour, capital, skills, and foreign exchange and in economic systems and types of government. Each case must therefore be considered individually.
(2) Microelectronics and information technology are not homogeneous but have many facets with different characteristics. All these aspects should be taken into account differentiating, for example, between component manufacture, new products, and processes.
(3) The incorporation of microelectronics into certain products is essential if underdeveloped countries are to survive in world markets. However, the decision to introduce microelectronics into the actual production processes depends on the characteristics and degree of specialisation of each individual country.
(4) The high cost of entering the production of microelectronic components and the risks involved suggest that almost all developing countries should continue to buy components from overseas.
(5) The creation of a complete science and technology infrastructure is essential for developing countries to get on the science/technology spiral and this can only be achieved by training and the acquisition of the necessary skills.
(6) In particular, shortage of skills is a major barrier to the successful introduction of microelectronic-based technology.
(7) The introduction of microelectronics into the information sector and its supporting infrastructure in developing countries is clearly desirable. However, in both the agriculture and manufacturing sectors the case is less obvious.
(8) It was generally concluded that, despite the all pervasiveness of microelectronics, the impact on employment would only be marginal. This may not be true for structural changes within employment levels and such structural changes and the adjustment process need to be addressed.
(9) Overall, the introduction of education and training programmes in all the skills relevant to microelectronics, organised at the regional level and funded independently of the microelectronics suppliers, is of utmost importance for the successful assimilation of this technology by developing countries.

## COMPUTERISATION IN DIFFERENT INDUSTRIAL SECTORS

(1) The wide differences in the characteristics of developing countries and the heterogeneous nature of societies within any specific country preclude the making of generalisations about applications before specific country and society case studies are carried out.
(2) A wide variety of structural and cultural constraints—as well as economic ones—hinder the effective implementation of even the simplest informatic applications. The relative cheapness of current micro-computers does not reduce the importance of this multiplicity of non-economic constraints.
(3) The major limitation on the application rate of informatics in any size of firm is the human factor; this is especially true in small and medium sized firms. In fact an effective informatics policy is people oriented: one that seeks to improve existing skills rather than displace them.
(4) This limitation results in: poor system design, lack of awareness on the part of line management, and lack of dedicated and co-ordinated effort by those who use or are affected by the applications.
(5) A need exists for detailed comparative case studies to examine the rate and dynamics of industrial innovation in specific industries and for the wide dissemination of results of these studies.
(6) Incremental forms of innovation and the avoidance of sweeping change have the advantage of maximising available skills, minimising risk, and providing financial payback even in the short-term.
(7) Sectoral studies to identify the skill requirements for specified technology should be performed by a central national body in each country.
(8) A greater emphasis is required on the managerial aspects of informatics.
(9) It is important that a sound information base be available and used for guiding policy.
(10) Case studies should be carried out to evaluate the economic usefulness of data bases such as ISIS in specific situations in developing countries.
(11) It is important that the alternative ways of implementing computer based applications, such as the sharing of equipment, the hiring of machine time externally, or the installation of in-house machines, be carefully evaluated in every case.
(12) Many computers, especially in the production and process fields, are under-utilised even in very large organisations.
(13) The use of inexpensive micro-computers can bring production flexibility even to small organisations.
(14) Computer aided design applications facilitate manufacturers in offering a wide range of products specially designed for local conditions.
(15) The incorporation and use of micro-processor based devices in vehicles can achieve both a reduction in national energy costs and encourage the development of valuable skills.

(16) The development of standards by large organisations can simplify the selection of equipment and introduction of informatics systems by limiting the choice of options to an appropriate subset.

## GENERAL DISCUSSION

THE CASE STUDIES of computerisation in different industrial sectors presented in the conference papers provided participants with a good opportunity to gauge the pulse of actual industrial development and the role that informatics is playing. It was concluded that there was no such thing as a readymade portfolio of informatic packages that could be transferred or transplanted to or developed on a standard basis in any developing country. It was also noted that there was no standard mechanism of implementing informatics which could be generally recommended. It was realised that there were heterogeneous developing countries working at different levels of development: while for some countries the basic need was for sensitisation before any beginning could be made, other countries had already built up a substantial pool of knowledge on informatics which could be shared with other developing countries and their skills and experience made available.

The specific microelectronic devices and support systems which could play a role in the industrial development of developing countries should be studied in depth, bearing in mind that management culture had a very important role to play in the use of informatics. While the importance of informatics in the general development process was acknowledged, a word of caution was sounded by Dr. Banerjee regarding the distinction between the mechanism of informatics and the management of informatics. In this connection a need for a number of evaluation studies in relation to developing countries was identified including industrial case studies, the management aspects of applying informatics, the cost-effectiveness of micro-computer based heuristic models, national initiatives in the development of software for micro-computer based systems, and the economic usefulness of some of the large databases now available, including packages like ISIS. Attention was drawn to the existence in India of SENDOC, a body specifically concerned with the informatics of small and medium-scale industries. Reference was made to the need for a clearing house which could direct one developing country's need to another and try to match the capabilities that have already been created in a number of developing countries, and reduce to that extent the dependence on developed countries. Small and medium-sized industries are very people oriented and for this reason it was felt that informatics applications in these industries should be kept at the level where it augments skill and does not replace it, keeping in view the wide varieties of development levels and cultures existing in different developing countries.

Dr. Kimbel observed that the discussion on the implications of microelectronics for productivity and employment reflected different schools in economics and that a certain polarisation emerged. In a country where there is a high degree of centralisation and planning the impact most likely will be different from that in a country where the diffusion speed of these technologies is a result of the interplay of market forces. Professor Braun, in his paper, presented what he called a constellation of circumstances as a prerequisite for the introduction of any new manufacturing equipment into an established plant producing an established product. It was considered that this was very useful, both from the point of view of getting onto the learning spiral and also from the point of view of translating the learning process into appropriate policies. Regarding the barriers to the spread of microelectronics, again and again the shortage of skilled manpower was considered as a major barrier to the successful introduction of microelectronic-based technology. An attempt was made to find out whether it would be possible to proceed in a sectoral manner through the advantages which microelectronics held for LDCs. However it seemed to be very difficult to generalise the various findings and indeed the elements could not be found for such generalisation. The discussion on how to reconcile the all-pervasiveness of these technologies with their impact led to certain doubts. It was again and again suggested that, although they are all pervasive, the impact of these technologies on employment levels seems to be marginal. This may not be true for structural changes within employment levels and such structural changes and the adjustment process need to be addressed.

Mr. Peter Vero, International Computer Education and Information Centre, Hungary, stated that the Centre to which he is attached was founded as a joint venture of the Hungarian Government and the UN Development Programme. He drew attention to courses organised at the Centre, for participants from developing countries, in the areas of data base management, online systems, distributed systems and programming languages.

Dr. Juan Rada, Centre for Education in International Management, Geneva, felt that in this decade the value of investment in informatic goods, including both data processing and telecommunications, would grow more-or-less by a factor of two between developed and developing countries and that therefore it was crucial that developing countries react very quickly to this challenge. A prospective assessment by developing countries themselves would be required as well as a surge for a long-term comparative advantage. He stressed the importance for all developing countries to stop reacting once the thing had happened and to start taking a much more prospective view including the introduction of a technological element into planning. He felt that the die was already cast in the case of microelectronics and that many other advanced technologies are now coming up. The experience of microelectronics should be taken, he added, to tell policy makers that elements of technological prospection and forecasting should be included in the planning process.

Mr. John Page, International Institute for Applied Systems Analysis, stressed the importance of telecommunications in all kinds of computer technologies including the newer upcoming technologies. Telecommunications was one area in which developing countries need a good deal of help, he believed, particularly in regard to the costs of international networking.

Dr. Gopal Gupta, Asian Institute of Technology, Bangkok, drew attention to the number of conferences on informatics which have been held in various parts of the world and to the wide range of international organisations concerned in some way or other with this matter. He detected a striking degree of repetition in the recommendations emanating from all these sources. He felt that more action was now needed and emphasised that educational training programmes should be accorded top priority in order that developing countries would be in a position to decide for themselves in relation to major informatic questions.

Mr. Sean Cooney, Irish Agricultural Institute, believed that society may be "living on a volcano" and that an important question was whether informatics will be effective as a means for transferring information on new technology and for promoting economic growth or whether fundamentally new insights are required in this area.

SECTION 9

# Conference Concluding Addresses

**Dr. Abdel Rahman**

My first observation relates to the overall situation in the advanced countries as regards the subject matter of our conference. I think in the leading advanced countries there is very intensive interest in, and support for, the development of microelectronics and its application in different forms. This intensive interest is almost taking a competitive form between the leading countries, United States, Japan, Germany, U.K. and others. There have been official announcements that this is heralding a major change, not only in technology but also in society as a whole. We are reading and hearing every day statements, very imaginative and speculative, maybe rather optimistic, of the importance of the subject of information. We are being told that we are becoming an information-rich society with certain forms. So we cannot really neglect all these indications. There are indications by business, and by heads of government, by presidents of republics. It cannot really be overlooked so easily. The business community is moving into an expanding volume of advances in the technology, expansion in sales, and it is a feverish situation.

In this conference we never mentioned the military basis of this matter and I think it cannot be overlooked. In connection with the military applications and the intensive military work, a lot of hardware and software has been developed. I do not know if from this military side there is a spin off to the non-military side, but definitely the involvement of military or para-military applications in this regard is quite important.

As has been pointed out in the discussion, the conjunction between microprocessors and telecommunications as a whole widens the subject of informatics to include information transmittable through the communications system, and this rather intensive work in the leading advanced countries is also accompanied, as we know, by very intensive work in three or four areas of scientific advance. Biology is one of them. Molecular biology, genetic engineering, a deeper understanding of the biological processes in different forms including the transmission of information for genetic control and growth, these seem to be equally under very intensive development.

We can add to biology and micro-processors one or two other areas, including those relating to energy and petrochemicals. There are people who think that what we are witnessing now in the last quarter of the twentieth century is something more or less similar to what happened in the last quarter of the nineteenth century when there were a series of wider applications of steam, electricity, telephone wiring, wireless, a change from solid fuel to liquid fuel, and a number of other major changes. We may actually be undergoing or entering into a phase that represents a wider jump into technological advances the consequences of which may actually extend beyond the scientific and technical community. I do not think I really share the pessimism expressed at this conference, that we are sitting on a volcano—I do not think it is a volcano. It is at best a revolution of some form and we hope it will be a peaceful and happy one.

**Emerging Problems**

The second remark is that these same advanced countries that are taking part in this big scientific and technological revolution are themselves in difficult positions. They have inflation, unemployment, balance of payments difficulties and urban disaster areas. They do not seem to be so good though they are the most advanced. As a matter of fact the policies which are emerging from these leading countries now are policies of trade protectionism and of rather regressive economic actions, in the face of increasing unemployment. The whole world maybe is facing serious problems of environment both at national and regional level. There is also the problem of energy.

How are the developing countries faring? The developing countries until now, relatively speaking, were feeling the international economic difficulties to a lesser degree than the advanced countries, but these are now being transmitted to the developing countries. Many of them now have large debts, they have difficulties in placing exports, there is expansion of consumption beyond their resources and they are paying more for oil: these difficulties are of an economic nature which exist in the advanced countries and are now overflowing to most of the developing countries.

**Impacts on Developing Countries**

The question for this conference to examine was whether there is an overflow or an impact, from the advancing countries, in their technological advances, to the developing countries or not. We have all realised that when we speak about the developing countries we really do not speak about one uniform group. It is recognised and agreed by everybody that there are as many developing country types as countries, and for any simplification you have to classify them into four or five groups according to searching criteria. Therefore the position of developing countries cannot be taken as uniform, but for the purpose of this conference one can classify them, as was being done in many of the discussions, as firstly the large, relatively advanced developing countries with a scientific and technological base; the OPEC countries are another; and then the rest, at different levels. Therefore any reference to the developing countries would have to take into consideration their diversified nature and situation.

Yet, when we come to the developing countries as users of technology, we find that, for example, all the developing countries are now using the technology of aviation. They are not producers of aeroplanes but they are users of aeroplanes. They are not producers of cars but they are users of cars, which means that the technology, especially what they call the universally spreading technology, of transport, communication and other facilities, does spread although the manufacturing capacity does not spread equally. This differentiation between the spreading of utilisation and the spreading of manufacture is not only limited to the developing countries. Amongst the advanced countries themselves, which are making aeroplanes? Two or three countries only are making them. Many advanced countries are not making aeroplanes, which means that there is a concentration in production facilities in contrast to the spreading of utilisation within the group of advanced countries, let alone the group of developing countries.

**The spread of industry**

Now, let us look at the second part of the subject of this conference: will industry spread with the utilisation of the technology? Will the manufacturing capacity of components or parts related to telecommunication, to the hardware, to the software, to microelectronics as a whole, spread and to what degree should it spread to the developing countries from the big centres where production is more and more being concentrated? We know that the spreading of utilisation is done by Asians who are not exactly the same Asians who would be effective in the spreading of manufacture. The role of telecommunication has been actually mentioned in some of the discussions of the workshops. This is point number three: the differentiation between the spreading of manufacture and spreading of utilisation. We have very clear current examples of past technologies and we should not really be worrying very much if the spreading of utilisation is much faster than the spreading of manufacture, as has happened in many other technologies.

Yet there is some hope here, and this is point number four, that in the case of microelectronics and allied activities, there may be a shift, though a relative shift, towards decentralised production units. This has been mentioned. Would this new technology be favourable to the economic functioning of smaller units, small and medium industries or decentralised industries? Or even applications in rural districts on a small scale, because you are adding precision and skill through the machine and through the system, and not necessarily losing economic viability through the size of the average cost of unit service. I think for the developing countries, as also for the advanced countries, this application of bringing the ability to smaller and medium industries and to smaller and medium countries, would be a very important point to follow and to examine. As a matter of fact, we have witnessed recently, in the last ten years, that from the management point of view some large industrial concerns and large business concerns find it more profitable to break down into units rather than manage one big complex. Therefore this question of decentralisation and smaller management units could really be quite important when we speak about the spreading of the technology and its uses to the developing countries and to the less developed among the industrial countries.

## The role of software

My fifth point is related to the question of, software. There are three legs, so to speak, of the subject we are speaking about, the hardware and the software and telecommunications as has been pointed out during the conference. These are the three legs on which the subject stands. In the developing countries telecommunications itself is at the lower level and this is a disadvangage, while in the advanced countries the telecommunications system, both by wire, wireless and such like and other means of communications, is already there. So you can superimpose an information system on top of it very easily and transmit information from one place to another. This is one leg. The second leg is the hardware, and we know that the hardware is made up of core units and peripheral units. It seems that the discussions have indicated that the developing countries may have a chance, as regards the peripheral units but nobody has advised that the developing countries can compete with Japan and Silicon Valley in California, in trying to produce chips. So apart from the chips and some key questions, there is an opportunity, which has been mentioned, for the developing countries in regard to the peripheral auxiliary units connected with the hardware. But when we come to the third leg, the software, I think everybody, more or less, in this Conference has agreed that the developing countries have a chance, and maybe the only chance: they have some good prospects in the area of software development. Software development requires, as has been said, both training in a specific sense and education and general cultural infrastructure as a whole. This could be a major policy point for the developing countries, the question of training and building of centres and all that is related to software; especially the software related to decentralised units and small production units and to the mini-computers which are likely to be more amenable for distribution and availability in the developing countries.

## Economic implications

Point number six covers the question of the economic implications of all of this. There is a general view that the economic implications were worked out by Ricardo more than one hundred and fifty years ago, that you replace the worker by the machine and this creates unemployment. We know that this has happened but, on the other hand there has been greater production and greater utilisation of workers. So there is a displacement of workers, yes, but this displacement of workers is not an absolute one: it is a partial one and with general economic growth and general employment growth the displacement could be easily absorbed, noting that in the early nineteenth century we are speaking about a population of the world which was 10 per cent of the present population. Therefore in the short term, there may be labour problems but noting the historical experience and also some of the economic analysis which has been put forward here, it does not seem that microelectronics, in the developing countries, is going to be a serious problem as regards unemployment and this has been put in the conclusions of one of the workshops and also mentioned by others. It may be that in the advanced countries there are more sensitivities because the labour movement is rather alert, and they may use this as a bargaining point, but, as I said in my previous address, it seems that the labour movement so far is living cautiously with the question of microelectronic developments. They have not taken up a fight against it, they are just keeping quiet about it to a certain extent.

But the economic implications not only relate to the question of employment: there is the question of production and productivity. If productivity could increase, that in itself is of great economic importance. Productivity may increase through the precision that would be introduced in the functions of production and industrial control by the new equipment and the new techniques of information which may replace the partial skills of the workers by more perfect skills of the machine in certain key functions. This is what they call the relation between skill and performance. Skill and performance in the very key functions in industry have been traditionally built on key workers, but now, gradually, you can get some machines to do this in a faster and more precise manner and hence the skill required in numbers may be less and you would have to change the skill pattern of employment at the higher level.

## Attitudes to information

But again economics would not stop at the question of the skill. I refer to something which I do not think has been discussed in the conference, and this is point number seven, the relation between information and attitude. The availability of larger amounts of information is supposed to produce a change of attitude in those who are receiving or taking part in the information, and this information/attitude interdependence may eventually be the most important consequence of microelectronics and not the machines that play around in front of you. This means that there may be direct implications, both economic and social, to microelectronics but I would dare to say that the indirect implications are going to be much larger. The indirect ones relate to what generally has been heralded as a new society. If we take the transformation from steam to electricity up to the internal combustion engine in the last century, the transformation was not only a matter of engineering efficiency, it was a complete change of society, which was depending on the steam engine and the railway, to a society which gradually became dependent on the motor car. The direct implication is the comparison between the motor car and the locomotive, which is a very small thing compared with the change between the society of the motor car which was created in the following fifty years or so, and the society of the railway engine.

These are the indirect implications which are more important eventually and go beyond the mere comparison between the two technologies of steam and the technology of diesel engines or electricity. It has been said that the tertiary sector in the economy of the advanced countries is now being split into two sectors. We know that economists are fond of speaking about the primary sector which is agriculture, the secondary sector which is transformation industry, and the tertiary sector which they used to call services. Now in the modern analysis, this services sector is divided into two: what you might call normal services and information as a fourth sector. There are analyses in the US and elsewhere to show that this information sector now represents forty to fifty per cent of the GNP of these countries. These still are concepts, but what concerns me here is to say that it should not be taken, the growth of the third or fourth sector in the economies of the advanced countries, that this is what the developing countries want. Because the third sector in the developing countries is really a redundancy sector. It is not an activity sector. Most of the small traders and shopkeepers and hawkers and the household services and all of this are put, technically, in the third sector in the developing countries. So in the developing countries there would still be need to develop a transformation from agriculture to industry in relative and absolute terms before we can speak about the advantage in having a fourth sector, like in the case of the developed countries.

## The information revolution

It may be that, as has been said, and this is the next point, number eight, that the advanced countries are coming into a new industrial revolution in which information, and everything related to information, would be the predominant feature. This is the second or the new industrial revolution. We know very little about it. But let us speak about something we know a little better, which is the first industrial revolution of the last century. The industrial revolution at that time, a hundred years ago, spread from the U.K. to Europe and later to the U.S., Japan and the Soviet Union, and this became *the* industrial revolution. Unfortunately with that industrial revolution of the last century, colonialism turned from a lower stage into a more intensified stage and the big empires and the great colonial structures were actually supported by the first industrial revolution. Later, through the industrial structures that developed in the advanced countries, more humane developments, which we call in general socialism, developed within the industrial structures and therefore we could say that the industrial revolution of the nineteenth century intensified colonialism and imperialism first and later gave birth to socialism and humanism in general. Now, are we going to have a second industrial revolution which would intensify new forms of imperialism? Or better forms? I would hope, and this is only a hope, and a speculation, that whatever you call it, this huge advance in technology in several fields not only microelectronics, would intensify and deepen the sense of interdependence between parts of the world, between countries rather than domination and would eventually lead to some kind of prosperity and rationalism in the world of management rather than the hectic, wasteful attitudes which we have now, even as advanced countries, towards natural resources and the environment. So we should say we hope for greater interdependence and greater possibilities of peace and prosperity rather than imperialism and eventual fight for socialism.

**Human aspects of informatics**

My ninth point is to speak about the human aspect. I think the human aspect has been touched on in the discussions of this conference and some of the workshops but we should really not leave this place considering that we are speaking about machines. I think we are still speaking about human beings, not only the human beings handling the machines but the human beings feeding the machines with information and the human beings receiving the information and reacting to it. And the effects, as I say, looking beyond the machine, beyond the technology to the wider social implications, may take another ten, twenty or thirty years to appear. It has happened before in previous ways, so I think, keeping in mind the humanistic attitude towards this question, that maybe the policies which the developing countries, and also the developed ones, envisage will lead to a benevolent and useful attitude.

My last remark relates to the conference. This is a meeting of talking, not a meeting of action. None of us has a decision-making process in his pocket to put on the table. If one wants to look to action one should look for action where the decision-makers are. This conference is an exchange of information which should be useful to everybody who takes part in it. If you want action you should go to the decision-makers and see, if you are one of them or in contact with them, that they take the proper decisions, and the talking which we do here would be helpful in building up their minds for decision and action. Therefore in this respect, I consider this conference has been a useful exercise and a very pleasant one for which we thank the sponsors.

**Mr. Didier Lecerf, ITM**

I have three suggestions. The first one is a very humble one. It is about a newsletter. This was suggested by a meeting we had a couple of days ago and I strongly believe that a link should be created between all those who attended this conference, who provided such bright and useful material and intellectual contributions. These contributions have to be a continuous stream to be offered both to governments and to international organisations such as IBI and those of the UN system, in the first instance to UNIDO with its special responsibilities regarding industrial development. And also, of course to UNESCO, UNDP, and other organisations at regional level. I believe that this collective effort has to start with the individuals who want to offer their capacities and their experience. There have been many people here of different walks of life and they have all contributed extremely useful papers and I would suggest that this newsletter be fed by contributions and articles that you would write and we could assemble this and I believe that Professor Foster, who has acted as a marvellous conductor of the orchestra of this meeting, could perhaps continue in this function and I believe that this material could be assembled and put together in consultation with a network of senior officials responsible for informatics policies and also of those responsible for industrial development policies, so as to prepare the publication of a white paper, a white paper giving the views of responsible people from developing countries on such key issues as the computerisation of production processes which is going to be one of the key issues for the eighties. I believe strongly that, as pointed out by Professor Bernasconi at the first SPIN Conference, that those countries who missed the industrial revolution in the past, are now the underdeveloped countries, and that those countries who miss the telematics revolution will be the underdeveloped countries of the future, and this should be avoided by all means. Therefore TCDC has a special function in the sharing of expertise and in increasing the spin-off of any experience, any advance, in view of the extreme inequality in the present distribution of the communication and information means between North and South. I believe that this white paper could be prepared by you with your contribution, with the leadership of a few words and I would suggest that I should be getting Professor Foster, and also perhaps Mr. Bouarfa and Mr. Mahdi ElMandjra who expressed readiness to contribute to streamlining and organising this material. I believe that this could help in the preparations for the SPIN II Conference, for which ElMandjra and Mustapha Bouarfa are already serving IBI's preparations. But the precondition for this, an essential one, perhaps the only one, is that each of you should continue to feed into both a newsletter and the white paper material, a continuous stream of contributions, such intellectual contributions as you have made here and this in conjunction with other experts who were unable to attend this conference.

My second proposal is very simple. I call it "UNispin". Let us have this immense reservoir of expertise as an instrument at the disposal of UNIDO, of IBI, as well as of government, to provide independent advisory services to governments, with particular reference to the policies of the informatisation of industries. I think developing countries have stressed time and again their strong desire to have independent advisory services in the sense that they should be unbiased and therefore independent from the manufacturers. And since such advisory services should be professional services, they should also be provided in a professional structure which could be a sort of multinational endeavour, a multinational structure. The function of UNispin could be to provide advisory services independently and also to facilitate joint ventures for innovative application systems for utilisation and even for manufacturing of novel and innovative application systems. Perhaps this could contribute to the elaboration of an international strategy on a worldwide scale.

I believe, and this is my last proposal, that incentives are needed to encourage innovation capacities in developing countries to combine with similar innovation capacity in industrialised countries. Therefore we believe that the communication function of the network structure should be emphasised in a variety of ways. One of the ways is to facilitate the combination of external assistance flows and this is what we are trying to do in action, trying to channel additional resources to those projects which have the support of UNIDO or IBI and by so doing remove barriers to the mechanism of external aid. In this way we can continue to act as a catalyst. For example, if a project for a micro-informatics club is presented at the Common Market Commission for financing, if this is presented by a NGO, non-associated countries as well as associated have equal footing for access to this financing. I believe that a network of micro-informatics clubs, telecommunicating micro-informatics clubs, can be an initial step for a number of initiatives and among those initiatives there should be room for, say, production and exchange of audio-visual sensitisation materials such as the film ITM has prepared just for the purpose of making suggestions for follow-up action. This can be an initiative of a common endeavour of Third World countries and this can be done in such a way that all revenue raised by the sensitisation campaigns of the governments comes back to the developing countries. This is the first incentive. The second incentive, as we thought in consultation with the various agencies, can be fiscal immunity and liberty of transfers for such revenues. I call for initiative and further action by developing countries in developing for themselves a network of workshops for manufacturing such materials; also in developing skills for maintenance centres.

## Mr. F. Piera Gomez, IBI

As the first conference to deal with the subject of the use of informatics in industry and the informatics industry itself in the widest possible sense, it has been very important to us who have been following it very carefully, and watching how the different issues were being raised. I can identify three factors that were put forward: (1) the innovation factor, (2) the information factor, (3) the human factor. These three elements I can assure you will be picked up carefully and studied in future activities. But the merit of this conference has been to raise the series of issues affecting the use and applications of informatics. These issues now need further in-depth research. This conference has covered two subsets of a larger objective, namely the informatisation of emerging nations. And it is going to be one of the principal inputs to the preparation of the Second Intergovernmental Conference on the Strategies and Policies for Informatics in 1983. IBI, of course, proposes not to stop here, but to continue. Already in this area we have been working for several years through one of our working groups. And we propose to continue in this area. What the exact programme of follow-up will be is yet to be designed, but one which we intend to examine in October will be the meeting of the IDUSPIN which will deal with the problems related to the topics of this conference. We look forward very much to co-operation with other international organisations in this field and even if we agree with previous speakers that there is an epidemic of international conferences, at the same time this is because everyone wants this type of disease because people like communication and want to speak to each other: it is the only way to keep up to date on the evolution both of thinking and of technology.

**Dr. G.S. Gouri, UNIDO**

We in UNIDO have already been influenced and affected by the delineation of a number of very interesting problems, and I think that Dr. Abdel Rahman has outlined some ten major areas of which probably we would touch on four or five. It has been emphasised in the conference that the strategies and policies are the key and perhaps it is here that ways and means of clarifying the strategies and policies for the developing countries and ways and means of sensitising the decision makers are really very important. In this process we would endeavour very much to concentrate on a number of recommendations, particularly recommendations relating to applications, experiences in applications, cross-national studies, so that one would be able to know more about the interface that is already with us.

It is our intention in UNIDO to approach you individually and collectively to participate in our endeavour. We would also search for the translation of the concept that has been indicated here, namely how do we approach the whole subject of micro-electronics, or informatics, as a tool for decentralisation. What are the packages? Do we have them? Can we develop them and then take them and see how they can be applied? It would also be useful to make an analysis of what the actual training programmes look like, beginning at the secondary level and leading up to the specialised level. How can they be adjusted? If a country wants to do something can we have a reference to all this material and some evaluation? I think this is a task which is well worthwhile undertaking.

# Appendix

# List of Conference Participants

## Algeria

M. Bouharfa,
Commissariat National en Informatique.
El Harrach, Algiers

## Brazil

Marcos R. da Silva Borges,
Universidade de Federal Rio de Janeiro,
Nucleo de Computacao Electronica,
Cx Postal 232Y, Rio de Janeiro

## China

Qin Linzheng,
The Chinese Society for Future Studies,
No. 8 West Street Den Shi Ko.,
Peking

## Denmark

Kjeld Klintoe,
Dansk Teknisk Oplysningstjeneste,
H.C. Andersens Boulev. 18,
Copenhagen

## Ecuador

S. Carrasco,
Banco Central del Ecuador,
P.O. Box 339, Quito

J. Ceuva,
Banco Central del Ecuador,
Calle Guanguiltagua 537,
Quito

## Egypt

Sayed Abdel Kader,
General Organisation for
Industrialisation,
6 Kalil Agha Street,
Garden City, Cairo

Ibrahim H. Abdel Rahman,
Advisor to the Prime Minister,
9 Talaat Harb Street, Cairo

## Ethiopia

Ahmed M. Sheikh,
Ethiopian Airlines,
P.O. Box 1755,
Addis Ababa
*S.D.P. Study Fellow at Trinity College
Dublin*

## France

M. Bosse,
Ministry of Industry,
Paris

James Clayson,
James Clayson & Associates,
42 Rue de Bruxelles,
75009 Paris

M. Coffin,
Ministry of Industry,
Paris

M. Terrisse,
Ministry of Industry,
Paris

## Ghana

F.K.A. Allotey,
University of Science and Technology,
Kumasi

Richard Q. Young-Gyampo,
Ghana Institute of Clinical Genetics,
P.O. Box 150, Korlebu
*S.D.P. Study Fellow at Trinity College Dublin*

## Hong Kong

R.C. Barquin,
IBM World Trade Asia Corporation

## Hungary

Robert E. Kalman,
Hungarian Academy of Sciences,
Budapest

Albert Kiss,
Hungarian Central Statistical Office,
H-1525 Budapest

Peter Vero,
Hungarian Central Statistical Office,
H-1525 Budapest

## India

Utpal Banerjee,
Computronics India,
53 Lodi Estate,
New Delhi 100 003

S.V. Deodhar,
Ballarpur Industries,
124 Janpath,
New Delhi 110 001

P.P. Gupta,
Computer Maintenance Corporation,
The Arcade,
World Trade Centre,
Cuffe Parade,
Bombay 400 005

P.D. Jain,
Computer Maintenance Corporation,
The Arcade,
World Trade Centre,
Cuffe Parade, Bombay 400 005

R. Narashimhan,
Tata Institute of Fundamental Research,
Bombay 400 005

## Indonesia

Harsono Taroepratjeka,
Bandung Institute of Technology,
Jalan Ganesha 10, Bandung

## Iraq

Mannam J. Zamzeer,
State Company for Sugar Production
Baghdad
*S.D.P. Study Fellow at Trinity College Dublin*

## Ireland

Sean P. Bedford,
IMPC (Ireland) Ltd.,
20 Upper Fitzwilliam Street,
Dublin

N.B. Browne,
Computer Centre,
Bank of Ireland, Dublin

J. Butler,
Industrial Development Authority,
Lansdowne House, Dublin

T. Byrnes,
An Bord Telecom,
Merrion House,
Dublin

Robert Cochran,
National Board for Science and Technology, Dublin

S. Cooney,
Agricultural Institute,
Dublin

Ross Cooper,
National Board for Science and Technology, Dublin

P.V. Cremin,
Computer Centre,
Bank of Ireland,
Dublin

Tevfik Dalgiç,
Dublin Institute for Technology,
Dublin

Peter Daly,
National Board for Science and Technology, Dublin

Seamus Doherty,
Coopers & Lybrand Associates,
Fitzwilton House,
Wilton Place, Dublin

Jim Doorley,
Stokes Kennedy Crowley,
Management Consultants,
Harcourt House, Dublin

F.G. Foster,
Trinity College,
Dublin

Roger Fox,
AnCO,
P.O. Box 456,
Upper Baggot Street, Dublin

David Hanna,
Industrial Development Authority,
Lansdowne House, Dublin

Mel Healy,
National Board for Science and Technology, Dublin

Tom Higgins,
National Board for Science and Technology, Dublin

Roy Johnston,
Applied Research and Consultancy,
Trinity College, Dublin

Richard Kavanagh,
National Board for Science and Technology, Dublin

Eamon Lalor,
National Board for Science and Technology, Dublin

Cahal Lennon,
Stokes Kennedy Crowley,
Management Consultants,
Harcourt House, Dublin

Barry McGovern,
Industrial Development Authority,
Lansdowne House, Dublin

Bernard Moran,
University College, Belfield, Dublin

Frank Moran,
Institute for Industrial Research and Standards, Dublin

John Moriarty,
Trinity College,
Dublin

D. Murphy,
National Board for Science and Technology, Dublin

Richard J. Nolan,
Department of the Public Service,
Dublin

Patrick O'Beirne,
5 Coolgreaney Park,
Arklow, Co. Wicklow

T.D. O'Connor,
Institute for Industrial Research and Standards, Dublin

Dermot O'Doherty,
National Board for Science and Technology, Dublin

Eugene O'Neill,
Coopers & Lybrand Associates,
Fitzwilton House,
Wilton Place, Dublin

Jim Ryan,
Telephone House,
Marlborough Street,
Dublin

Kevin Ryan,
Trinity College,
Dublin

Lilly Sheehan,
National Board for Science and Technology, Dublin

Michael Sherwood-Smith,
Unilever Computer Services Ireland,
12 Lower Hatch Street, Dublin

Denis Toomey,
Institute for Industrial Research and Standards, Dublin

Norman Wood,
National Board for Science and Technology, Dublin

## Ivory Coast

Raymond Balliet,
7 Ter Rue d'Alesia,
75014 Paris, France

## Kenya

N.K. Mwara,
Government Computer Centre,
Ministry of Finance,
P.O. Box 30266, Nairobi

Robert J.P. Scott,
Institute of Computer Science,
University of Nairobi, Nairobi

## Korea

Young-Hwan Choi,
Bureau of Information and Systems Development, Seoul

Sang Joon Hahn,
Korea Institute of Electronics Technology,
P.O. Box 6,
Yang Jae, Seoul

## Libya

I.S. El Miligi,
Department of Statistics,
University of Garyounis,
Benghazi

## Malawi

Phillip K.S. Kaliati,
Agricultural Ministry,
P.O. Box 30134,
Capital City,
Lilongwe 3.
*S.D.P. Study Fellow at Trinity College Dublin*

## Malaysia

Chan-Onn Fong,
University of Malaya,
Faculty of Economics and Administration,
Kuala Lumpur

## Mexico

Carlos E. Acaraz,
Director General de Política Informática,
Balderas 71—4º piso, Mexico City

Carlos Enriquez,
Director General de Política Informática,
Balderas 71—4º piso, Mexico City

Antonio Garcia,
Secretaria de Programmacion,
Balderas 71—4º piso, Mexico City

## Morocco

Chaouki Benazzou,
Office of the Prime Minister,
Rabat

## Netherlands

Paul Flapper,
DAF Trucks B.V.,
Eindhoven

## Nigeria

C.C. Agunwamba,
University of Nigeria,
Nsukka

J. Anamaleze,
Governor's Office,
Information Department,
Government House,
Publications Unit,
Anambra State

Nweke Anyigor,
Commissioner for Information,
Governor's Office,
Government House,
Anambra State

G.M. Chikwendu,
University of Nigeria,
Nsukka

Nduka Isaac Eya,
Government Ministry,
Information Department,
Governor's Office,
Anambra State

## Norway

Øyvind Kvale,
Resource Policy Group,
Sagveien 21, Oslo 4

## Poland

Konrad Fialkowski,
Instytut Informacji NTE,
A.L. Niepodlegeosci 188,
00950 Warsaw 252809

## Portugal

Joaquim Jose Pestana,
Foreign Investment Institute,
Av. de Liberdada 258-4,
Lisbon

**Qatar**

Ali Al-Jaber,
Gulf Organisation for Industrial Consulting,
P.O. Box 5114, Doha

Abdulilah Dewachi,
Gulf Organisation for Industrial Consulting,
P.O. Box 5114, Doha

**Saudi Arabia**

Sharaf A. Eldin,
Chemical Engineering Department,
College of Engineering,
Riyad University,
P.O. Box 800, Riyad

M.S. Medani,
Chemical Engineering Department,
College of Engineering,
Riyad University,
P.O. Box 800, Riyad

A. Nouh,
Chemical Engineering Department,
College of Engineering,
Riyad University,
P.O. Box 800, Riyad

**Sierra Leone**

Samba Gbawuru-Mansaray,
Bank of Sierra Leone,
Freetown
*S.D.P. Study Fellow at Trinity College Dublin*

**Singapore**

Leon V. Chico,
Technonet Asia

Robert Iau,
Singapore Computer Society,
Shenton Way

Lim Tew Wee,
Singapore Computer Society,
Shenton Way

**Spain**

R. Portaencaza Baeza,
Dean, Faculty of Informatics,
Polytechnic University,
Madrid 31

De La Torre A. Prabos,
Ministry of Industry,
Madrid

**Sri Lanka**

Ira Unamboowe,
Industrial Development Board,
615 Galle Rd., Colombo

**Swaziland**

Andreas M. Fakudze,
Swaziland Government,
Treasury Department,
P.O. Box 38, Mbabane

**Sweden**

S. Dedijer,
Lund University

**Switzerland**

Dieter Altenpohl,
Alusuisse,
Postfach 495,
8034 Zurich

Juan F. Rada,
Centre for Education in International Management,
4 Chemin de Conches,
1231 Conches, Geneva

## Syria

A. Mufti,
Centre d'Études et Recherches Scientifiques,
B.P. 4470, Damas

## Tanzania

Edward Mazula,
Small Industries Development Organisation (SIDO),
P.O. Box 2476, Dar-es-Salaam
*S.D.P. Study Fellow at Trinity College Dublin*

Veremund H. Mwingira,
Institute of Rural Development Planning,
P.O. Box 643, Morogoro
*S.D.P. Study Fellow at Trinity College Dublin*

Daniel K. Rulagora,
Small Industries Development Organisation (SIDO),
P.O. Box 2476, Dar-es-Salaam
*S.D.P. Study Fellow at Trinity College Dublin*

## Thailand

Gopal K. Gupta,
Asian Institute of Technology,
Bangkok

## Turkey

Ural Sözen,
Ankara Academy of Economic and Commercial Sciences,
Kirçicegi Sok, No. 8,
Gaziosmanpaşa, Ankara

Çevik Uraz,
Işikyolu Sokak 3/2,
Paris CAD, Ankara

## United Kingdom

Ernest Braun,
Technology Policy Unit,
University of Aston,
Gosta Green,
Birmingham B4 7ET

T.R. Crossley,
University of Salford,
Salford M5 4WT

Maurice Goldsmith,
Science Policy Foundation,
Benjamin Franklin House,
36 Craven St.,
London WC2N 5NG

H.K. Hoffman,
Science Policy Research Unit,
University of Sussex,
Falmer, Brighton,
Sussex BN1 9RF

Colin Palmer,
66 St. Helen's Road,
Solihull, West Midlands

Brinley Platts,
National Computing Centre,
Oxford Road,
Manchester M1 7ED

H.J Rush,
Science Policy Research Unit,
University of Sussex,
Falmer, Brighton,
Sussex BN1 9RF

Paul Stoneman,
Department of Economics,
University of Warwick,
Coventry

R.M.D. Ure,
Department of Industry,
London

### United States

Lloyd S. Kaufman,
CBEMA,
1828 L Street,
N.W. Washington D.C.

James R. Kreer,
U.S. Air Force,
European Office of Aerospace Research and Development,
223/231 Old Marylebone Road,
London NW1 5TH

George M. Sokol,
U.S. Army,
European Research Office,
223/231 Old Marylebone Road,
London NW1 5TH

### Yugoslavia

Andrez Jerman-Blazic,
Committee for Information of SR Slovenia,
Levstickova 10, Ljubljana

Tone Vahen,
Committee for Information of SR Slovenia,
Levstickova 10, Ljubljana

### Zambia

Passmore P.M. Hamukoma,
Mining Industry Manpower Service Unit,
P.O. Box 1699, Kitwe
*S.D.P. Study Fellow at Trinity College Dublin*

## INTERNATIONAL BODIES

### European Economic Community (EEC)

H.J. Helms,
Head, Informatics Programme,
Joint Research Centre,
ISPRA, Italy

D. Watson,
Head of Task Force for Innovation,
Brussels

### Informatique pour les Tiers Mondes (ITM)

Didier M. Lecerf,
13 Avenue Soeur-Rosalie,
75013 Paris

### Intergovernmental Bureau for Informatics (IBI)

L. Castelli Avolio,
Director, Department of Co-Operation,
Rome

C.S. Caduda,
Department of Operations, Rome

André Michel,
Head, Information Services, Rome

F. Piera Gomez,
Director of Operations, Rome

## International Institute for Applied Systems Analysis (IIASA)

John R.U. Page,
Laxenburg,
Austria

## Organisation for Economic Co-operation and Development (OECD)

Dieter Kimbel,
Directorate for Science Technology and Industry, Paris

## United Nations Development Programme (UNDP)

Erling Dessau,
Resident Representative,
Burma

## United Nations Educational, Scientific and Cultural Organisation (UNESCO)

E. A. Owolabi,
Informatics Section,
Division of Scientific Research and Higher Education, Paris

## United Nations Industrial Development Organisation (UNIDO)

A. Amato-Gauci,
Vienna

Roch T. de Mautort,
Head, Public Information Section,
Vienna

G.S. Gouri,
Director, Division for Industrial Studies,
Vienna

## United Nations University (UNU)

I. Wesley-Tanasković
c/o National Commission UNESCO,
Mose Piyade 8/VI,
1100 Belgrade,
Yugoslavia